AWS Observability Handbook

Monitor, trace, and alert your cloud applications with AWS' myriad observability tools

Phani Kumar Lingamallu

Fabio Braga de Oliveira

BIRMINGHAM—MUMBAI

AWS Observability Handbook

Group Product Manager: Mohd Riyan Khan

Publishing Product Manager: Surbhi Suman

Senior Content Development Editor: Adrija Mitra

Technical Editor: Irfa Ansari

Copy Editor: Safis Editing

Project Coordinator: Prajakta Naik

Proofreader: Safis Editing

Indexer: Pratik Shirodkar

Production Designer: Prashant Ghare

Marketing Coordinator: Agnes D'souza

First published: April 2023

Production reference: 1190423

Published by Packt Publishing Ltd.
Livery Place
35 Livery Street
Birmingham
B3 2PB, UK.

978-1-80461-671-0

www.packtpub.com

I would like to take this opportunity to express my heartfelt gratitude to two very special people in my life. To my parents, Lakshmi and Mohan Rao, your unwavering support and guidance throughout my life have been a source of strength and inspiration. Your love and sacrifices have shaped me into the person I am today, and I am forever grateful for all that you have done for me. And to my wife, Usha, my loving partner throughout our joint life journey; you have been my rock during difficult times and my partner in every adventure. Your love has given me strength, and your friendship has brought me endless joy. Thank you for both being an integral part of my life and for making it truly special.

– Phani Kumar Lingamallu

My parents had a difficult life; they migrated from an impoverished region in Brazil to try for a better life. They found each other, fell in love, and brought up a family, doing the best they could with the knowledge they had. All three of their kids attended the best universities in our country. Thanks to their struggle and sacrifices, and despite all the bumps in the road, I was able to experience an international career, migrating to Germany with my small family. We may be far away, but I can't start to think about any of my accomplishments without feeling thankful for everything they provided for me.

– Fabio Braga de Oliveira

Contributors

About the authors

Phani Kumar Lingamallu works as a senior partner solution architect at **Amazon Web Services** (**AWS**). With around 19 years of IT experience, he previously served as a consultant for several well-known companies, such as Microsoft, HCL Technologies, and Harsco. He has worked on projects such as the large-scale migration of workloads to AWS and the Azure cloud. He has hands-on experience with the setup of monitoring/management for over 45,000 servers, and the design and implementation of large-scale AIOps transformations for clients across Europe, the US, and APAC, covering monitoring, automation, reporting, and analytics. He holds a Master of Science in electronics and possesses certifications including AWS Solution Architect Professional and Microsoft Certified Azure Solution Architect Expert.

I am immensely grateful to my fellow colleagues, both those I have worked with in my current role and those from my previous roles. Your unwavering dedication and passion for the work we do have been a constant source of inspiration to me.

Fabio Braga de Oliveira works as a senior partner solution architect at AWS. He carries a wealth of experience from various industries – automotive, industrial, and financial services, working in the last 19 years as a software engineer/team lead/solutions architect. His professional interests range from big to small: he loves event-driven architectures, helping build complex, highly efficient systems, and also working on small devices, building devices fleet to collect data and support companies to drive new insights, using analytics techniques and machine learning. He majored in electronics and has a BS in computer science, an MBA in project management, and a series of IT certifications, among them AWS Certified Solution Architect – Professional. Nowadays, he supports AWS partners in the DACH/CEE region with application modernization (serverless and containers) and IoT workloads.

I would love to be as brilliant and smart as many of my colleagues. I am standing on the shoulders of giants, definitely. Every example, every code excerpt, and every concept is the result of the accumulated knowledge of practitioners and the computer science community as a whole. So, to all of you, my humble thank you; without all of you, I wouldn't be half of what I am.

About the reviewers

Anand Rajanala's expertise in **Application Performance Management** (**APM**) and AIOps observability. As a product manager, he has created roadmaps and identified opportunities to enhance product offerings to ensure that products meet customer needs and are aligned with business goals.

Working with companies such as CA, HCL Technologies, Broadcom, ConnX, and Rakuten has given Anand the opportunity to develop a deep understanding of the technology industry and the challenges that businesses face. He has strong communication and collaboration skills, which are essential for managing cross-functional teams and building relationships with stakeholders.

Anand Rajanala has a supportive family behind him – his lovely kids, Srisubodh and Paanya SriSisira, and his wife, Vasavi. To my family, my brothers and sister – I cannot thank you enough for everything you do for me and our family. Words cannot express how much your support means to me. From the bottom of my heart, thank you for being there for me every step of the way.

Peter Gergely Marczis brings 15 years of industry experience to his role as the leader of the DevOps platforms practice at Nordcloud. Starting his career as an embedded programmer, he quickly developed a passion for cutting-edge technologies, and he now works with industry-leading companies on their DevOps strategies. His unique expertise and insights have made him the ideal candidate to review technical content for its quality and relevance to the field.

Table of Contents

3

Gathering Operational Data and Alerting Using Amazon CloudWatch   49

4

Implementing Distributed Tracing Using AWS X-Ray   91

Part 2: Automated and Machine Learning-Powered Observability on AWS

5

6

7

Observability for Serverless Applications on AWS   197

8

End User Experience Monitoring on AWS   231

Part 3: Open Source Managed Services on AWS

9

Collecting Metrics and Traces Using OpenTelemetry 273

10

Deploying and Configuring an Amazon Managed Service for Prometheus 297

13

Observability Best Practices at Scale   401

14

Be Well-Architected for Operational Excellence   419

15

The Role of Observability in the Cloud Adoption Framework   439

Preface

Observability refers to the ability to gain insights into the internal state of a system by analyzing the external outputs or data produced by the system. Achieving observability is complex in modern application architectures due to their distributed nature.

While talking to customers and builders, we realized the information required to leverage observability benefits using AWS's native tools and services is spread across many service-specific documents without a concise view and practical examples. That's why we decided to write this book for practitioners looking for a straightforward, hands-on source.

In this book, we will explore how to configure and use various AWS services to achieve full-stack observability for your workloads running on AWS. The guide covers key concepts such as understanding the need for observability for different architectures, such as monolith, microservices, and serverless computing, on AWS. The book also highlights how *Site Reliability Engineers* (*SREs*) can benefit from AWS's automated and machine learning offerings to achieve more with less management overhead. We will also look into how developers can achieve observability for their applications and roll out changes confidently with the help of observability. Furthermore, we will dive into the open source observability options available on AWS.

Then, we will look into the architecture best practice recommendations for your observability workloads, the importance of observability in achieving faster adoption of the cloud, and the approach to observability in a large organization.

Who this book is for

This book is intended for SREs, Cloud Developers, DevOps engineers, and Solution Architects who are looking to use AWS's native services and open source managed services on AWS to achieve the required observability targets. Solution architects seeking to achieve operational excellence by implementing cloud observability solutions for their workloads will also find guidance in this book. You are expected to have a basic understanding of AWS cloud fundamentals and the different service offerings available on the AWS cloud to run applications, such as EC2, storage solutions such as S3, and container solutions such as ECS and EKS.

What this book covers

Chapter 1, Observability 101, will go through the fundamentals of observability and discuss its building blocks and concepts. It provides you with the required terminology and introduces the vocabulary and concepts that you need to know relating to observability in a modern distributed application environment.

Chapter 2, Overview of the Observability Landscape on AWS, will help you understand the basic, foundational services and infrastructure-, application-, and machine learning-based tools available in AWS in terms of cloud-native observability and managed open source observability solutions.

Chapter 3, Gathering Operational Data and Alerting Using Amazon CloudWatch, helps you navigate the fundamentals of CloudWatch metrics, CloudWatch Logs, CloudWatch alarms, and CloudWatch dashboards. It provides hands-on experience in the installation of a unified agent and ingesting metrics and logs from EC2 instances and provides an overview of how to visualize them on a unified dashboard. It also introduces the requirement of the EventBridge service and event rules and how they would be used for fault monitoring.

Chapter 4, Implementing Distributed Tracing Using AWS X-Ray, will take you through what the requirement for distributed tracing is in modern applications and the fundamentals of the services offered by AWS relating to performance monitoring and distributed tracing.

Chapter 5, Insights into Operational Data with CloudWatch, will deep-dive into CloudWatch metrics and CloudWatch dashboards. We will see how to do more with less using CloudWatch Log Insights, CloudWatch Contributor Insights, and CloudWatch Application Insights, deriving operational intelligence automatically from log data and metrics and allowing for faster troubleshooting during operations.

Chapter 6, Observability for Containerized Applications on AWS, enables you to understand the setup of end-to-end containerized applications running on ECS and EKS to achieve observability.

Chapter 7, Observability for Serverless Applications on AWS, gives an overview of Lambda Insights and explores the data generated from it. You will understand how to gather metrics, logs, and traces from the serverless Lambda application and how they can be visualized as a unified dashboard for end-to-end operational visibility.

Chapter 8, End User Experience Monitoring on AWS, will take you through the importance of user experience monitoring. It provides an overview of how synthetic canaries can be implemented in understanding the user experience for a web application. We will provide an overview of how to collect metrics to capture real user behavior while interacting with a web application.

Chapter 9, Collecting Metrics and Traces Using OpenTelemetry, will discuss the existing SDKs, APIs, and AWS services that support organizations looking for ways to implement observability but using the open source ecosystem. It shows how AWS services can easily integrate with existing practices, helping to reduce much of the heavy lifting of deploying and managing those open source tools done by your own infrastructure team.

Chapter 10, Deploying and Configuring an Amazon Managed Service for Prometheus, enables you to understand the foundation of Amazon Managed Grafana and Prometheus and guides you in setting up the services, ingesting metrics, logs, and traces from the cloud-native observability services, and setting up advanced dashboards for operational visibility. It also discusses how to set up Prometheus monitoring for containerized workloads on AWS.

Chapter 11, Deploying the Elasticsearch, Logstash, and Kibana Stack Using Amazon OpenSearch Service, enables you to understand the foundation of Amazon OSS and guides you on how to set up the services and ingest logs and traces from your application workloads and set up dashboards for operational visibility.

Chapter 12, Augmenting the Human Operator with Amazon DevOps Guru, looks at AWS DevOps Guru, which is a service powered by machine learning that automatically extracts the relevant metrics about workloads and detects anomalies before they impact end users. In this chapter, you will learn how to use it to enrich the already deployed set of tools and use it as an advisor to detect issues and recommend remediations.

Chapter 13, Observability Best Practices at Scale, covers some patterns and recommendations on how to scale the observability of applications in complex organizations for workloads distributed in multiple accounts and regions.

Chapter 14, Be Well-Architected for Operational Excellence, looks at the AWS Well-Architected Framework, which provides guidelines on how to apply best practices of the design, delivery, and operations of AWS environments. Its Operational Excellence pillar and Management and Governance Lens include guidance on how to run workloads effectively and continuously improve operations. In this chapter, we discuss some of those principles and how they are interconnected with observability best practices.

Chapter 15, The Role of Observability in the Cloud Adoption Framework, looks at **the Cloud Adoption Framework (CAF)**, which helps customers and users to digitally transform their businesses by leveraging the AWS experience and best practices. Among the CAF pillars are Management, Governance, and Operations. This chapter will discuss the role of observability in an organization's transformation journey.

To get the most out of this book

To get the most out of the book, we recommend you have an AWS account to practice the concepts discussed in the book. We have used quick-start templates where applicable to make your exercises as practical as possible. If you would like to understand the code and the CloudFormation templates used in detail, we suggest you access the book's GitHub repository (a link is available in the next section).

Software/hardware covered in the book	Operating system requirements
Python 3.9	Windows, macOS, or Linux
Node.js 14/Node.js 16	
JSON	

If you are using the digital version of this book, we advise you to type the code yourself or access the code from the book's GitHub repository (a link is available in the next section). Doing so will help you avoid any potential errors related to the copying and pasting of code.

Download the example code files

You can download the example code files for this book from GitHub at `https://github.com/PacktPublishing/AWS-Observability-Handbook`. If there's an update to the code, it will be updated in the GitHub repository.

We also have other code bundles from our rich catalog of books and videos available at `https://github.com/PacktPublishing/`. Check them out!

Download the color images

We also provide a PDF file that has color images of the screenshots and diagrams used in this book. You can download it here: `https://packt.link/n7E68`.

Conventions used

There are a number of text conventions used throughout this book.

`Code in text`: Indicates code words in text, database table names, folder names, filenames, file extensions, pathnames, dummy URLs, user input, and Twitter handles. Here is an example: "Set the dataset name to `my-dataset1`."

A block of code is set as follows:

```
Function:
  Runtime: nodejs16.x
  Timeout: 100
  Layers:
    - !Sub "arn:aws:lambda:${AWS::Region}:580247275435:layer:
LambdaInsightsExtension:21"
  TracingConfig:
      Mode: Active
```

Any command-line input or output is written as follows:

```
python sendAPIRequest.py
```

Bold: Indicates a new term, an important word, or words that you see onscreen. For instance, words in menus or dialog boxes appear in **bold**. Here is an example: "For the next step, let's go ahead and decrease the table capacity in DynamoDB for both **Read Capacity** and **Write Capacity** to **1**."

> Tips or important notes
> Appear like this.

Get in touch

Feedback from our readers is always welcome.

General feedback: If you have questions about any aspect of this book, email us at `customercare@packtpub.com` and mention the book title in the subject of your message.

Errata: Although we have taken every care to ensure the accuracy of our content, mistakes do happen. If you have found a mistake in this book, we would be grateful if you would report this to us. Please visit `www.packtpub.com/support/errata` and fill in the form.

Piracy: If you come across any illegal copies of our works in any form on the internet, we would be grateful if you would provide us with the location address or website name. Please contact us at `copyright@packt.com` with a link to the material.

If you are interested in becoming an author: If there is a topic that you have expertise in and you are interested in either writing or contributing to a book, please visit `authors.packtpub.com`.

Share Your Thoughts

Once you've read *AWS Observability Handbook*, we'd love to hear your thoughts! Scan the QR code below to go straight to the Amazon review page for this book and share your feedback.

https://packt.link/r/1804616710

Your review is important to us and the tech community and will help us make sure we're delivering excellent quality content.

Download a free PDF copy of this book

Thanks for purchasing this book!

Do you like to read on the go but are unable to carry your print books everywhere? Is your eBook purchase not compatible with the device of your choice?

Don't worry, now with every Packt book you get a DRM-free PDF version of that book at no cost.

Read anywhere, any place, on any device. Search, copy, and paste code from your favorite technical books directly into your application.

The perks don't stop there, you can get exclusive access to discounts, newsletters, and great free content in your inbox daily

Follow these simple steps to get the benefits:

1. Scan the QR code or visit the link below

https://packt.link/free-ebook/9781804616710

2. Submit your proof of purchase
3. That's it! We'll send your free PDF and other benefits to your email directly

Part 1: Getting Started with Observability on AWS

This part provides an overview of observability and a discussion about the building blocks of observability. Additionally, it provides a review of the different services available in AWS to achieve observability in a modern distributed application environment.

This section has the following chapters:

- *Chapter 1, Observability 101*
- *Chapter 2, Overview of the Observability Landscape on AWS*
- *Chapter 3, Gathering Operational Data and Alerting Using Amazon CloudWatch*
- *Chapter 4, Implementing Distributed Tracing Using AWS X-Ray*

1
Observability 101

Observability is the hot new tech buzzword. Observability is confused with many other practices, such as monitoring, tracing, logging, telemetry, and instrumentation. But observability is a superset of all these, and all are required to achieve observability. It includes measuring your infrastructure, application, and user experience to understand how they are doing and then acting on the findings with predictive or reactive solutions.

One of the benefits of working with older technologies was the limited set of defined failure modes. Yes, things broke, but you would know what went wrong at any given time, or you could find out quickly because many older systems repeatedly failed in the same ways. As systems became more complex, the possible failures became more abundant. To address the possible failures of these complex systems, monitoring tools were created. We kept track of our application performance with monitoring, data collection, and time-series analytics. This process was manageable for a while but quickly got out of hand.

Modern systems are extraordinarily complex, with everything depending on open source libraries and turning into cloud-native microservices running on Kubernetes clusters. Further, we develop them faster than ever, and the possible failure modes multiply as we implement and deploy these distributed systems more quickly.

When something fails, it's no longer obvious what caused it. Nothing is perfect; every software system will fail at some point, and the best thing we can do as developers is to make sure that when our software fails, it's as easy as possible for us to fix it. Standard monitoring, which is always reactive, cannot fix this problem, and it can only track known unknowns. The new unknowns mean that we have to do more work to figure out what's going on. Observability goes beyond mere monitoring (even of very complicated infrastructures) and is instead about building visibility into every layer of your business. Increased visibility gives everyone invested in the business more significant insight into issues and user experience, and creates more time for more strategic initiatives, instead of firefighting issues.

In this chapter, we are going to cover the following topics:

- What is observability?
- The need for observability in a distributed application environment
- Building blocks of observability
- Benefits of observability

Technical requirements

For this chapter, you must have a basic understanding of *application deployment* and *operations.*

Some basic coding skills are also required. We will use some code samples to illustrate concepts, but we will keep it simple and focus less on the code and more on the ideas explained.

Finally, we will use the Python language for all the code samples if not explicitly stated otherwise.

What is observability?

If you are reading this book, the odds are you have already read about or heard the term observability elsewhere, and have decided to apply it to your AWS workloads. You are in the right place. But even being a book for the practitioner, we can't start this book without defining some terms. They will become our guide for the rest of this book, helping us drive our discussions. Let's start with the main one: **observability**.

The engineer Rudolf E. Kálmán coined the term observability (abbreviated as **o11y**) in 1960.

In his 1960 paper, Kálmán describes what he calls observability in the field of *control theory*: the measure of how well someone can infer a system's internal states from knowledge of its external signals/outputs.

Observability is another borrowed term, in the same way as **software architecture**, **software engineering**, and **design patterns**. We borrow a complex, mathematical term from an older, more mature field and make it ours in our younger computing field. And to do that, we need to make it softer to make it usable.

So, in this book, we will say an application has observability if the following is true:

- You can read any variable that affects the application state
- You can understand how the application reached that state
- You can execute both the aforementioned points without deploying any new code

So, your application is observable if you can answer questions that you knew you should ask, but you can also answer questions that you didn't know you needed to ask.

So far, we have defined what observability is. But if you are like me, the first time I saw a description of observability like the one provided here, it didn't help me understand it or even what made it different from our old friend: *monitoring*. But I like examples, so let me try to do a better job to help you. In the next section, we will see a small application example, we will apply monitoring practices to keep our application up and running, and we will fail. Let's discuss why we failed and how observability principles can improve the situation in our sample scenario.

The need for observability in a distributed application environment

Let's suppose you want to create the definitive Hello World program so that no other developer will need to implement it again. But you want to add a minor new feature: the users can give their names, and the application should remember them, all based on modern REST APIs. So, you implement something as follows:

```
from flask import Flask, request
import os.path
app = Flask(__name__)

@app.route("/")
def hello_world():

    name = request.args.get('name')
    if name:
        with open("name.txt", "w") as text_file:
            text_file.write(name)

    name_file = None

    if os.path.exists("name.txt"):
        with open("name.txt") as text_file:
            name_file = text_file.read()

    if name_file:
        return {
            "msg" : f"Hello, {name_file}!"
        }
```

```
return {
    "msg": "Hello, World!"
}
```

In this small example, written in the Python (`https://www.python.org`) language and using the Flask (`https://flask.palletsprojects.com/en/2.0.x/`) web framework, we have an optional name query parameter, which, if we receive it, we store in a file. Anyway, we always read from the file, and if there's something in it, we return a friendly hello to our old, returning friend. Otherwise, we return an also friendly but generic `Hello, World!` message.

We can see an example of user interaction with our REST API here:

```
> curl http://127.0.0.1:5000/
{"msg":"Hello, World!"}
> curl http://127.0.0.1:5000/?name=User
{"msg":"Hello, User!"}
> curl http://127.0.0.1:5000/
{"msg":"Hello, User!"}
```

Our local tests show the implementation works as intended, so we are ready to shock and revolutionize the world. Our organization follows best practices, so we need to define and monitor key application metrics before we deploy our application in production. After years of deploying and monitoring applications, we, as software engineers, start to understand what can go wrong and what to keep an eye on. Usually, applications can be CPU-, memory-, or I/O-intensive. Given that our application writes and reads data to/from a file, we decided a key metric is **input/output operations per second** (**IOPS**). We add the necessary tools to monitor it and the CPU and memory just in case. We also create dashboards to have visual clues of our current state, and we implement alarms to notify us when we think we are reaching any system limits. This all looks good, so let's open the gates for our beloved users!

But after a few users start to use our application, reports of unexpected behaviors begin to pour into our issue system. Some users sent their names, but the application failed to store them. Or even worse, some users received the names of other users in a significant data privacy leak. Nobody wants to be in the news because of that.

What happened to our perfect, simple, little application? During the deployment, our operations teams used a typical deployment pattern to increase the application's scalability and availability, as shown in the following diagram:

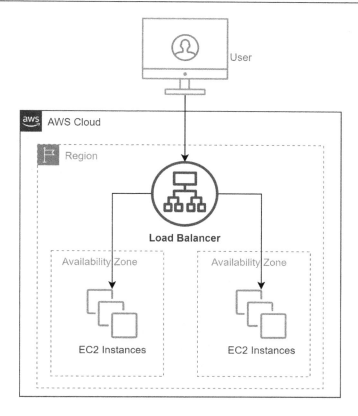

Figure 1.1 – Load balancing requests to multiple servers

Many of you may recognize the pattern described in this diagram. For many years, even on-premises operations teams have deployed multiple nodes of the same application behind a load balancer, which distributes incoming requests in a round-robin fashion to all of them. In this way, you can quickly scale the number of requests the application can handle by the number of nodes, and if a node fails, the load balancer automatically redirects new requests to the yet-available nodes.

We look at our configured metrics and we are clueless. None of our metrics helps us solve the problem. We deploy new metrics. We watch the problem occur a couple of times again (with new, angry users). And after debugging a bit, we find that the users who could not see their names after sending them received responses from servers that did not have their names stored in the local storage. Even worse, the users receiving other users' names received responses from servers that stored names from other users. What a mess!

Postmortem time: what happened, and how can we prevent it from happening again? When our operations team deployed our application behind a load balancer, we had multiple nodes, not just one anymore. New nodes could appear and disappear. This failure of nodes, combined with the fact we keep the application state in the individual nodes, causes the issue.

This is a simplistic, even silly, example of the jump in complexity from the local, single-user development environment to a distributed, multi-node, auto-scaling production environment. Our code is simple, and because of that, we thought nothing could go wrong. But there are many things outside our application code we don't understand entirely. Still, we take them for granted: the CPU run queue, the kernel multi-threading, the language virtual machine, the network stack, the load balancing strategy... and many more. They all contain the application state and the potential root cause for an issue.

This simple example shows that an initially observable application, deployed as a standalone process, as many monoliths are, no longer remains observable as soon as we use modern techniques such as multiple nodes and load balancing. Those components added more complexity and issues we didn't expect. As our user base grows and we split our monolithic application into many related services, what was the right observability tool before may not be the right tool now. This mismatch can catch us off guard because the complexity jump is exponential. As a terrifying example, see the following graph:

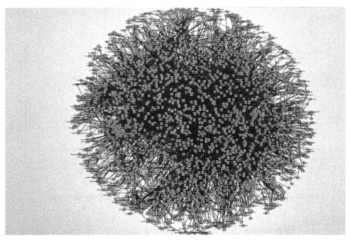

Figure 1.2 – Real-time graph of microservice dependencies at http://amazon.com in 2008

In our small example, we applied the usual techniques under the monitoring umbrella. The practice of monitoring is good enough for monolithic and small-scale distributed applications. And in this book, we will start with them, and we will progress, showing you the right tools for the job. With some experience, operations teams can reduce the potential failure space from hundreds, maybe thousands, of possibilities to a few. But we expect our businesses to grow, and with it, the supporting applications. The number of possible application and error states grows exponentially. As soon as our application reaches a specific size, at any moment, a call in the middle of the night can quickly become a sleepless night while we try to navigate the maze of our metrics to find the right set of inputs that have caused a new, unforeseen issue.

Modern applications have gotten good at accounting for failures that can be caught by tests and use established techniques such as autoscaling and failovers to make the application more resilient. As we catch up on known variables and take action to monitor them, the unknown unknowns are left. The issues we often see in modern applications are emergent failure modes, which happen when many unlikely events line up to degrade the performance of the system or even take it down. These scenarios are challenging to debug, which entails the need for observability.

If we want to understand any application state without deploying new code, we need to collect as much context as possible and store it all. We need mechanisms to query, slice, and summarize this data in new ways. Some of this complexity may not fit in our human brains anymore, so the support of machine learning tools is a must. Dashboards and alarms continue to be necessary for the well-known failure states, but to reach the next step, we need new tools in our tool belt.

So far, we have seen what observability is and how it evolved from more traditional monitoring practices to support more complex systems. We saw the need to collect more data and answer questions we didn't know we should answer. In the next section, we will see the basic observability components and how they relate.

Building blocks of observability

There are three fundamental building blocks of observability: **metrics**, **logs**, and **traces**. Each plays a specific role in infrastructure and application monitoring, so you need to understand what they bring to the table. They can be called the golden triangle of observability, as depicted in the following figure:

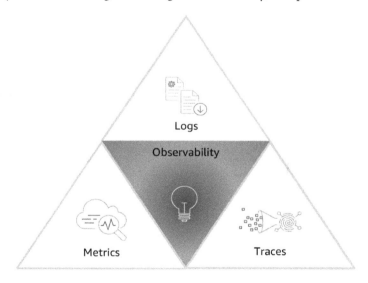

Figure 1.3 – Observability building blocks

Now, let's try to understand the three building blocks.

Metrics

Metrics are measurements of resource usage or behavior of your system over time. They might be low-level measurements of system resources, such as the CPU, memory utilization, disk space, or the number of I/O operations per second. They could also be high-level indicators, such as how the user interacts with your system – for example, how many customer requests, the number of clicks on a web page, the number of products added to the shopping cart, and so on.

Everything from the operating system to the application can generate metrics, and a metric is composed of a name, a timestamp, a field representing some value, and potentially a unit. Metrics are a prominent place to start observability.

For many years, metrics have been the starting point to measure a system's health, representing the data on which monitoring systems are built to give a holistic view of your environment, automate responses to events, and alert humans when something needs their attention. In the following figure, you can see a simple example of a CPU utilization metric:

Figure 1.4 – A CloudWatch metric

When a solution expands to hundreds or thousands of microservices, the risk of false positives and false negatives increases, causing *alarm fatigue*. The root cause of this alarm fatigue is twofold.

First, we are keeping old habits from the monolithic times, when we had a single system to care for, and operations engineers did their best to keep it up all the time. The objective was to avoid failures entirely. We collect metrics and establish healthy/unhealthy thresholds for many of them. And on every unexpected outage, a postmortem evaluation of the causes will point out which metrics/alarms were missing in a rinse-and-repeat fashion.

Second, for any highly distributed and scalable system:

Everything Fails All the Time

– Werner Vogels, AWS CTO

The mechanisms and controls we use on monolithic or small-scale applications are not the right choices on higher scales because failures are expected. The question now is whether the issues are or aren't affecting our end customer experience or business processes and not whether a single service is up and running.

That's why we see a change in the metrics being used to notify operation engineers that something is wrong, from low-level metrics (CPU, memory utilization, and disk space), to aggregated metrics related to the user experience and business outcomes (web page time to interact, error rate, and conversion rate).

We will look at different tools for collecting and analyzing metrics in this book.

Logs

Event logs, or simply logs, are probably the oldest and simplest way to expose the internal state of an application. A log is a file or collection of files that contains the history of all the clues the application developers decided to leave to someone else. In case of issues, they could read it and understand the application's steps until the failure. See the following example:

```python
import logging

logging.basicConfig(format='%(asctime)s - %(name)s - %(levelname)s -
%(message)s',filename='example.log', encoding='utf-8', level=logging.
DEBUG)

logging.info('Store input numbers')
num1 = input('Enter first number: ')
num2 = input('Enter second number: ')

logging.debug('First number entered: %s', num1)
logging.debug('Second number entered: %s', num2)

logging.info('Add two numbers')
sum = float(num1) + float(num2)

logging.debug('Sum of the two numbers: %d', sum)

logging.info('Displaying the sum')
msg = 'The sum of {0} and {1} is {2}'.num1

logging.debug('Rendered message: %s', msg)

print(msg)
```

After executing this program, the resulting log file looks like this:

```
2022-03-20 17:21:40,886 - root - INFO - Store input numbers
2022-03-20 17:21:43,758 - root - DEBUG - First number entered: 1
2022-03-20 17:21:43,758 - root - DEBUG - Second number entered: 2
```

```
2022-03-20 17:21:43,758 - root - INFO - Add two numbers
2022-03-20 17:21:43,758 - root - DEBUG - Sum of the two numbers: 3
2022-03-20 17:21:43,759 - root - INFO - Displaying the sum
2022-03-20 17:21:43,759 - root - DEBUG - Rendered message: The sum of
1 and 2 is 3.0
```

As we can see, logs initially used an unstructured format because they were meant to be readable by humans. And initially, they were written on the local disk of the machine running the application.

We can quickly see how the jump from a single, monolithic application to a distributed system, or even a collection of distributed systems, can affect how we use or process log files. I used SSH to connect to a machine and check the server logs. Today, we have applications dynamically coming online because of a scale-out event or terminated because they failed a health check. We can't store the logs on the local machine anymore; otherwise, they would be lost sooner or later. We need a place to send them and keep them.

Another substantial improvement is to make them machine-readable. In our investigation to understand what happened with our application, we need to collect as much context as possible and make it available in a system where we can query, slice, and aggregate it in new and unexpected ways. We can't simply connect to a single machine and read a single log file anymore. Instead, we need to understand the execution steps of potentially hundreds of servers.

Check out the same log example here, but now using structured logs:

```
import logging
import structlog

logging.basicConfig(format='%(message)s',filename='example.log',
encoding='utf-8', level=logging.DEBUG)



structlog.configure(
    processors=[
        structlog.stdlib.filter_by_level,
        structlog.stdlib.add_logger_name,
        structlog.stdlib.add_log_level,
        structlog.stdlib.PositionalArgumentsFormatter(),
        structlog.processors.TimeStamper(fmt="iso"),
        structlog.processors.StackInfoRenderer(),
        structlog.processors.format_exc_info,
```

```
            structlog.processors.UnicodeDecoder(),
            structlog.processors.JSONRenderer()
        ],
        wrapper_class=structlog.stdlib.BoundLogger,
        logger_factory=structlog.stdlib.LoggerFactory(),
        cache_logger_on_first_use=True,
    )

    log = structlog.get_logger()

    num1 = input('Enter first number: ')
    num2 = input('Enter second number: ')

    log = log.bind(num1=num1)
    log = log.bind(num2=num2)

    sum = float(num1) + float(num2)

    log = log.bind(sum=sum)

    msg = 'The sum of {0} and {1} is {2}'.num1

    log.debug('Rendered message', msg=msg)

    print(msg)
```

The resulting logs are as follows:

```
{"num1": "1", "num2": "2", "sum": 3.0, "msg": "The sum of 1 and 2 is
3.0", "event": "Rendered message", "logger": "__main__", "level":
"debug", "timestamp": "2022-03-22T07:43:11.694537Z"}
```

As you can see, the structured logs contain key-value pairs with the relevant data. To make it easier for machine consumption, we can use a semi-structured format such as JSON. And also, instead of multiple lines that tell us what happened, the logs are structured to represent a unit of work, so you can aggregate more data in a single context.

We can also see a profound shift in how we debug issues in our production system. Initially, it was reactive: we collected metrics and defined healthy thresholds for some of them. As soon as one of those thresholds was crossed, the monitoring system would send an alert via an SMS or pager to the engineer of that shift to go and investigate further. So, the engineer would check the metric that raised the alarm, as well as all the other metrics, create a hypothesis of what could be the problem, and only then use logs to prove or refute the hypothesis. So, in this case, if the metrics show that the system is malfunctioning, logs show why it is malfunctioning.

With the explosion in the number of servers and services a team must handle, we see a shift toward the proactive use of observability tools, where the engineers don't just use them when there's an issue but all the time. When doing a new release or when activating a new feature using a feature flag, we need to check not only the 99.9% satisfied end users but the other 0.1%. And to collect all the necessary data, structured logs are a fundamental tool, and the path for the investigation starts with them instead. We see engineers using analytic tools to make complex queries against the data generated by structured logs first and checking some other auxiliary data second to confirm the issue.

Throughout this book, we will look at tools for collecting and analyzing data for systems of any size so that you can decide which one fits your case best.

Traces

Last but not least in the observability triangle is application trace data. Trace and logs are sometimes difficult to differentiate, but the main difference is in nature and intent. While logs are discrete events that localize issues and errors, traces are continuous. They understand the application flow while processing a single task/event or request.

Traces are more verbose. They include information such as which methods/functions were called, with which parameters, how long a method took to return a value, the call order, information about the thread context, and more. Because of that, tracing is often implemented using instrumentation, utilizing the programing language runtime reflection mechanism to introduce hooks and automatically collect this information.

Traces add the critical visibility of the application end to end. Traces typically focus on the application layer and provide limited visibility into the underlying infrastructure's health. So, metrics and traces complement each other to give you complete visibility into the end-to-end application environment.

But more interesting than just tracing is distributed tracing. Distributed tracing is the capability of a tracing solution to track and observe service requests as they flow through multiple systems. The tracing process starts at one of the application's entry points (for example, a user request on the web application), which generates a unique identifier. This identifier is carried along while traversing the local method calls, using techniques such as attaching it to the thread context. When a request is made to an external system, the request carries this unique ID as part of the request metadata (for example, part of the HTTP headers in an HTTP-based REST call). The recipient system unpacks the ID and carries it along similarly.

In this way, when we aggregate the data generated by different systems, we can see the request flow from application to application, the time it took to process locally, or how much time it took to call external data sources.

A distributed tracing map will look like this:

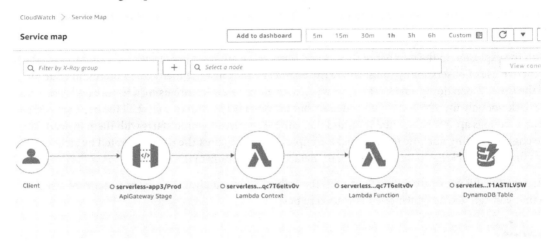

Figure 1.5 – A service map on X-Ray

Later in this book, you will learn how to add distributed tracing capabilities to your application.

What is the relationship between the three pillars?

When a user request occurs, and a delay has occurred for the request, metrics provide the data to demonstrate data quantitatively, such as the number of requests. At the same time, it can also record the number of services the request passes through when it occurs using the trace data. If you would like to record detailed information when an error occurs, you can do so using the log data.

As we can see, it is easy for us to see metrics, tracing, and logging and the connection between these three kinds of data.

Will I need to adapt all three pillars?

The simpler your environment and the more tolerant you are of performance degradation and outages, the fewer tools are required to keep it running and simple metrics will be able to work fine for you.

If the environment becomes complex and has to be up and running all the time or needs to be fixed as quickly as possible, you will require a mix of tools to understand where it is broken. Metrics and logs will support you with this requirement.

If your environment consists of a lot of microservices, then adding traces will save you effort when it comes to troubleshooting problems across the environment.

In this section, we saw the basic observability building blocks, a few of their historical origins, and how they evolved. We also briefly saw the need to connect all three to create a holistic view. In the next section, we will see why we should invest in improving our system's observability.

Benefits of observability

Adopting observability to analyze system performance used to be the job of sysadmins and ops teams, who cared most about the **mean time to detect** (**MTTD**) and **mean time to resolve** (**MTTR**). Today, more job roles than ever need to use observability data. With the rise of DevOps, CI/CD, and Agile methods, developers are often directly responsible for the performance of their apps in production. SREs and DevOps staff care about meeting **service-level indicators** (**SLIs**) and **service-level objectives** (**SLOs**). Information about systems and workloads is also used by business leaders in making decisions about capacity, spending, risk, and end user experience. Each stakeholder in an organization has different needs for what is monitored and how the resulting data is analyzed, reported, and displayed. Let's try to understand the benefits of observability in the real world for different personas.

Understanding application health and performance to improve customer experience

The main observability goal is to know what is going on anywhere in your system to ensure the best possible experience for your end users. You want to detect problems quickly, investigate them efficiently, and remediate them as soon as possible to minimize downtime and other disruptions to your customers.

Improving developer productivity

Traditional debugging by analyzing logs or instrumenting breakpoints into code is tedious, repetitive, and time-consuming. It doesn't scale well for production applications or those built using microservice or serverless architectures. To analyze performance across distributed applications, developers need to correlate metrics and traces to identify user impact from any source and to find broken or expensive code paths as quickly as possible. And they need to do all this without having to re-instrument their code when they want to add new observability tools to their kit.

Getting more insight with visualizations

Observability, especially at scale, can generate huge volumes of data that become difficult for humans to parse. Visualization tools help humans make sense of data by correlating observability data into intuitive graphic displays. However, having a bunch of graphs, charts, and more scattered across multiple tools and displays becomes a problem. It's essential to centralize visual data into a single dashboard, giving you a unified view of your system's critical information and performance.

Digital eperience monitoring

Digital Experience Monitoring (DEM) correlates infrastructure and operations metrics with business outcomes by focusing on the end user experience. It seeks to reduce the MTTR in the event of client-side performance issues by monitoring the client-side performance on web and mobile applications in real time. Resolution is assisted by the relevant debugging data such as error messages, stack traces, and user sessions to fix performance issues such as JavaScript errors, crashes, and latencies.

Controlling cost and planning capacity

A key advantage to operating in the cloud is that you can scale quickly to meet demand during peak load times. However, unplanned and uncontrolled growth can result in unexpected costs. Observability can help you find performance improvements, such as reducing the CPU footprint. Across a fleet of thousands or hundreds of thousands of instances, a slight percentage performance improvement in how much CPU an application uses can save millions of dollars. Similarly, by using observability to understand and predict your future capacity needs, you can take advantage of the cost savings available from reserve and spot pricing and avoid cost surprises.

Summary

In this chapter, we saw what observability means in the context of software applications and what makes it different from monitoring. We saw increased observability complexity, from more straightforward, monolithic applications to more complex, distributed applications. We discussed the observability building blocks and how they evolved. Finally, we saw some critical use cases where observability principles bring attractive business advantages.

Now, you can more easily discuss the differences between monitoring and observability, and when to adopt one of them. You can also advocate for observability principles in your organization, clearly understanding the requirements and advantages.

In the next chapter, we will map the different AWS services we can use to make applications observable.

Questions

Answer the following questions to test your knowledge of this chapter:

1. Which characteristics must a solution have to make it observable?

2. What's the difference between monitoring and observability?

3. Why is observability important for complex, distributed applications?

4. What is alarm fatigue and what are its root causes?

5. What's the difference between unstructured and structured logs? What makes structured logs better for more complex use cases?

6. What's the difference between tracing and distributed tracing?

7. Can you cite three use cases for observability?

2

Overview of the Observability Landscape on AWS

We spent the previous chapter understanding the requirement of observability in the modern application landscape and the building blocks of observability. In this chapter, we will go through various services that can be used to observe and manage your application landscape available on **Amazon Web Services (AWS)**.

AWS offers several services that can be used to observe and manage your overall application landscape:

- **Infrastructure monitoring**: Amazon CloudWatch, or simply CloudWatch, metrics and logs will support infrastructure monitoring for components such as VMs, containers, **operating systems (OSes)**, and applications.

- **Distributed tracing**: AWS X-Ray provides support in distributed tracing and profiling for your application. AWS X-Ray support distributed tracing for applications written in the .NET, Java, Node.js, Python, Ruby, and Go programming languages.

- **AWS services vended monitoring**: AWS services natively send metrics and logs to Amazon CloudWatch. These metrics and logs are configurable or can be use without much configuration to manage your infrastructure.

- **Digital experience monitoring**: Digital experience monitoring adds the outside, end user perspective to ensure applications and services are available and functional across all user interfaces or devices. Digital experience monitoring tools combine application performance data, real user behavior, and synthetic monitoring to help you gain deeper experience insights, such as via session replays, understand the impact of changes, and identify bottlenecks. AWS provides three services to help you understand your application's digital experience, namely CloudWatch **Real User Monitoring (RUM)**, CloudWatch Synthetics canaries, and CloudWatch Evidently.

In this chapter, we are going to cover the following topics:

- Overview of observability tools in AWS
- Overview of native observability services in AWS
- Overview of AWS-managed open source observability services in AWS
- Adoption of observability services in AWS

Technical requirements

To engage in the technical section of this chapter, you need to have an AWS account. You can quickly sign up for the AWS free tier if you do not have one.

Check out the following link to learn how to sign up for an AWS account: `https://aws.amazon.com/premiumsupport/knowledge-center/createand-activate-aws-account/`.

Overview of observability tools in AWS

We can divide the monitoring/observability tools available in AWS into two categories:

- AWS-native services
- Open source managed services

The following figures show a high-level representation of the AWS services available to you:

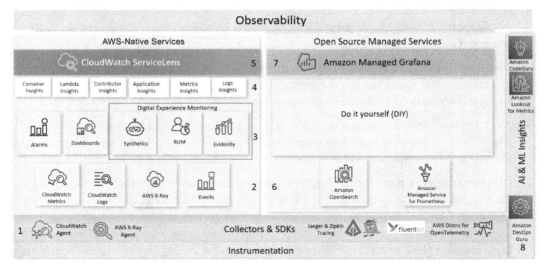

Figure 2.1 – Overview of observability services on AWS

Observing the layers, you can see that similar functionality is provided natively or using the open source managed services available on the AWS platform. We will discuss each layer and the different options available.

Let's try to navigate each layer to understand the AWS services available to you.

The first layer, which is the instrumentation layer, provides the tools required to collect metrics, logs, and traces from your applications. AWS provides **CloudWatch** and **X-Ray** agents to install and collect metrics, logs, and trace data from both operating systems and applications, instrument them appropriately, and send the data to AWS-native services. Similarly, the open source **AWS Distro for Open Telemetry** (**ADOT**) supports collecting traces and application metrics, and **Fluent Bit** supports collecting metrics and logs for container services. We can also use Jaeger or Zipkin tracing to collect distributed application tracing based on the requirements.

In the second layer, metrics (**M**), events (**E**), logs (**L**), and traces (**T**) are stored and processed and can be referred to as **MELT** for easy recollection. Once the raw data is collected, the data metrics, logs, and traces are processed and aggregated in the **CloudWatch Metrics**, **CloudWatch Logs**, and **AWS X-Ray** services, respectively. Additionally, AWS events communicate changes in the state of the services as events through **EventBridge**, which supports forwarding them to third-party systems for additional engagement.

The third layer adds visibility to the data generated and received by **CloudWatch Metrics** and **Logs** with **CloudWatch Dashboards**. **CloudWatch Alarms** provides a mechanism to notify you and activate any actions required to be taken on the data. The third layer also consists of services related to measuring the digital experience of the end users. **CloudWatch Synthetics** supports mimicking the end user behavior and understanding the issue before impacting the real users. **RUM** captures the real user data and understands any issues caused to the users. **CloudWatch Evidently** supports carrying out *blue/green deployments* and understanding the performance impact of the new code.

The fourth layer consists of the **Insight** services to help you build intelligence on top of the data and support gathering additional information about managed compute services. **Container Insights** provides a method to collect, aggregate metrics and logs from container services such as **Amazon Elastic Container Service** (**ECS**) and **Amazon Elastic Kubernetes Service** (**EKS**), Lambda Insights for serverless services such as **AWS Lambda** functions and **Application Insights** to automate the onboarding of applications, eliminating the overhead of manual setup and also configuring the data collection. **Metric Insights** provides intelligence to query metrics using SQL and **Logs Insights** provides interactive search features and allows you to analyze the log data in real time.

The fifth layer acts as an umbrella in providing a bird's-eye view of **Application Health** using **CloudWatch ServiceLens** and provides a navigation experience to help you drill down and understand root cause issues.

The sixth layer consists of the open source managed services, powered by **Open Search** and **Amazon Managed Service for Prometheus** (**AMP**), and provides services to capture metrics, logs, and application traces using **Open Source Managed** services.

The seventh layer, which is powered by **AMG**, provides *dashboarding* services on the data gathered about metrics, logs, and traces from different observability services and provides a top-down view of the application, as well as business health metrics.

The eighth layer provides machine learning services such as **Amazon DevOps Guru** to help you identify application health issues. **Amazon CodeGuru** delivers a deep dive into code-level issues and provides a way to detect code that has the longest execution times. **Amazon Lookout for Metrics** provides anomaly detection in metrics gathered from different data sources.

In this section, we learned about the services offered by AWS for achieving observability and how they can be segregated into different layers. We also learned about the relationship between them in both AWS cloud-native services and AWS open source managed observability services. In the next section, we will understand what the purpose of each service is and the functionality it provides for achieving observability for your application workloads.

Overview of native observability services in AWS

Let's start by understanding the AWS-native observability services in AWS, which are powered by **Amazon CloudWatch**. Amazon CloudWatch is a cloud-native monitoring and observability solution that provides comprehensive functionality in terms of metrics, logs, traces, dashboarding, alarms, real user monitoring, synthetic monitoring, and observability tools for over 200 AWS services, including serverless technologies such as AWS Lambda and AWS Fargate. Amazon CloudWatch is used to monitor more than 6,000 trillion metric observations, ingests more than 3.5 Exabytes of logs, and when combined with EventBridge, triggers more than 32 trillion events per month.

CloudWatch delivers actionable insights by collecting logs, metrics, and events, providing a single pane of glass for all applications and services on AWS and on-premises. You can configure alarms and be notified when metrics are outside the desired range, visualize logs and metrics side by side, trigger automated actions, troubleshoot issues, and discover insights to help you fix your applications before your users notice. In this section, we will understand the components of Amazon CloudWatch and look into how they can be leveraged to fulfill the golden triangle of observability discussed in The previous chapter. Let's start with layer 2 in the AWS-native observability services and the first component of the golden triangle, known as *metrics*.

Amazon CloudWatch Metrics

Amazon CloudWatch Metrics provides data points about the performance of your systems. By default, many AWS services provide built-in metrics (for example, **Amazon Elastic Compute Cloud (EC2)** CPU utilization or DynamoDB consumed read capacity) with a frequency of 5-minute intervals. If you are looking to gather at a frequency of fewer than 5 minutes for high-critical workloads running on AWS, you could enable detailed monitoring for some resources, such as your Amazon EC2 instances, which publish the metrics at a frequency of 1 minute. Amazon CloudWatch will load all the metrics, aggregate them, and index them for you to search, graph, and configure alarms.

Metric data is kept for up to 15 months and is aggregated in rolling windows as follows:

- CloudWatch keeps metrics with a period of fewer than 60 seconds for 3 hours. CloudWatch aggregates them into 1-minute groups after 3 hours.

- Metrics with a period of 1 minute are available for 15 days. After that, CloudWatch aggregates them into 5-minute groups.

- Metrics with a period of 5 minutes are available for 63 days. After that, CloudWatch aggregates them into 1-hour groups.

- Metrics with a period of 1 hour are available for 15 months.

You can visualize the **Metrics** section in AWS CloudWatch in the following screenshot:

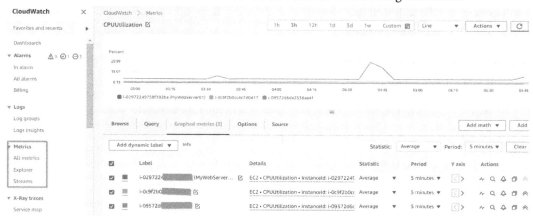

Figure 2.2 – CloudWatch Metrics

CloudWatch allows you to publish custom metrics, which will help you to publish application-related metrics such as the number of processors used by a Java application, business application metrics, such as the number of orders, and more. We will discuss the custom metric functionality in *Chapter 7, Observability for Serverless Applications on AWS* **Metrics Explorer** is a tool that allows you to explore the data in CloudWatch Metrics. It is easy to select the metric you are interested in, for example, by viewing it by a specific Lambda tag and then splitting it by errors in each group. Its collection of features immediately gives a bird's-eye view of all the crawlers and errors in each group. You can visualize the **Metrics Explorer** view of EC2 instances by type in the following screenshot:

Figure 2.3 – EC2 Instances by type view in CloudWatch Metrics Explorer

We will understand the technical components of metrics and learn how to gather default and custom metrics from EC2 in the next chapter.

Amazon CloudWatch Logs

CloudWatch Logs is a highly scalable service that can be used to ingest, store, and access logs from your applications and AWS services. CloudWatch Logs provides useful functionalities such as the following:

- Real-time access to your logs
- Aggregation of logs from different hosts into a central place
- Allowing search functionality on logs to identify root cause issues quickly
- Long-term retention of logs for investigation
- Easy integration with storage services such as S3

CloudWatch Logs and **Log groups** can be seen in the following screenshot:

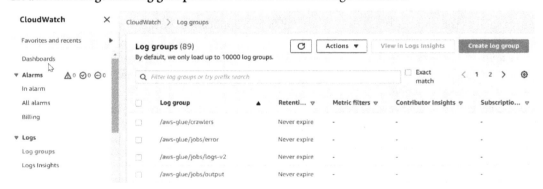

Figure 2.4 – CloudWatch Log groups

We will dive deep into CloudWatch logs and gathering logs from various services in *Chapter 3, Gathering Operational Data and Alerting Using Amazon CloudWatch, Chapter 6, Observability for Containerized Applications on AWS* and *Chapter 7, Observability for Serverless Applications on AWS,* of this book.

AWS X-Ray

AWS X-Ray helps developers analyze and debug distributed applications built using microservices or serverless architectures. Tracing can help developers and architects find latency in a chain of calls to fulfill a request. A trace records the execution path across multiple services/applications, together with metrics such as the execution time. AWS X-Ray provides a daemon that runs on-premises on Linux, macOS, and Windows, in a Docker container, or on AWS compute resources to forward data from your instrumented application to the X-Ray service.

Traces generated from a sample application with high latency can be used to identify bottlenecks in your application, as shown in the following screenshot:

Figure 2.5 – CloudWatch X-Ray traces with high latency and the Sample queries view

We will look into how to instrument applications using AWS X-Ray in *Chapter 4, Implementing Distributed Tracing Using AWS X-Ray, Chapter 6, Observability for Containerized Applications on AWS,* and *Chapter 7, Observability for Serverless Applications on AWS,* of this book.

Amazon EventBridge

Amazon EventBridge delivers a near real-time stream of events that describe changes in your AWS infrastructure. When resources change state, they automatically emit events in the stream. Then, you can write declarative rules to match events of interest and route them to targets to take action.

From an observability practice standpoint, Amazon EventBridge will help you do the following:

- Reduce polling and associated cost and complexity
- Create a uniform interface for events across AWS services
- Provide near real-time notifications of resource changes
- Provide an end-to-end solution, from event detection to automated remediation
- Provide a scheduled execution of tasks in a fully managed service
- Provide a unified experience for event notifications and actions

The **Rules** view of Amazon EventBridge can be seen in the following screenshot:

Figure 2.6 – CloudWatch EventBridge

Additionally, with EventBridge, you can build event-driven computing architectures for business applications using serverless applications such as AWS Lambda.

CloudWatch Alarms

Amazon CloudWatch Alarms allows you to define thresholds around CloudWatch Metrics and receive notifications when the metrics fall outside a certain range. Each metric can trigger multiple alarms, and each alarm can have many actions associated with it.

A CloudWatch Alarm is always in one of three possible states: **OK**, **ALARM**, or **INSUFFICIENT_DATA**. You can see the different states of the alarms in the CloudWatch **Alarms** console:

Figure 2.7 – State of CloudWatch Alarms

CloudWatch Dashboards

CloudWatch Dashboards is a customizable collection of visual displays of your metrics that you or your team deem relevant to see side by side. CloudWatch Dashboards helps you visualize the system performance and interpret your AWS services and workloads metrics. You can use cross-account dashboards to share multiple AWS regions and account data. This cross-account/region sharing is beneficial when working with cloud applications and workflows that involve numerous accounts and metrics you or your team need to monitor/view in a centralized manner.

An automated *dashboard* generated for an Amazon **EKS Cluster** produced from a sample application is shown in the following screenshot:

Figure 2.8 – Automated EKS performance dashboard

We will learn how to set up and configure CloudWatch Events, CloudWatch Alarms, and CloudWatch Dashboards in the next chapter of this book.

CloudWatch Synthetics

Amazon CloudWatch Synthetics is part of the digital experience monitoring set of services designed to help you understand your customer experience, even when you don't have any customer traffic on your application. As a result, you can identify and fix the issues before they impact your customer experience.

CloudWatch Synthetics lets you monitor web applications using modular, lightweight canary tests and allows you to do the following:

- **Website and API endpoint monitoring**: Amazon CloudWatch Synthetics monitors your website and API endpoint for latency, availability, and performance monitoring

- **Outside-in monitoring**: Client-side to transaction monitoring

- **Continuous monitoring**: Acts like a user 24/7 and continuously runs health checks

- **Visual monitoring**: Visual monitoring will let you detect defects in your application by comparing the screenshots with an established baseline

CloudWatch Synthetics can test your endpoints every minute, 24/7, and notify you when something deviates from the expected behavior. You can customize those tests to check for a list of metrics, such as the following:

- Availability

- Latency

- Transactions

- Broken or dead links

- Step-by-step task completions

- Page load errors

- Load latencies for UI assets

- Complex wizard flows

- Checkout flows

A dashboard view of website availability as measured using CloudWatch Synthetic canaries can be seen in the following screenshot:

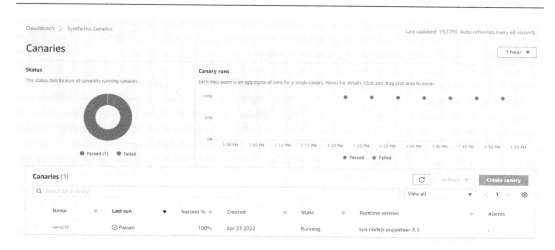

Figure 2.9 – Website availability using CloudWatch Synthetics canaries

Real User Monitoring (RUM)

CloudWatch RUM is another service that's part of digital experience monitoring. It enables your applications to send telemetry data, which will enable application developers and DevOps engineers to provide quicker resolution and optimize the end user experience. CloudWatch RUM helps in identifying client-side performance issues, debugging client-side errors, and collecting client-side metrics on the web application performance in real time.

CloudWatch RUM detects anomalies in performance and aggregates debugging data such as stack traces, error messages, and user sessions, helping to diagnose performance issues such as JavaScript crashes, latencies, and errors. Customers can also visualize the impact of the problem on the end customer population, including the browsers affected, the number of users, and their geolocations. RUM helps developers prioritize features and bug fixes by aggregating user behaviors and click-stream paths, providing information such as bounced user sessions.

You can see the data generated by CloudWatch RUM in the following screenshot providing information about **Page loads and load time**, **Apdex by country**, and more:

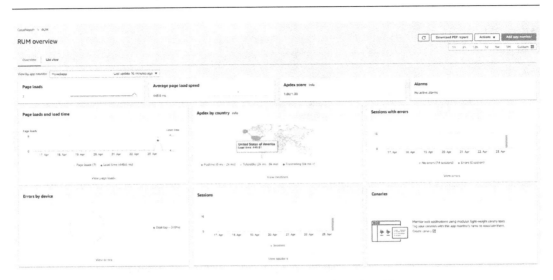

Figure 2.10 – RUM metrics in CloudWatch RUM

CloudWatch Evidently

CloudWatch Evidently is another service that's part of digital experience monitoring. It helps application developers safely validate new features via A/B testing and experimentation to do safe launches across the entire application stack, covering user-facing and backend features. Application developers can use Evidently to run experiments on new application features to identify unintended consequences, thus reducing risks. When releasing new features, developers can publish the features to a smaller population of users, monitor vital metrics such as conversions and page load times, and safely publish the feature to a broader audience if the team is satisfied with the results. It also allows developers to collect user data, experiment with different designs, and release the best one to production.

CloudWatch Evidently helps you remove the guesswork when deciding which features are the best for your business, whether it's a new user experience, machine learning recommendations model, or server-side implementation. Experimental results are described clearly, so you don't need advanced statistical knowledge to interpret them. While an experiment is running, anytime p-value and confidence intervals allow you to see when there is statistical significance so that you can end the experiment. It also has a granular scheduling capability to dial up traffic in a controlled manner so that you can launch your new application changes with confidence while monitoring key business and performance metrics for the new feature. You can define alarms to roll back to a safe state if there are issues with the launch. CloudWatch Evidently also integrates with CloudWatch RUM, adding client-side application monitoring so that you can use RUM metrics directly in Evidently.

We will go through the configuration, management, and usage of Digital Experience Monitoring in *Chapter 8, End User Experience Monitoring on AWS,* of this book.

CloudWatch Container Insights

CloudWatch Container Insights is a fully managed service that summarizes and traces correlations between metrics and logs from your containerized applications. Container Insights collects data as performance log events in a structured JSON while leveraging **Embedded Metric Format (EMF)** in CloudWatch. CloudWatch EMF automatically extracts custom metrics from the log data and uses them to provide visualizations and create alarms. CloudWatch Container Insights is available for AWS services such as **EKS, ECS**, and **AWS Fargate**. CloudWatch Container Insights also provides diagnostic information, such as the number of times a container restarted (for example, the infamous **CrashLoopBackOff** in Kubernetes), to help you isolate issues and resolve them quickly.

The following screenshot shows an automatic dashboard generated by CloudWatch Container Insights:

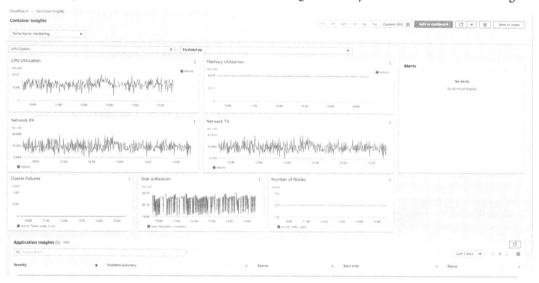

Figure 2.11 – Performance view of an EKS cluster in CloudWatch Container Insights

CloudWatch Lambda Insights

Similar to CloudWatch Container Insights, **CloudWatch Lambda Insights** also summarizes and correlates metrics and logs for Lambda functions. It processes metrics for the CPU time, memory, disk, and network. It helps you find the root cause of cold starts and worker shutdowns to help you fix them quickly.

Lambda Insights has use cases other than fundamental performance understanding, including identifying functions that are having issues with a memory leak, identifying functions that have high costs compared to the rest of the Lambda functions in the account, and understanding latency drivers in specific functions.

The following screenshot shows the multifunction comparison view provided by CloudWatch Lambda Insights:

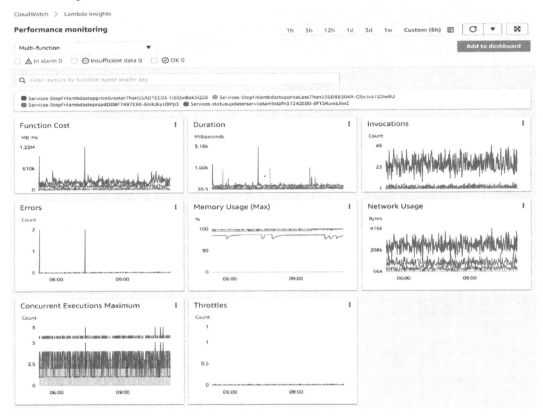

Figure 2.12 – A multifunction view of a Lambda function in CloudWatch Lambda Insights

CloudWatch Contributor Insights

CloudWatch Contributor Insights allows you to set up a real-time analysis of time-series data quickly and easily in Amazon CloudWatch Logs, to understand which process impacts the system or application performance the most. Today, you can use contributor insights for two different use cases:

- **Contributor Insights for CloudWatch Logs**: You can evaluate patterns in structured log events in real time
- **Contributor Insights for DynamoDB**: You can view the most accessed and throttled items

When analyzing CloudWatch Logs, you can see statistical summaries as the top-N contributors and the total number of unique contributors. Those summaries help you understand what impacts the system's performance. For example, you can identify the users who consume your network bandwidth the most, find bad hosts, or find the URLs that generate the most errors.

The following screenshot shows the top IP traffic patterns analyzed from the **VPCFlowLogs** area in Contributor Insights:

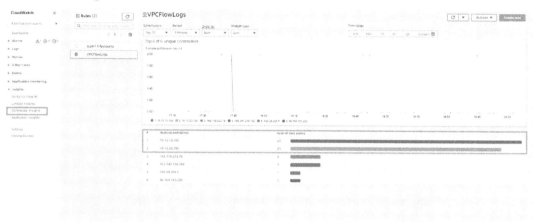

Figure 2.13 – Top IP traffic patterns in VPCFlowLogs in Contributor Insights

When analyzing DynamoDB, you can understand what the hotkeys are, identify table access patterns over time, and identify the most frequent keys. As the contributor insights are asynchronous, there is no performance impact on the DynamoDB database.

CloudWatch Application Insights

Generally, when setting up observability using CloudWatch, we need to take time to identify and set up monitoring, detect and correlate anomalies, and diagnose and troubleshoot issues. Typically, 60 to 70% of our time is spent on the setup. CloudWatch Application Insights provides an easy way to automate the setup of monitors for the application resources via application discovery. It performs intelligent problem detection by correlating observations using algorithms and built-in rules. It provides visualization using CloudWatch automatic dashboards with additional insights to pinpoint the potential root cause.

General use cases of CloudWatch Application Insights include, but are not limited to, the following:

- Monitoring for .NET applications and MS SQL Server running on an EC2 instance
- Monitoring Microsoft Active Directory and SharePoint
- Observability for SAP HANA running on an EC2 instance

A typical view of the Application Insights dashboard that's generated for the monitored components can be seen in the following screenshot:

Figure 2.14 – CloudWatch Application Insights summary view

CloudWatch Metric Insights

CloudWatch Metrics Insights is a flexible, SQL-based query engine that you can use to analyze your metrics at scale. Based on your use cases and business requirements, you can group and aggregate your custom and AWS built-in metrics in real time. A single query can process up to 10,000 metrics to identify trends and patterns.

When using the CloudWatch console, you can choose one of the pre-built sample queries or create one yourself. You can build queries in two different ways:

- **Builder view**: This lets you browse your existing metrics and dimensions to build a query
- **Query editor view**: This lets you write SQL queries manually

A query showcasing the average CPU utilization of instances for EC2 instances grouped by **InstanceID** is shown in the following screenshot:

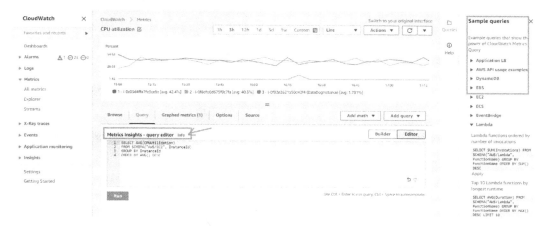

Figure 2.15 – Average CPU utilization of EC2 instances by InstanceID
in the CloudWatch Metric Insights query editor view

CloudWatch Logs Insights

CloudWatch Logs Insights is a fully managed, highly scalable, cost-efficient, and interactive log analytics solution for CloudWatch Logs. It offers log analytics capabilities, such as support for aggregations, regular expressions, and time-series visualizations. You can execute ad hoc analytics queries to help you identify the root cause of operational issues.

CloudWatch Logs Insights implements a purpose-built query language with the key benefits of fast execution, query auto-completion, and log field discovery.

The following screenshot shows a query showcasing the source and destination traffic from the VPC Flow logs using CloudWatch Logs Insights:

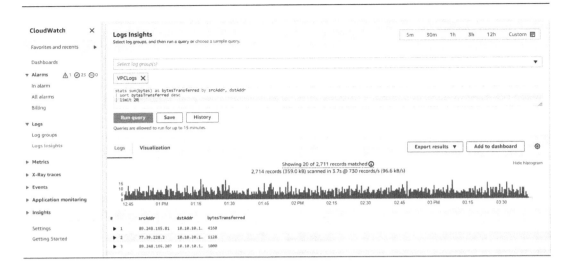

Figure 2.16 – Querying VPC Flow logs from CloudWatch Logs Insights

CloudWatch ServiceLens

CloudWatch ServiceLens gives you unified access to metrics, logs, traces, and canaries, enabling performance monitoring from end user interaction to infrastructure layer insights. ServiceLens integrates CloudWatch and AWS X-Ray to provide a holistic view of your application to help you pinpoint performance bottlenecks and identify impacted users.

A ServiceLens map will provide an overview of the application's health, with the ability to drill down into specific components to view metrics, logs, and traces and view end-to-end customer transactions and impact.

With the introduction of the **ServiceLens** map, the customer's observability journey will change from tracking individual components to changing a single view of the entire application into a single view. From the ServiceLens map, you can do the following:

- **Observe**: Get an overview of all resources and their health in a single view. Visualize dependencies and contextual linking of resources. Reduce the MTTR by correlating latencies and requests at each node and edge.

- **Inspect**: Know when your end user experience has degraded. Tie end user experience back to infrastructure-level insights.

- **Isolate**: Identify performance bottlenecks and issues. View the top resources with issues by sorting and filtering.

- **Diagnose**: Dive deep into node-level insights using pre-canned dashboards. Investigate correlated metrics, alarms, traces, and logs.

An application traceability view from the CloudWatch ServiceLens map is shown in the following screenshot:

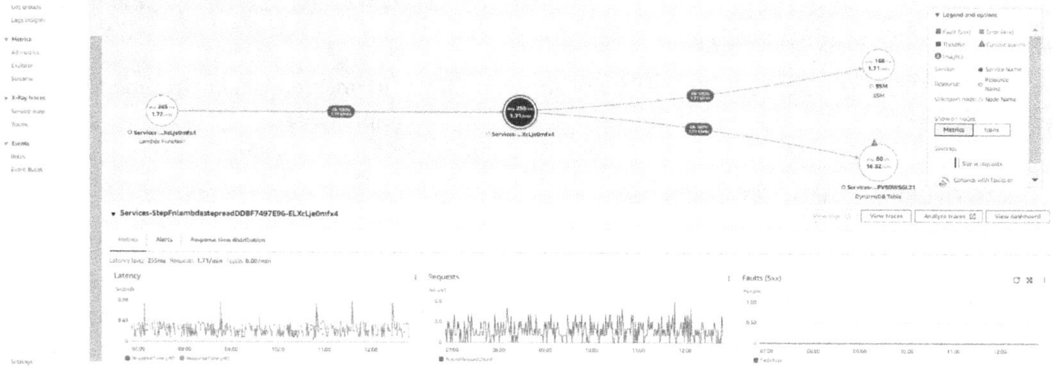

Figure 2.17 – Representation of traffic, latency, and errors on a ServiceLens map

In this section, we explored what is offered by AWS's cloud-native services to provide observability for your applications. In the next section, we'll understand the open source managed offerings from AWS that support observability for your application and understand the functionality of each.

Overview of AWS-managed open source observability services in AWS

When managing open source solutions for observability, it could be challenging to keep the software updated, secured, patched, and distributed across the company. AWS helps you by providing a range of managed services, fully compatible with popular open source observability software. They will support you in using the tools of choice that you love and have invested in while avoiding the burden of managing them yourself. We will explore the offerings from AWS in open source managed services and the functionalities provided by them in the following subsections.

Amazon Managed Service for Prometheus

Amazon Managed Service for Prometheus (**AMP**) is a fully managed, secure, and highly available metric monitoring solution that makes it easy to monitor containerized applications at scale. You can use this service to monitor workloads from AWS environments and non-AWS environments, including on-premises services. Prometheus is easy to use and has a fantastic query language called PromoQL, which can provide support for high-cardinality metric data and has a powerful alerting feature that offers alarm groups, inhibition, and silencing.

It scales up and down based on your workload's requirements and integrates with AWS security services to enable secure access to data. It integrates well with Kubernetes-native service discovery of resources and provides support for dozens of exporters that help you capture performance and health metrics of various workloads with minimal effort.

You can use Prometheus with Amazon ECS and Amazon EKS or in on-premises environments using AWS Distro for Open Telemetry or Prometheus servers as collection agents.

Amazon OpenSearch Service

Amazon OpenSearch Service is a fully managed, secure, and highly available service that makes it easy to deploy, operate, and scale OpenSearch clusters in the AWS cloud. Amazon OpenSearch Service makes it easy for you to perform log analytics, real-time application monitoring, clickstream analysis, and more. It also automatically detects and replaces unhealthy OpenSearch nodes, reducing the overhead of managing the required infrastructure yourself. There are three major components from OpenSearch that are of interest from an observability standpoint:

- **Amazon OpenSearch Service Log Analytics**: Log analytics involves searching, analyzing, and visualizing machine data generated by your IT systems and technology infrastructure to gain operational insights.

- **Amazon OpenSearch Dashboards**: OpenSearch Dashboards is a lightweight, real-time visualization tool.

- **Amazon OpenSearch Service Trace Analytics**: The addition of trace data to OpenSearch Log Analytics makes it easy to use the same service to isolate the source of performance problems and also diagnose their root cause. End-to-end insights are possible when a trace is added to traditional logs and metrics in OpenSearch.

Amazon Managed Grafana

Amazon Managed Grafana (**AMG**) is a fully managed service developed by Grafana Labs and AWS based on Grafana, a popular open source analytics platform. It allows you to create a rich, single-pane-of-glass view for all your disparate data sources. It also allows you to create alerts within the dashboard and dispatch notifications to different destinations, such as Amazon SNS, PagerDuty, Slack, Opsgenie, and others. AMG also enables you to analyze, monitor, and alarm across multiple data sources such as AMP, Amazon CloudWatch, AWS X-Ray, Amazon Time Stream, Amazon OpenSearch service, and third-party ISVs such as Datadog and Splunk, as well as self-managed data sources such as Influx DB. AMG reduces the operational management of Grafana by automatically scaling compute and storage infrastructure as demands increase with automated version updates and security patching.

With AMG Enterprise, you can access even more Enterprise plugins, which give customers more flexibility and options to unify data visualization on the managed Grafana service.

A view of the AMG dashboard for the ingested CloudWatch Logs can be seen in the following screenshot:

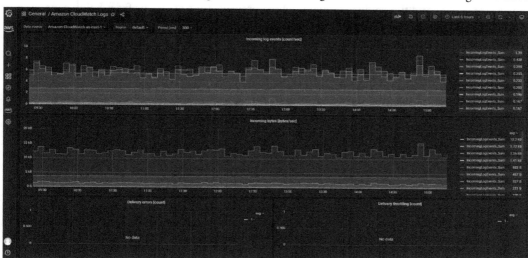

Figure 2.18 – CloudWatch log ingestion statistics on an AMG dashboard

With that, we understand the open source managed observability services offered on AWS. Now, let's explore the **artificial intelligence** (**AI**) and **machine learning** (**ML**) services offered by AWS to support observability for your application and transition from manual operations to AI for IT operations.

AI and ML insights

AIOPs is the process of using AI or ML techniques to solve operational problems. The goal of AIOPs is to reduce human intervention in the IT operations process. AWS offers multiple AI and ML services to help you automate your IT operations. We will review some of the AIOPs services (Amazon DevOps Guru, Amazon CodeGuru, and Amazon Lookout for Metrics) offered by AWS and their role in observability in the next few subsections.

Amazon DevOps Guru

The key challenges operators face today when an issue occurs are due to large and disparate data volumes, lots of time and effort in manually correlating across data sources and tools, alarms and notifications from multiple tools resulting in alarm fatigue, and the inability to identify the most critical issue. That's where **Amazon DevOps Guru** comes into the picture.

Amazon DevOps Guru is an ML-powered service that is designed to improve an application or service's operational performance and availability by reducing expensive downtime with no ML experience. When DevOps Guru identifies a critical issue by detecting behaviors that deviate from normal operating patterns, such as increased latency, high error rates, resource constraints, and so on, it automatically sends an alert and provides a summary of related anomalies, along with the likely root cause and context for when and where the issue occurred. It prevents the operations team from having to set alarms and thresholds manually, which prevents tons of alarms from being generated. When an issue occurs, it guides operators via Insights, hence reducing the MTTR issue. When possible, DevOps Guru also helps provide recommendations on how to remediate the issue.

Additionally, Amazon DevOps Guru for RDS combines the database depth of **Performance Insights** with ML detection. Amazon DevOps Guru for Serverless services that use AWS Lambda detects anomalous behaviors at the function and correlates anomalies across resources into a single issue.

Amazon CodeGuru

Historically, if a developer were to write code, it would be reviewed by a senior developer to understand any specific issues with the code. That's where Amazon CodeGuru helps you. Amazon CodeGuru is a developer tool that provides automated code reviews and identifies the most expensive lines of code that affect application performance. Amazon CodeGuru is an ML service for automated code review and provides application performance profiling. Furthermore, it accepts user feedback and enhances the recommendations based on user feedback.

Amazon CodeGuru has two different services:

- **Amazon CodeGuru Reviewer**
- **Amazon CodeGuru Profiler**

Amazon CodeGuru Reviewer can integrate with GitHub and AWS CodeCommit to scan the code, provide automated code review comments, and addresses code areas such as concurrency, resource leaks, sensitive information leaks, and more. Amazon CodeGuru needs read-only access and will not store the code. Amazon CodeGuru Reviewer supports the Java and Python languages.

Amazon CodeGuru Profiler finds your most expensive lines of code. It is trained to find methods with high potential for performance optimization and provides recommendations on how to fix the code. CodeGuru Profiler supports Amazon EC2 instances, Amazon ECS, EKS, AWS Fargate containers, and AWS Lambda functions. Amazon CodeGuru Profiler supports Java, Python, and JVM languages such as Scala and Kotlin.

The following screenshot shows the heap size anomalies in an application captured by Amazon CodeGuru Profiler:

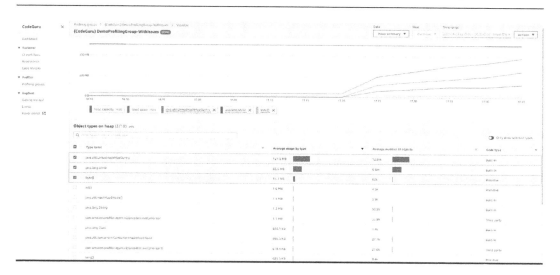

Figure 2.19 – Visualization of heap size anomalies in a sample app in CodeGuru

Amazon Lookout for Metrics

Amazon Lookout for Metrics is a fully managed ML service that detects anomalies in any time-series metrics and determines their root cause with no ML experience. You can connect your data to the service via any supported data sources, ranging from Amazon S3, Amazon Redshift, and Amazon CloudWatch to name a few. From an observability standpoint, it can detect anomalies in operational metrics and business metrics, and accurately detect data points that are outside the norm. Then, it creates an impact summary to help you diagnose the criticality and root cause of the anomalies quickly. Once anomalies have been detected, you can investigate the results on the AWS console, consume them via APIs, and create custom notifications and actions via an output channel. Amazon Lookout for Metrics also allows you to provide feedback and tune the system for continuous improvement.

The use cases of Amazon Lookout for Metrics are beyond operational IT metrics and could be expanded to enhance customer engagement across the customer journey, marketing metrics to understand the overall traffic volume, revenue churn, and more.

You can see the anomaly detection that is generated by Amazon Lookout for Metrics in the following screenshot:

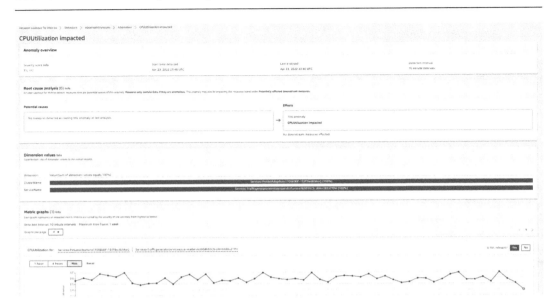

Figure 2.20 – CPU utilization anomaly detection in Amazon Lookout for Metrics

Now that we have understood the services offered on AWS for observability, let's try to understand the methodologies available for gathering metrics, logs, and traces from your application running on EC2, and containerized applications running on EKS, ECS, and others.

Instrumentation

Fundamentally, **instrumentation** involves measuring an application. It is the very foundation on which we can monitor, troubleshoot, debug, profile, and understand how applications function and why they function in a certain way. AWS provides different agents and SDKs to instrument infrastructure and applications. These agents are used to collect metrics, logs, and traces. These metrics are collected by AWS services and are then used to provide insight into the application.

CloudWatch Agent

CloudWatch Agent is an open source software package that autonomously and continuously runs on EC2, containers, hybrid, and on-premises servers running both Linux and Windows. CloudWatch Agent sends metrics and logs to Amazon CloudWatch for storage and analysis, allowing you to create a unified dashboard view of metrics and logs across your hybrid environment.

X-Ray Agent

X-Ray Agent is a software package that runs on EC2, hybrid, and on-premises servers running both Linux and Windows. X-Ray Agent sends traces of applications to AWS X-Ray for storage and analysis, allowing you to create a unified dashboard view of traces across your hybrid environment.

The X-Ray SDK is a library that provides a high-level API for AWS X-Ray. It also provides a way to interact with AWS X-Ray and generate and consume traces. The X-Ray SDK is available in different programming languages, such as Java, Python, and Go.

AWS Distro for OpenTelemetry

AWS Distro for OpenTelemetry (**ADOT**) is a secure, production-ready, AWS-supported distribution of the OpenTelemetry project. OpenTelemetry provides an agnostic instrumentation library for multiple runtimes and platforms with automatic and manual instrumentation support to collect distributed traces and metrics. With AWS Distro for OpenTelemetry, you can instrument your applications to send correlated metrics and traces to multiple AWS and Partner monitoring solutions. You can use ADOT to instrument your applications running on Amazon ECS, **Amazon EC2**, Amazon EKS on EC2, AWS Fargate, and AWS Lambda, as well as on-premises.

ADOT consists of SDKs, auto-instrumentation agents, collectors, and exporters to send data to backend services. OpenTelemetry's components include the following:

- **The OpenTelemetry library/SDK**: The OpenTelemetry SDK allows you to collect AWS resource-specific metadata, such as container IDs, Lambda function versions, tasks, and Pod IDs. It allows you to correlate ingested trace and metrics data from AWS X-Ray and CloudWatch.

- **An auto-instrumentation agent**: AWS has added support in the OpenTelemetry Java auto-instrumentation agent for AWS SDK and AWS X-Ray traces.

- **OpenTelemetry collector**: AWS has added AWS-specific exporters to the distro upstream collector to send data to AWS services, including AWS X-Ray, Amazon CloudWatch, and AMP.

Jaeger and Zipkin tracing

Jaeger is an open source distributed tracing system initially developed by Uber. It stores traces and spans (in a storage backend) and hosts a UI that gives visibility to these traces and spans. **Zipkin** is a distributed tracing system, and it helps gather the timing data needed to troubleshoot latency problems in service architectures.

The new ADOT releases include the Jaeger receiver and Zipkin receiver as a part of the AWS Distro. Based on the requirements, the backends where the traces and logs are sent can be either X-Ray, OpenSearch, or others, based on the observability services being adopted.

Fluentbit

Fluentbit is a super-fast, lightweight, highly scalable logging and metrics processor and forwarder. AWS provides a fluent bit image with CloudWatch Logs and Kinesis Data Firehose plugins. If you are looking to capture application logs from EKS, fluentbit and **fluentd** are the options to publish the data to CloudWatch Logs. With the latest release of 1.19, fluentbit also supports publishing logs to OpenSearch Service.

With that, we've explored all the blocks of the observability services provided by AWS in native, open source managed, and AI and ML services. We've also understood the tools available to instrument applications and gather metrics, logs, and traces. Next, we'll provide guidance on how you should choose various services to achieve observability for your applications.

Adoption of observability services in AWS

Observability covers many concepts and tools that are used by different organizational roles with widely varying needs. AWS has a broad, deep, and growing portfolio of solutions. Observability solutions are tending to converge, in part because of the industry-wide adoption of OpenTelemetry as a standard.

For customers looking for an all-in-one solution, that's Amazon CloudWatch, with a large and growing number of features across a wide range of observability needs, and more on the way. However, other services, even some that are not primarily observability tools, also have multiple observability features, and when a customer can meet a need within a familiar tool, they often prefer to use that. For example, logs can be analyzed using Amazon CloudWatch Logs, but also with Amazon OpenSearch Service. Customers who are already familiar with Elasticsearch and use it in other contexts may use OpenSearch for log analysis. Similarly, distributed traces can be analyzed with AWS X-Ray, but also with Amazon OpenSearch Service's new Trace Analytics feature.

In this reality, any customer's choice of tool may be less about the availability of a specific feature than about their preferences across several axes, such as the following:

- The level of sophistication
- Preference for open source
- Need to integrate with third-party solutions
- Organization-wise versus individual/team choices
- Bias toward familiarity
- All-in-one AWS versus hybrid versus multi-cloud
- Architecture – standard stack versus containers versus serverless

We will look into the details of these topics throughout this book to understand them in greater detail.

Summary

In this chapter, we saw what options are available for adopting observability for applications running on AWS and in Hybrid mode. We understood what native AWS services are available to support observability needs in a distributed application environment and where they would be useful. We also covered what AWS-managed open source services are available on AWS. Then, we understood high-level ways to choose the AWS and open source managed services dimensions when adopting the observability requirements for an organization.

In the next chapter, we will start understanding each of the services in detail, starting with CloudWatch Metrics and CloudWatch Logs.

Questions

Answer the following questions to test your knowledge of this chapter:

1. Which AWS-native services are available to support observability needs in a distributed application environment?

2. What AWS-managed open source services are available on AWS to support observability needs in a distributed application environment?

3. What instrumentation tools are available on AWS to support observability needs in a distributed application environment?

4. What is the difference between AWS-native and AWS-managed open source services?

5. What are the different ways to choose the AWS-native and AWS-managed open source services?

3

Gathering Operational Data and Alerting Using Amazon CloudWatch

We spent the whole of the last chapter understanding the various services available on AWS to observe our applications, using cloud-native and open source observability solutions to observe and monitor the compute, microservices, and serverless components. We categorized the services into layers and provided a brief description of the services in each layer. Also, we gained an understanding of the functionalities of those services and when you should adopt those services in general. Those services will be discussed as we progress, and we will practically demonstrate them.

In this chapter, we are going to cover the following topics:

- Overview of CloudWatch metrics and logs
- Deployment and configuration of the CloudWatch agent in an EC2 instance
- Overview of CloudWatch alarms and dashboard
- Setup of CloudWatch alarms and dashboard
- Overview of Amazon EventBridge
- Setup rules for state change events in Amazon EventBridge

Technical requirements

To be able to accomplish the technical tasks in the chapter, you will need to have the following technical prerequisites:

- A working **AWS account** (you can opt for the free tier, which will cost $0/month for 1 year)
- A **Windows EC2 instance set up** in the AWS account

- A fundamental understanding of **AWS IAM** (*Roles*, *Users*, *Policies*, and *Permissions*)

- Fundamental knowledge of **AWS Systems Manager** (**SSM**)

- Basic PowerShell and shell script execution knowledge

- An understanding of **AWS Simple Notification Service** (**SNS**)

The code files for this chapter can be downloaded from the Chapter03 folder at https://github.com/PacktPublishing/An-Insider-s-Guide-to-Observability-on-AWS.

Overview of CloudWatch metrics and logs

In *Chapter 2, Overview of the Observability Landscape on AWS*, you gained an understanding of the functionalities provided by **CloudWatch metrics**, **CloudWatch logs**, and the **dashboard**. Now let's dive into the technical terms related to these services.

CloudWatch metrics have three main terms and they are as follows:

- **Metric**: A metric is a variable to monitor and is a time-ordered set of data points.

- **Metric namespace**: A CloudWatch metric namespace is a logical store for storing different CloudWatch metrics. Metrics from different namespaces are isolated from each other using metric namespaces. There are two main types of namespaces, namely **AWS namespaces** and **custom namespaces**. All the default **AWS vended metrics** will be published to AWS namespaces with the AWS/Service. Metrics published from the **CloudWatch agent** and metrics published from applications are published to custom namespaces. You can see the namespaces available in the CloudWatch metrics in the following screenshot.

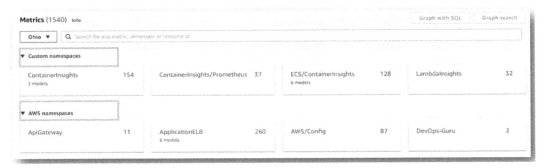

Figure 3.1 – Metric namespaces in CloudWatch

- **Metric dimensions**: A dimension is a *name/value* pair that is part of a metric's identity. By using dimensions, you are creating a new variation of a metric. For example, you can publish a metric to a dimension called *InstanceId*, and another dimension called *InstanceType*. This functionality allows you to see how a specific *InstanceType* is being utilized from a grouping perspective.

CloudWatch retains metric data for 15 months. Data points that are initially published within a shorter period are aggregated together for long-term storage. For example, if you collect data for 1 minute, the data remains available for 15 days with a 1-minute resolution. After 15 days, this data is still available, but is aggregated and is retrievable only with a resolution of 5 minutes. After 63 days, the data is further aggregated and is available with a resolution of 1 hour.

Let's try to understand the technical terms related to CloudWatch logs:

- **Log event**: A record of an activity generated by an application or a resource being monitored. The log event generated in CloudWatch has two properties – *timestamp* and *raw event message*.
- **Log stream**: A log stream is a sequence of log events that share the same source. A log stream is generally intended to represent the sequence of events coming from the application instance or the resource being monitored.
- **Log group**: A log group is a collection of log streams that share the same retention, monitoring, and access control settings. You can define log groups and specify which streams to put into each group. There is no limit on the number of log streams that can belong to one log group. You can also define a log group with a retention policy that specifies how long to retain the log events in the group.

Let's proceed with the installation of the CloudWatch agent and gather custom metrics and logs from an EC2 instance.

Deployment and configuration of the CloudWatch agent in an EC2 instance

When you launch an EC2 instance, whether Linux or Windows, CloudWatch will provide metrics related to the instance, such as *CPU*, *network*, and *disk* metrics, by default. The automated dashboard in *Figure 3.2* provides the default AWS vended metrics provided by CloudWatch as part of the EC2 instance built with no additional configuration.

You can view the automatic dashboard generated by AWS CloudWatch by navigating to **CloudWatch** –> **Dashboard** –> **Automatic Dashboard** –> **EC2**. This should display the default AWS vended metrics provided by CloudWatch, as seen in the following screenshot:

Figure 3.2 – Automated dashboard generated by CloudWatch for default AWS vended EC2 metrics

Building upon the default vended metrics provided by AWS, there are certain metrics that are only visible to the guest operating system, such as memory metrics and page file utilization. In some cases, you also need to monitor application-level metrics, **Operating System (OS)** logs, or application logs. You can start monitoring these additional levels of detail by utilizing the **unified CloudWatch agent**.

The unified CloudWatch agent

As discussed in *Chapter 2, Overview of the Observability Landscape on AWS*, the unified CloudWatch agent is an open source package available on GitHub: `https://github.com/aws/amazon-cloudwatch-agent`. The stable version of the CloudWatch agent can be downloaded from the AWS repository, based on the OS and the region of the instance, from the following link: `https://docs.aws.amazon.com/AmazonCloudWatch/latest/monitoring/download-cloudwatch-agent-commandline.html`.

You can gather the OS metrics and log files from an EC2 instance using the CloudWatch agent. Additionally, you can also enable the **Statsd** and **Collectd protocols**, and the **procstat** plugin to gather application metrics, network metrics, and individual process level metrics respectively, and send them to the CloudWatch metrics namespace. The following diagram provides a high-level representation of the CloudWatch agent and the supported plugins:

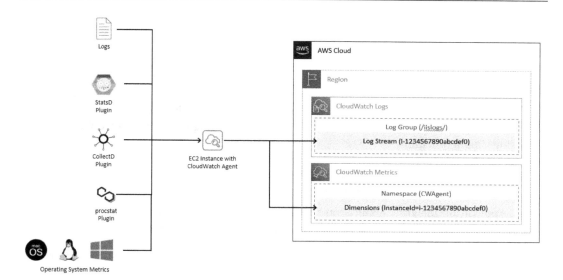

Figure 3.3 – CloudWatch agent and plugins

Next, let's try to set up the unified CloudWatch agent on an **EC2 Windows instance**.

EC2 Windows instance monitoring with the unified CloudWatch agent

Currently, the CloudWatch agent for Windows can only be used on 64-bit versions of the operating system. Now Let's take a look at how unified CloudWatch agent works in Windows. Once you install the unified CloudWatch agent, there are three main components that will provide the required functionality in a Windows instance. The CloudWatch agent installation creates our **Windows Service** that reads the given **Configuration File** and sends the data to the **Amazon CloudWatch** service, as shown in the following diagram:

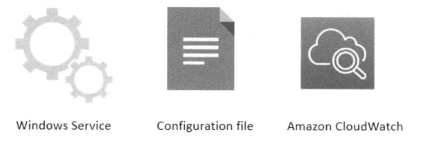

Windows Service   Configuration file   Amazon CloudWatch

Figure 3.4 – The components of CloudWatch agent in the Windows OS

You can install the CloudWatch agent in two different ways:

- **Manual installation**: Download the agent from the AWS repository and install it manually on your EC2 instance.

- **Automated installation**: If you are looking to roll out the CloudWatch agent to multiple EC2 instances, it would be preferred to go with automated installation. Automated installation can be carried out in two different ways other than leveraging **Infrastructure as Code (IaC)** using CloudFormation:

  - **AWS Systems Manager**: AWS Systems Manager can be used to automate the process of agent installation in EC2 instances along with applying the standard templates.

  - **Application Insights**: Application Insights can be used to automate agent installation and discovery and monitor resources, and eliminate the need for manual instrumentation.

Let's build the Windows EC2 instance along with the **Internet Information Services (IIS)** web server. The prerequisites are as follows:

1. Set up an EC2 Windows instance with the IIS web server. To set up the EC2 Windows instance, follow the instructions at `https://docs.aws.amazon.com/AWSEC2/latest/WindowsGuide/EC2_GetStarted.html`.

2. As we launch of the EC2 instance, we will use **User data** to install the IIS web server, as depicted in *Figure 3.5*. Let's configure the security group to allow port `80` to allow `http` requests, along with **Remote Desktop Protocol (RDP)** port `3389`, for the EC2 instance to allow the web application to be accessible from the internet. This will allow you to generate user access logs and capture them using the CloudWatch agent when you browse the web application.

3. To install the IIS web server, copy the following script to **User data**:

```powershell
<powershell>
Install-WindowsFeature -name Web-Server
-IncludeManagementTools
</powershell>
```

Once you've copied the **User data** script, it should look like the following:

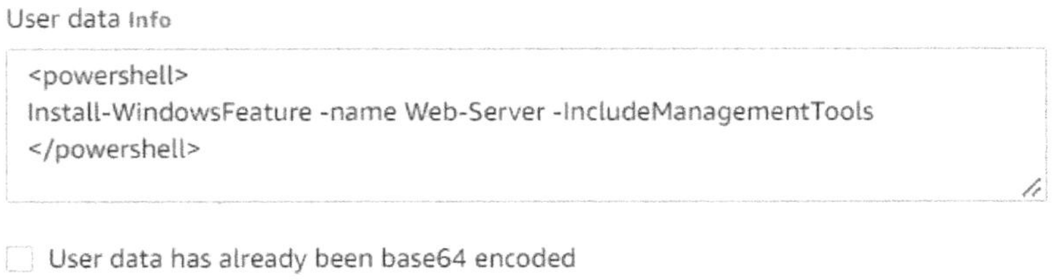

User data Info

```
<powershell>
Install-WindowsFeature -name Web-Server -IncludeManagementTools
</powershell>
```

☐ User data has already been base64 encoded

Figure 3.5 – IIS web server installation using User data

As we have built the EC2 Windows instance, let's install the unified CloudWatch agent using the manual installation method and explore automation in the next section.

Manual installation and configuration of the CloudWatch agent

Let's look at the steps required for the installation and configuration of the unified CloudWatch agent:

1. *Set up IAM role*: Create the IAM role and attach the IAM role to an EC2 instance.
2. *Download and install the agent*: Download the CloudWatch agent from S3 or GitHub and install it on the EC2 instance.
3. *Create a configuration file*: Create a JSON configuration file using the wizard.
4. *Start the agent*: Start the CloudWatch agent using the configuration file.

Let's proceed with *step 1* – creating the IAM role and adding it to the EC2 instance!

Step 1 – set up IAM role: The IAM role provides the permission to publish the custom metrics gathered by the agent to the CloudWatch metrics and logs service:

1. Let's go ahead and create a new IAM role named **CWAgentAdminRole** with **CloudWatch AgentAdminPolicy**.

> **Note**
>
> We are using a more privileged policy, **CloudWatchAgentAdminPolicy**, as we are looking to store the configuration file generated in the AWS SSM Parameter Store. **CloudWatchAgentServerPolicy** will be sufficient to gather metrics using the CloudWatch agent and store them in the CloudWatch service.

2. Input IAM in the search engine, then select the **IAM** service, as shown in the following screenshot, and navigate to it:

Figure 3.6 – Navigate to IAM

3. Next, you will select **Roles | Create role**:

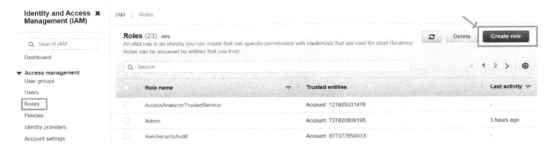

Figure 3.7 – Create role

4. Select **Trusted Entity | AWS service | EC2**, as shown in the following screenshot and select **Next**:

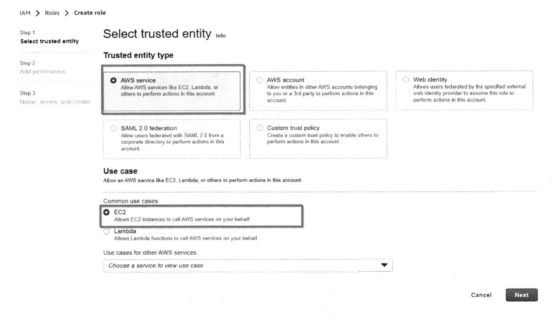

Figure 3.8 – Select EC2

5. Search for and select `CloudWatchAgentAdminPolicy` and select **Next**.

6. Provide the **Role name** as `CWAgentAdminRole` and click on the **Create Role** button.

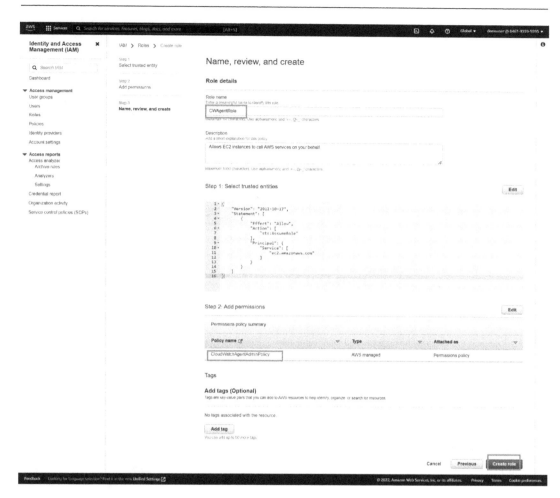

Figure 3.9 – Create a CloudWatch agent role

This completes the creation of the new IAM role with the required permissions. Let's proceed with attaching the newly created role to the EC2 instance we have created.

7. By attaching the IAM role to the E2 instance, we grant it the required permission to publish CloudWatch metrics and also save the configuration file in Parameter Store in AWS Systems Manager.

8. Right-click on the instance created, then navigate to **Security,** and select **Modify IAM role**:

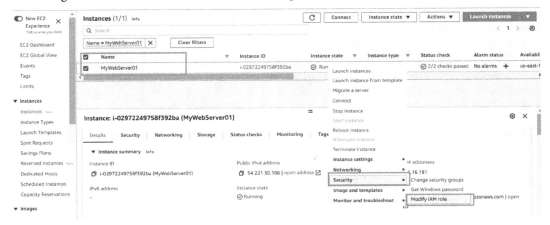

Figure 3.10 – Modify IAM role

9. From the dropdown, select the role named **CWAgentAdminRole** and click **Save**:

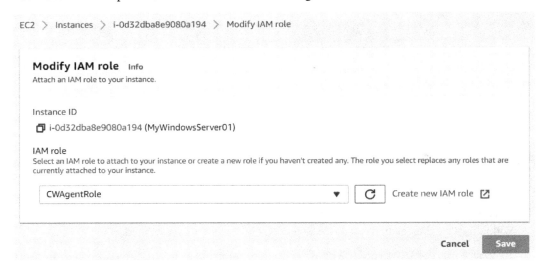

Figure 3.11 – Select CWAgentAdminRole from the dropdown

For further information about the IAM role and instructions on how to create one, refer to the following link: `https://docs.aws.amazon.com/IAM/latest/UserGuide/id_roles_create_for-service.html`.

Now, let's proceed with *step 2* of installing the CloudWatch agent!

Step 2 – installation of the CloudWatch agent: We need to download the agent files from the S3 location provided by AWS that contains the CloudWatch agent source files. This can be done using either the browser or **PowerShell**. We will use PowerShell for the installation and configuration.

Let's log in to the Windows EC2 instance created as a part of the prerequisites and download the CloudWatch Agent installation package to the user's desktop and follow these steps:

1. Make sure to open PowerShell in **Administrator** mode and run the following command to download the unified CloudWatch agent:

   ```
   Invoke-WebRequest -Uri https://s3.amazonaws.com/amazon-
   cloudwatch-agent/windows/amd64/latest/amazon-cloudwatch-
   agent.msi -OutFile $env:USERPROFILE\Desktop\amazon-
   cloudwatch-agent.msi
   ```

2. Verify that the Amazon CloudWatch agent was successfully downloaded using the following command and verify that it returns `True`:

   ```
   Test-Path -Path $env:USERPROFILE\Desktop\amazon-
   cloudwatch-agent.msi
   ```

3. Install the CloudWatch agent using the following PowerShell command. The command will execute silently and will install the agent on the EC2 instance:

   `msiexec /i $env:USERPROFILE\Desktop\amazon-cloudwatch-agent.msi`

   Executing all the three steps will result in the following:

Figure 3.12 – Executing agent download and agent installation from PowerShell

Step 3 – create a configuration file: Now, let's proceed with generating and saving the configuration file. To generate the configuration file, we need to execute a file named `amazon-cloudwatch-agent-config-wizard.exe`. The executable program is menu-driven and will provide configurable options within your PowerShell console session. The executable will create a `config.json` file and also provides you with the option to store the configuration file in AWS Systems Manager Parameter Store. Let's proceed with creating the configuration file:

1. Navigate to `%ProgramFiles%\Amazon\AmazonCloudWatchAgent`, launch PowerShell, and run the `amazon-cloudwatch-agent-config-wizard.exe` executable using PowerShell, which will help you generate the configuration file with the options for gathering

additional metrics from the operating system by executing the following code snippet (please make sure to navigate to the path %programfiles%\Amazon\AmazonCloudWatchAgent. This is the path where the config.json file should be stored to execute the exercise correctly):

```
& $env:ProgramFiles\Amazon\AmazonCloudWatchAgent\amazon-
cloudwatch-agent-config-wizard.exe
```

2. The first question, shown in *Figure 3.13*, asks you which OS you would like to install the agent for. As we are on Windows EC2, we simply press *Enter* as it is the default option.

Figure 3.13 – OS selection

3. The second question, shown in *Figure 3.14*, asks for the location of the instance. The CloudWatch agent wizard will query the EC2 instance metadata to check the details and provide the default option. As we are using an EC2 instance on AWS, let's proceed with the default selection.

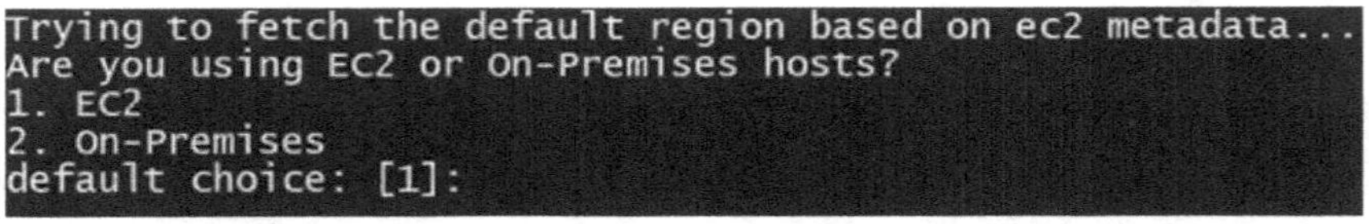

Figure 3.14 – Location selection

4. The third question, shown in *Figure 3.15*, is about enabling the StatsD plugin. We will not configure the plugin for this exercise as there are no additional application requirements. Let's provide option 2.

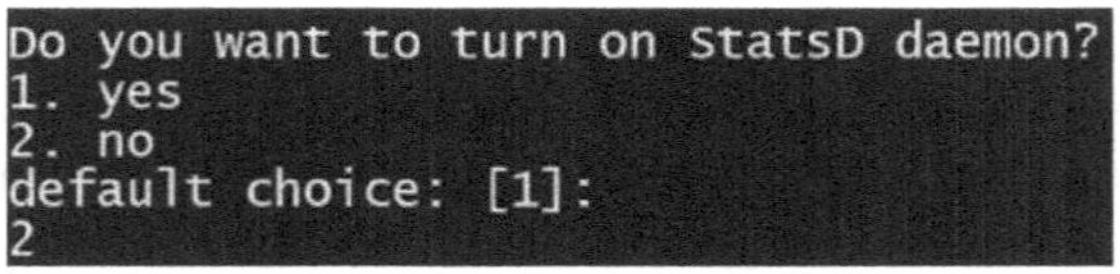

Figure 3.15 – StatsD plugin options

5. The fourth question is about the migration of any legacy agent configuration from the old CloudWatch Log Agent. CloudWatch Log Agent was a standalone agent before the introduction of a unified CloudWatch agent. Let's continue with option 2 as we are not transitioning any configuration from the legacy agent:

```
Do you have any existing CloudWatch Log Agent configuration file to import for migration?
1. yes
2. no
default choice: [2]:
2
```

Figure 3.16 – Migration of log configuration from the legacy Log Agent

6. The next four questions, shown in *Figure 3.17*, are about what type of metrics you would like to gather from the OS and the metric dimensions. We will go with the default options. Note that adding dimensions to the metrics will result in additional costs for CloudWatch metrics. Therefore, if specific metric dimensions are not required, it can be changed to no:

```
Do you want to monitor any host metrics? e.g. CPU, memory, etc.
1. yes
2. no
default choice: [1]:

Do you want to monitor cpu metrics per core?
1. yes
2. no
default choice: [1]:

Do you want to add ec2 dimensions (ImageId, InstanceId, InstanceType, AutoScalingGroupNam
ilable?
1. yes
2. no
default choice: [1]:

Do you want to aggregate ec2 dimensions (InstanceId)?
1. yes
2. no
default choice: [1]:
```

Figure 3.17 – Core metric and dimension options

7. The next question is about the resolution of the metrics. Based on the criticality of the application, you can gather metrics at a frequency of 1 second. For this exercise, we'll select **1s** (1 second):

```
would you like to collect your metrics at high resolution (sub-minute resolution)? This enables sub-minute resolution for all metrics,
but you can customize for specific metrics in the output json file.
1. 1s
2. 10s
3. 30s
4. 60s
default choice: [4]:
1
```

Figure 3.18 – Resolution selection for CloudWatch metrics

8. The next question is about the additional metrics that you would like to gather by default. To make it easy, CloudWatch has predefined metrics for each configuration. The standard option will provide **Memory**, **Paging**, **Processor**, **PhysicalDisk**, and **LogicalDisk** information. Details can be verified at this link: https://docs.aws.amazon.com/AmazonCloudWatch/latest/monitoring/create-cloudwatch-agent-configuration-file-wizard.html. Refer to the **CloudWatch agent predefined metric sets** section. For this exercise, we are going with option 2 (**Standard**):

```
Which default metrics config do you want?
1. Basic
2. Standard
3. Advanced
4. None
default choice: [1]:
2
```

Figure 3.19 – Selecting predefined metrics to be gathered

9. The next question, seen in *Figure 3.20*, is about confirming whether the configuration is fine or whether you would like to change it. The configuration generated as a JSON file can be customized at a later stage if required. We will confirm that the configuration is good.

```
Are you satisfied with the above config? Note: it can be manually customized after the wizard completes to add additional items.
1. yes
2. no
default choice: [1]:
1
```

Figure 3.20 – Confirming the configuration of options

10. The next four questions are about the logs that you would like to collect using the CloudWatch agent. We are going to collect IIS logs from the default path. We will select yes and provide the log path and the log group name to ingest the logs.

11. Next, you will need to provide the log file path to gather the IIS log files, as shown here:

    C:\inetpub\logs\LogFiles\W3SVC1*.log

12. We will be providing the name IISLogs as the log group name to store the logs sent by the CloudWatch agent for IIS:

    IISLogs

13. **Log Group Retention in days**: We will select option 5, which will set up the retention of logs for 7 days in the CloudWatch log group and specify any additional log files to monitor as no.

```
Do you want to monitor any customized log files?
1. yes
2. no
default choice: [1]:

Log file path:
C:\inetpub\logs\LogFiles\W3SVC1\*.log
Log group name:
default choice: [*.log]
IISLogs
Log stream name:
default choice: [{instance_id}]

Log Group Retention in days
1.  -1
2.  1
3.  3
4.  5
5.  7
6.  14
7.  30
8.  60
9.  90
10.  120
11.  150
12.  180
13.  365
14.  400
15.  545
16.  731
17.  1827
18.  3653
default choice: [1]:
5
Do you want to specify any additional log files to monitor?
1. yes
2. no
default choice: [1]:
2
```

Figure 3.21 – Custom log retention configuration

14. The next question is whether you would like to monitor any Windows event logs. This will be useful if you would like to gather OS or application events (such as ad hoc OS reboots, application backup-related events, etc.). We will select **no** for this exercise, as shown in the following screenshot:

```
Do you want to monitor any Windows event log?
1. yes
2. no
default choice: [1]:
2
```

Figure 3.22 – Windows event gathering using the CloudWatch agent

15. The configuration file will be saved locally. If you would like to automate the agent installation and configuration in the future, the config file can be saved to the AWS **SSM** Parameter Store. Let's go ahead and save it in Parameter Store so that we can look at automation for the next exercise. Select the default options for the next four questions, as shown in *Figure 3.23*. In these four options, we are providing the name for the parameter store, which will default to **AmazonCloudWatch-windows**, the region to store the configuration file in (AWS System Manager defaults to the region you are building the EC2 instance in), and the credentials to access Parameter Store (provided by the AWS IAM role attached to the EC2 instance provided in *step 1*).

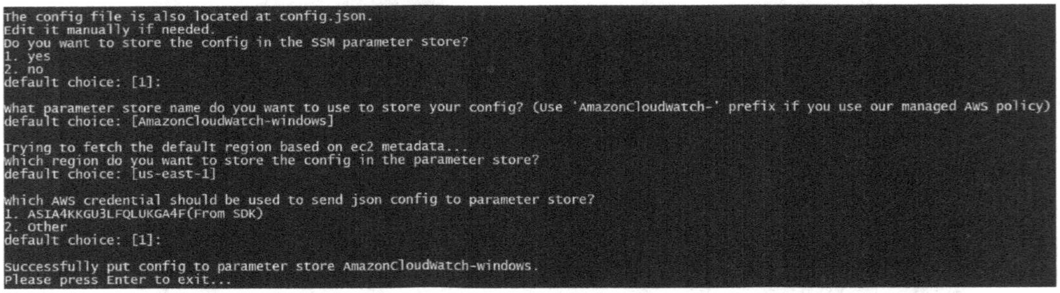

```
The config file is also located at config.json.
Edit it manually if needed.
Do you want to store the config in the SSM parameter store?
1. yes
2. no
default choice: [1]:

what parameter store name do you want to use to store your config? (Use 'AmazonCloudwatch-' prefix if you use our managed AWS policy)
default choice: [AmazonCloudwatch-windows]

Trying to fetch the default region based on ec2 metadata...
Which region do you want to store the config in the parameter store?
default choice: [us-east-1]

Which AWS credential should be used to send json config to parameter store?
1. ASIA4KKGU3LFQLUKGA4F(From SDK)
2. other
default choice: [1]:

Successfully put config to parameter store AmazonCloudwatch-windows.
Please press Enter to exit...
```

Figure 3.23 – Saving the configuration file to Parameter store

Let's proceed with *step 4*: starting the agent using the config file generated!

Step 4 – starting the agent: We need to run a PowerShell command to start the agent along with the configuration file generated as part of *step 3* to start the CloudWatch agent. Let's proceed with starting the agent:

1. Let's start the agent using `amazon-cloudwatch-agent-ctl.ps1` and using the configuration file generated in *step 3* by running the following command:

```
& $env:ProgramFiles\Amazon\AmazonCloudWatchAgent\
amazon-cloudwatch-agent-ctl.ps1 -a fetch-config -m ec2
-c file:$env:ProgramFiles\Amazon\AmazonCloudWatchAgent\
config.json -s
```

2. After running the command, you should see that the CloudWatch agent has been successfully started using the configuration file provided, as shown in *Figure 3.24*.

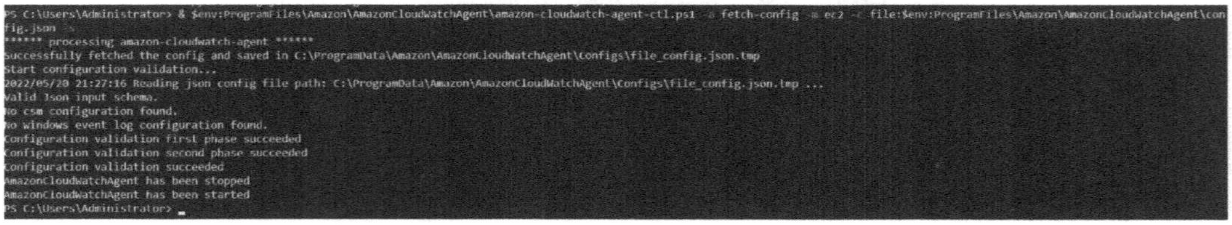

Figure 3.24 – Starting the CloudWatch agent using Powershell

3. You can check that the service has been successfully started and is running from **Services (Local)** in the Windows OS, as shown in *Figure 3.25*.

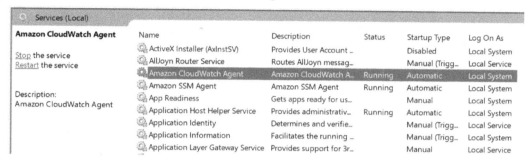

Figure 3.25 – The CloudWatch agent service in Windows services

Next, let's explore the data received by the CloudWatch metrics and logs from the CloudWatch agent:

1. Navigate to **CloudWatch -> Metrics -> All metrics** and observe that there is a new custom namespace created with the name **CWAgent**, as shown in the following screenshot:

Figure 3.26 – Custom namespace created for the CloudWatch agent

2. Click on the **CWAgent** metric namespace, then select **InstanceID**, which will provide the detailed metrics gathered for the EC2 instance, such as **Processor % Usertime**, **Physical % Disk Time**, **Pagefile Usage**, **MemoryCommitted**, and so on.

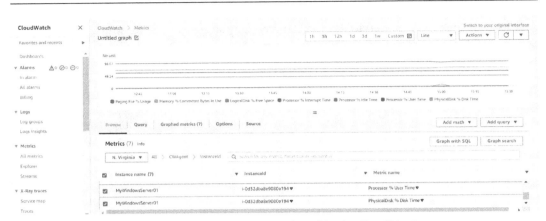

Figure 3.27 – Exploring the metrics gathered by the CloudWatch agent

3. Browse the public IP of the EC2 instance created, which will generate the IIS logs. To verify the logs gathered, navigate to **CloudWatch | Log groups**. You should see, as in *Figure 3.28*, that there's a new log group named **IISLogs**.

Figure 3.28 – Exploring the web server logs gathered by the CloudWatch agent

When you navigate the log group, you will observe the log streams in the CloudWatch Log groups that have been created by the instance ID for each EC2 instance.

This verifies that the manual agent installation was successful and we are able to succesfully gather additional metrics from the EC2 instance, in addition to collecting logs from the web server.

As a part of **Well-Architected** best practices, it would be good to leverage an automated agent rollout rather than a manual installation. We already generated the configuration file required for the configuration from the manual installation and stored it in AWS SSM Parameter Store. Let's continue with the installation and configuration of the CloudWatch agent using AWS Systems Manager.

Automated installation using AWS Systems Manager

To begin this section, please proceed with the new EC2 instance built along with the IIS installation, which was previously covered as part of the prerequisites. Next, let's see how to install the agent using Systems Manager:

1. *Set up IAM roles*: Create an IAM role with policies of **AmazonEC2RoleforSSM** and **CloudWatchAgentServerPolicy**. This will enable you to manage the EC2 instance using AWS Systems Manager and gather custom metrics and logs from the EC2 instances in CloudWatch.

2. *Installation of the agent*: We will run the `AWS-ConfigureAWSPackage` command and install the CloudWatch agent on the target EC2 instances.

   A high-level representation of configuring the role and the installation of the agent is depicted in the following diagram:

Figure 3.29 – Installation process of the CloudWatch agent using Systems Manager

3. *Create and store a configuration file*: If you have skipped manual installation, please generate the configuration file as described in *step 3* of the *Manual installation and configuration of the CloudWatch agent* section or download it from the repository and create it as a parameter in AWS Systems Manager's Parameter Store. You can find the configuration file (`AmazonCloudWatch-windows.json`) on Github: `https://github.com/PacktPublishing/AWS-Observability-Handbook/blob/main/Chapter03/AmazonCloudWatch-windows.json`.

4. *Starting the agent*: The process of starting the CloudWatch agent using AWS System manager can be accomplished by running the SSM `AmazonCloudWatch-ManageAgent` document with the saved configuration file in AWS SSM Parameter Store and starting the agent in the EC2 instances to be monitored. This process is illustrated in the following diagram:

Figure 3.30 – Process to start the CloudWatch agent using AWS Systems Manager

With a clear understanding of the procedure for installing the CloudWatch agent using AWS SSM, we can now move forward with installing the CloudWatch agent using Systems Manager:

1. IAM roles provide the permission to publish the custom metrics gathered by the agent to the CloudWatch metrics and logs service.

2. Let's go ahead and create a new IAM role named `CWAgentRole`. Please refer to the instructions in the manual installation on how to create a role and add the **AmazonEC2roleforSSM** and **CloudWatchAgentServerPolicy** policies and create a new IAM role na,ed `CWAgentRole`. Then attach it to the newly created instance.

3. To install the CloudWatch agent using Systems Manager, do the following:

    I.   Navigate to **Systems Manager** as follows:

Figure 3.31 – Navigate to Systems Manager

    II.   Select **Run Command** | **Run a Command**, as you can see in the following screenshot:

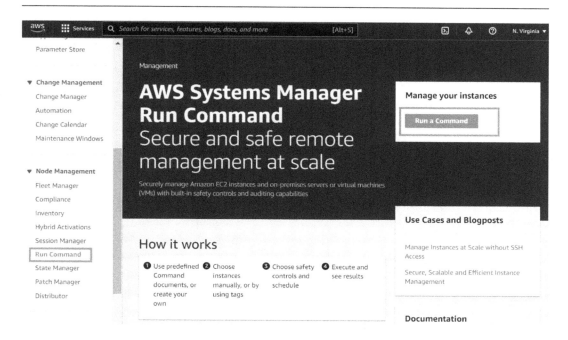

Figure 3.32 – Select Run a Command

III.   Select **Command document | AWS-ConfigureAWSPackage**, and leave the Document Version to Default:

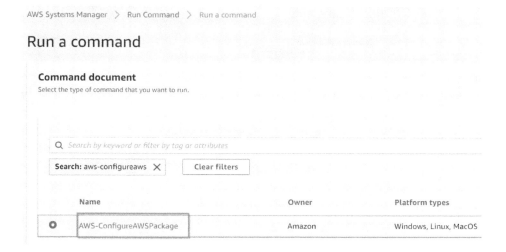

Figure 3.33 – CloudWatch agent install using AWS SSSM Run a ComV.nd

IV.   Provide the name of the package to Install as **AmazonCloudWatchAgent** and select the **Choose instances manually** option and then select the newly created AWS EC2 Instance and uncheck **Enable an S3 bucket**:

Command parameters

Action
(Required) Specify whether or not to install or uninstall the package.

Install	▼

Installation Type
(Optional) Specify the type of installation. Uninstall and reinstall: The application is taken offline until the reinstallation process completes. In-place update: The application is available installation.

Uninstall and reinstall	▼

Name
(Required) The package to install/uninstall.

AmazonCloudWatchAgent

Version
(Optional) The version of the package to install or uninstall. If you don't specify a version, the system installs the latest published version by default. The system will only attempt to uni If no version of the package is installed, the system returns an error.

Additional Arguments

Figure 3.34 – Provide package name as AmazonCloudWatchAgent

V.   Once submitted, verify that the installation is successful, which should look like the following:

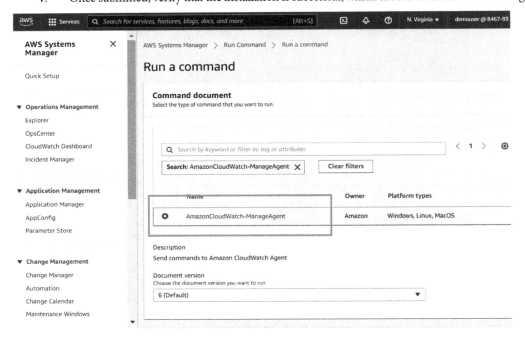

Figure 3.35 – CloudWatch agent installed successfully

4. Let's start the CloudWatch agent using the AWS SSM document:

   I.   Select **Run command** | **Command document** | **AmazonCloudWatch-ManageAgent**.

Command parameters

Action
The action CloudWatch Agent should take.

configure ▼

Mode
Controls platform-specific default behavior such as whether to include EC2 Metadata in metrics.

ec2 ▼

Optional Configuration Source
Only for 'configure' related actions. Use 'ssm' to apply a ssm parameter as config. Use 'default' to apply default config for amazon-cloudwatch-agent. Use 'all' with 'configure (remove)' to clean all configs for amazon-cloudwatch-a

ssm ▼

Optional Configuration Location
Only for 'configure' related actions. Only needed when Optional Configuration Source is set to 'ssm'. The value should be a ssm parameter name.

AmazonCloudWatch-windows

Optional Open Telemetry Collector Configuration Source
Only for 'configure' related actions. Use 'ssm' to apply a ssm parameter as config. Use 'default' to apply default config for amazon-cloudwatch-agent. Use 'all' with 'configure (remove)' to clean all configs for amazon-cloudwatch-a
not support MacOS instance.

ssm ▼

Optional Open Telemetry Collector Configuration Location
Only for 'configure' related actions. Only needed when Optional Configuration Source is set to 'ssm'. The value should be a ssm parameter name. It does not support MacOS instance.

Optional Restart
Only for 'configure' related actions. If 'yes', restarts the agent to use the new configuration. Otherwise the new config will only apply on the next agent restart.

yes ▼

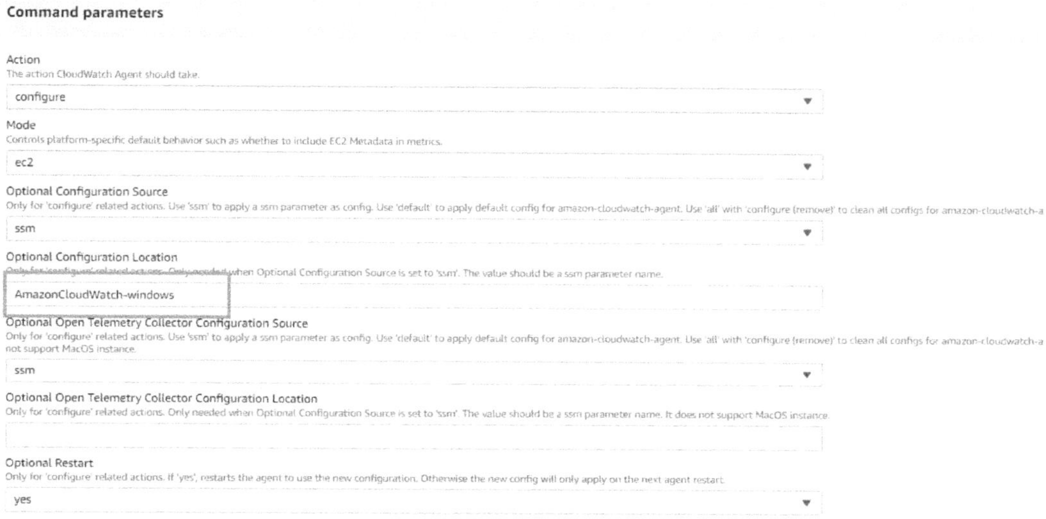

Figure 3.36 – Starting the CloudWatch agent using AmazonCloudWatch-ManageAgent

   II. Provide the `Parameter` file from Parameter Store as **AmazonCloudWatch-windows** in the **Optional Configuration Location** field.

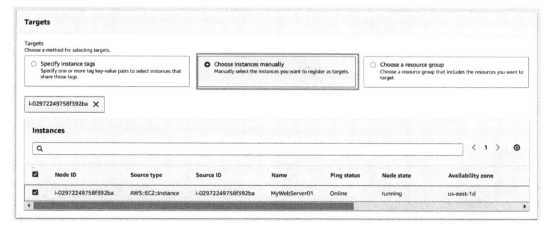

Figure 3.37 – Selecting the CloudWatch agent configuration file from Parameter Store

III.   Select **Choose instances manually,** as shown in *Figure 3.37*, select the newly created instance, and configure the agent with the parameters created as per the JSON file. Unselect **Enable an S3 bucket** and leave the remaining values at the default. Then click **Run.**

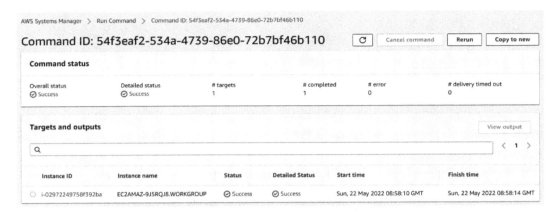

Figure 3.38 – Selecting the instances where the CloudWatch agent should be configured

IV.   Verify that the agent configuration is successful as follows.

Figure 3.39 – Successful rollout of the CloudWatch agent configuration using Systems Manager

V.   When you navigate to the CloudWatch metrics, you should be able to see the instance sending the custom CloudWatch metrics.

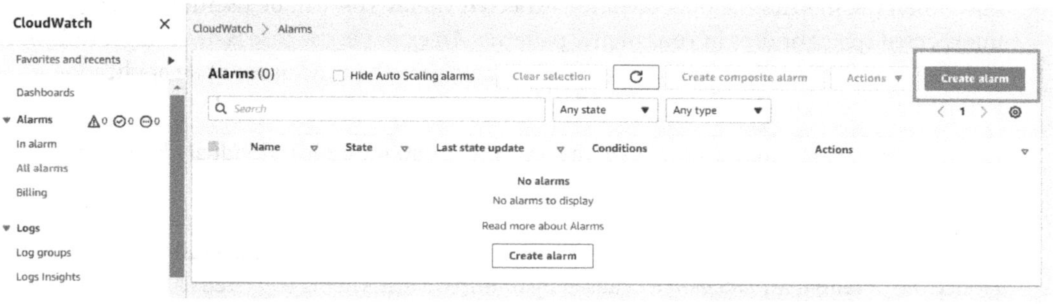

Figure 3.40 – Verification of the gathered CloudWatch agent metrics

We have successfully installed and configured the CloudWatch agent using AWS Systems Manager.

We will be covering the method to discover and install the CloudWatch agent using AWS Application Insights in *Chapter 5, Insights into Operational Data with CloudWatch*. Now, let's continue with configuring the alarms for the metrics being monitored.

Overview of CloudWatch alarms and dashboards

In the last section, we started gathering custom OS metrics and logs using the CloudWatch agent. Let's proceed with the technical concepts of CloudWatch alarms and configure CloudWatch for the gathered metrics.

CloudWatch alarms

We gained an understanding of what CloudWatch alarms are and the states in them in the previous chapter.

Let's understand what an alarm can be used for in observability. CloudWatch alarms will help you measure a defined data point and determine whether it is in an **alarm** state, **OK** state, or has **insufficient data**. Based on the state of the alarm, we can set up notifications and notify the team, group, or a third-party system that will be interested in the application's state. Additionally, an automated response can be set up to act on the alarm so that auto-remediation activities can be carried out.

Different types of alarms can be set up in CloudWatch. They are as follows:

- **Static threshold**: Static thresholds will be good when there are well-defined KPIs, for example, instance metrics that need to be alerted when they cross a threshold.

- **Anomaly detection**: Anomaly detection in CloudWatch applies a machine learning algorithm to define the upper limit and lower limit for each of the enabled metrics and generates an alarm only when the metrics fall outside of the expected values This can be useful for identifying unexpected spikes or dips in your metric patterns. An example use case is detecting a sudden increase in usage, which could be a result of organic growth in the usage of an application or a sign of a DDOS attack.

- **Composite alarms**: Composite alarms are a way to group multiple individual alarms. Instead of having to manage and respond to multiple alarms for different parts of the environment, composite alarms allow you to group related alarms together. This can help to reduce the number of alarms that need to be monitored and managed, which can improve overall efficiency and reduce the potential for *alarm noise* caused by multiple alarms being triggered at the same time.

- **Metric math expressions**: Metric math expressions can be used to build more meaningful KPIs and alarms for them. You can combine multiple metrics and create a combined utilization metric and alarm on the composite metric. For example, you would like to set up an alarm only for when all the pods in a particular cluster have high CPU utilization.

Let's go ahead and create a fundamental **CPU alarm** for the monitored EC2 instance and send notifications via email using AWS **SNS**:

1. Let's set up a static threshold alarm. Navigate to **CloudWatch | Alarms | All Alarms| Create alarm**. Let's see what that looks like in the following screenshot.

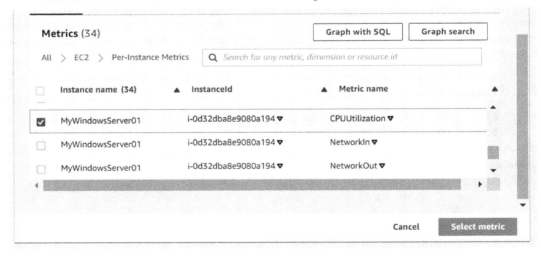

Figure 3.41 – Create alarm

2. Click **Select Metric | Browse AWS namespaces | EC2 | Per-Instance Metrics**. Now, select **CPU utilization** for the instance that you created in the earlier exercise in the *Deployment and configuration of the CloudWatch agent in an EC2 instance* section.

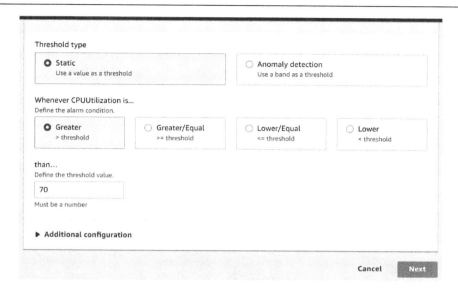

Figure 3.42 – CPU utilization metric for an EC2 instance

3. Select the **Greater** threshold and set the threshold value to 70%, as shown in *Figure 3.42*, then click on **Next.**

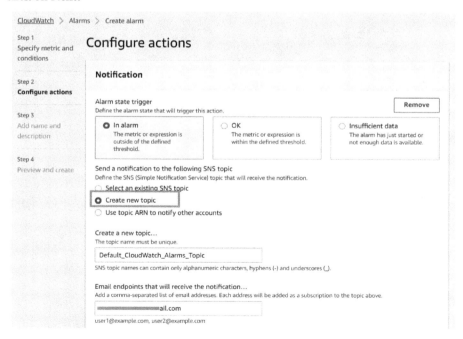

Figure 3.43 – Set the threshold to greater than 70%

4. Next, you will select **Create new topic** and provide your email address. Then select **Create topic** and click on **Next**.

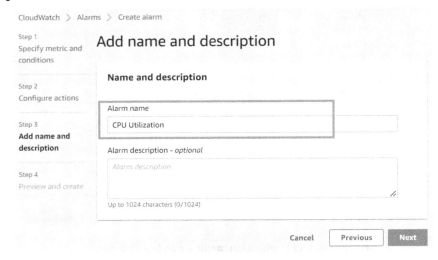

Figure 3.44 – Configure email notifications using an SNS topic

5. Provide the alarm name as CPU Utilization, then click **Next,** and create the alarm.

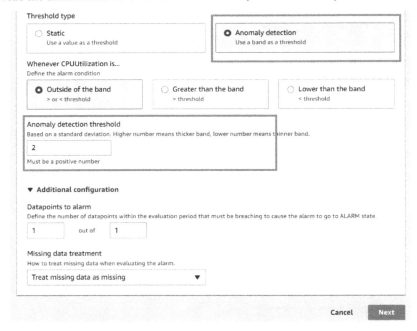

Figure 3.45 – Name the alarm

Repeat the process to create an alarm for the CPU credit balance. We will then create a composite alarm based on these two alarms, as follows:

1. **Anomaly detection**: You can create alarms based on anomaly detection, like the static threshold alarm, except that instead of setting up the alarms based on a threshold, you select **Anomaly detection** as shown in *Figure 3.45*. You can configure notifications as described in the steps for setting up a static threshold alarm.

Figure 3.46 – Configuring anomaly detection for a CloudWatch alarm

We can create an alarm based on three different categories as a part of anomaly detection:

- **Outside of the band**: An alarm will be generated when there is a deviation in the lower and upper thresholds

- **Greater than the band**: There will be an alarm only when the upper band shows anomalous behavior

- **Lower than the band**: There will be an alarm only when the lower band shows anomalous behavio.

2. **Composite alarms**: You can create composite alarms to reduce alarm fatigue and provide better visibility of the problems that are impacting your application. For example, if you consistently have high CPU usage and low CPU credit balance, it may be necessary to increase the instance size to handle high load and prevent application failures.

We have created two static threshold alarms, one for **CPU Utilization**, and the other for **CPU Credit Balance**. Let's create a composite alarm, combining them:

I.   Select the **CPU Utilization** and **CPU Credit Balance** alarms and click on **Create composite alarm** as shown here.

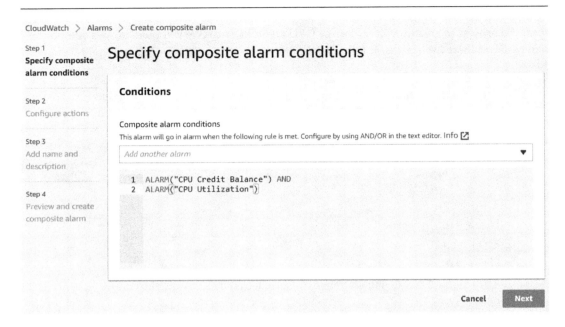

Figure 3.47 – Create composite alarm

II.   Change the condition to AND .

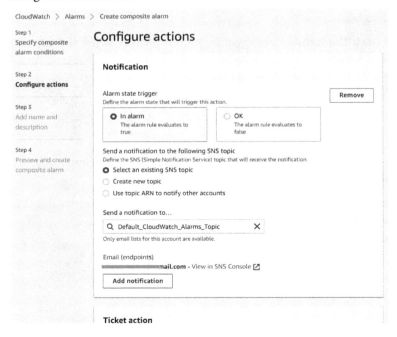

Figure 3.48 – Conditions to create a composite alarm

III.   Select **Action -> Next -> Create composite alarm.**

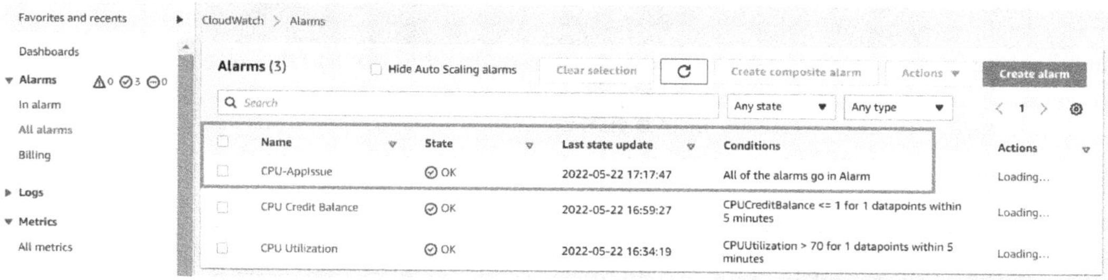

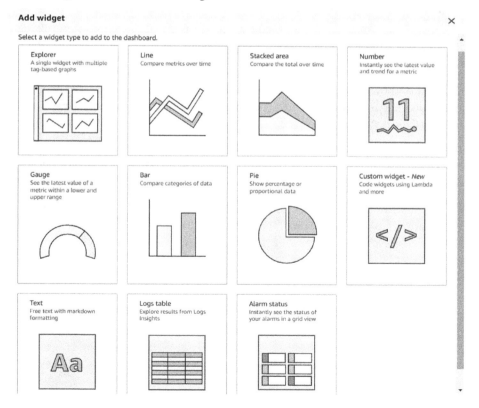

Figure 3.49 – Configuring notifications for the composite alarm

IV.   This creates a composite alarm and alert when both the alarms are in the **Alarm** state, which avoids alarms when CPU utilization is high, and also when CPU credit is not available for constant high CPU utilization to avoid application issues. You can see the composite alarm created in *Figure 3.49*.

Figure 3.50 – Notification about the composite alarm

We will look at the example of the metric math alarms in *Chapter 7, Observability for Serverless Applications on AWS*.

The following are some of the best practices to keep in mind when configuring alarms:

- Create alarms based on meaningful metrics and KPIs
- Test alarm actions
- Use composite alarms to eliminate noise
- Use alarms for automatic responses to events
- Iterate alarm definitions over time

We now understand what an alarm is, the different types of alarms that are available in CloudWatch, and how to configure them. We also know the best practices to keep in mind when configuring alarms to avoid alarm fatigue. Let's explore CloudWatch dashboards next.

CloudWatch dashboards

We learned what a CloudWatch dashboard is and about its functionalities in in the previous chapter. In this chapter, let's further explore the types of CloudWatch dashboards and how to create them. There are two different types of dashboards available in CloudWatch:

- **Automatic dashboards**: These are created automatically based on the metrics being gathered by the CloudWatch service and require no intervention from the user. We have already seen an automated dashboard created by CloudWatch for an EC2 instance as a part of *Figure 3.2*.
- **Custom dashboards**: These are created by the user and can be combined with different types of metrics, logs, events, and traces of information gathered from the services being used by the application.

Widgets are the foundation of custom dashboards and can be added based on the type of data being displayed. They are the main component of the dashboard display and come in various forms, such as **Line**, **Stacked area**, **Number**, **Gauge**, **Bar**, **Pie**, **Text**, **Logs table**, and **Alarm status**, as shown in *Figure 3.50*.

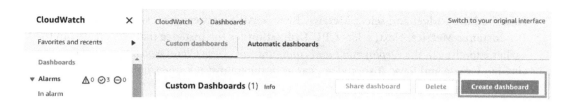

Figure 3.51 – Widgets available in the CloudWatch dashboard

Let's go ahead and create a custom dashboard for the data generated by the EC2 instance. We will be creating a custom dashboard with default metrics provided for an EC2 instance (CPU utilization and CPU credit balance), custom metrics generated by the CloudWatch agent (**Memory % Committed Bytes in Use**), logs generated by IIS, and the current alarm status. We will be using the **Line**, Number, **Logs table**, and **Alarm** status widgets as a part of this custom dashboard:

1. Navigate to **CloudWatch | Dashboards | Custom dashboards | Create dashboard.**

Figure 3.52 – Creating a custom dashboard

2. Provide a name for the dashboard, such as EC2_Dashboard, then click on **Create dashboard.**

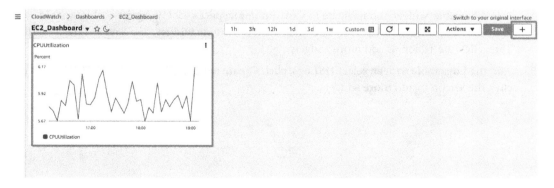

Figure 3.53 – Provide a name for the custom dashboard

3. Add the **Line** widget and select **Metrics**. Browse **AWS namespaces**, select **EC2**, and then **Per-instance Metrics**. Next, select **CPU Utilization** for the instance that you created in the earlier exercise (in the *Deployment and configuration of the CloudWatch agent* section) by clicking **Create** and **Save**. It should appear as in the following screenshot. Then, click the + icon to add more widgets.

Figure 3.54 – Adding CPU Utilization on a Line widget

4. Add a **Number** widget and navigate to **Custom namespaces** -> **CWAgent** -> **InstanceId** -> **Memory % Committed Bytes in Use**, then click on **Create widget,** and save the dashboard. Then click the + icon to add more widgets.

5. Add the **Logs table** widget, select **IISLogs**, click **Create widget**, and save the dashboard. Then click the + icon to add more widgets.

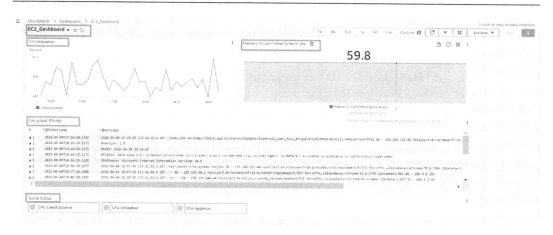

Figure 3.55 – Adding IISLogs to the CloudWatch dashboard

6. Add the **Alarm Status** widget, select all the alarms statuses, click **Add to Dashboard**, and provide the name `Alarm Status` before you save the dashboard.

7. The end result of the custom dashboard, **EC2_Dashboard**, will look as follows.

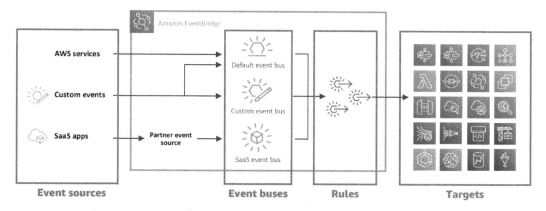

Figure 3.56 – End state of the custom EC2 dashboard

We now understand the different types of dashboards that can be created in CloudWatch and have learned how to create a custom dashboard to simplify the experience for **Site Reliability Engineers (SREs)**. Now let's learn about the fault monitoring provided by AWS, using Amazon EventBridge.

Overview of Amazon EventBridge

Amazon EventBridge is a serverless event bus service for AWS services, your applications, and **Software-as-a-Service (SaaS)** providers. Events are generated for any change in the state of resources. From an **observability** point of view, Amazon EventBridge is a service that allows you to subscribe to events and receive notifications when those events occur.

An event is a signal that a system's state has changed, such as an EC2 instance that has been shut down. To write code to react to events, you need to know the events' schemas, which include information such as the title, format, and validation rules for each piece of event data.

Let's take a look at how EventBridge works. EventBridge has four main components:

- **Event sources**: Event sources are the place where an event comes from. AWS services generate events when there is any change in the state of resources. For example, Amazon EC2 generates events when there is any change in the state of an instance. Event sources are of three types: AWS services, custom events, and SaaS applications. There are over 90 AWS services that will publish events to EventBridge.

- **Event buses**: An event bus is a pipeline that receives events from event sources. Traditionally, there is a default bus where AWS service-related events are published. In addition to that, there is now a **custom event bus** to publish your custom events and a **SaaS event bus** that is dedicated to ingesting **partner events**.

- **Rules**: You can associate rules to an event bus. Rules allow you to match values against metadata and payloads of events as they are ingested, and to determine which events are routed to which destinations.

- **Targets**: Targets are the place where events are sent. You can associate multiple targets with each rule. Targets can include built-in targets such as Lambda functions, Kinesis Data Streams, Amazon SNS topics for notifications, and custom targets. A target processes the events.

You can see the high-level architecture of EventBridge in the following diagram:

Figure 3.57 – EventBridge architecture

Beyond observability, Amazon EventBridge could be leveraged to build event-driven architecture for your business applications.

Now, let's look at subscribing to an event for an EC2 state change for events such as shut down, restart, stopped, and so on:

1. Navigate to **Service** –> **EventBridge**. Then click on **Create rule** as shown in the following screenshot.

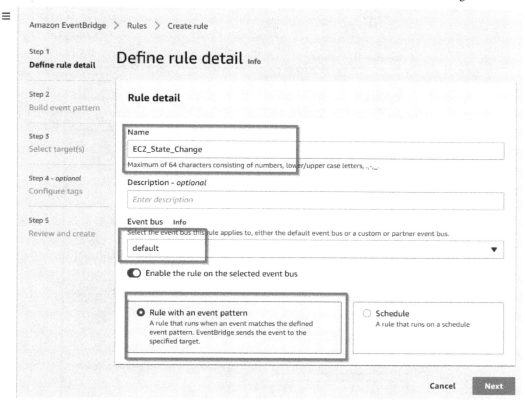

Figure 3.58 – Create rule

2. Provide a name for the rule, such as EC2_State_Change, then set **Event bus** as **default**. Set **Rule type** to **Rule with an event pattern**. All of this is shown in the following screenshot.

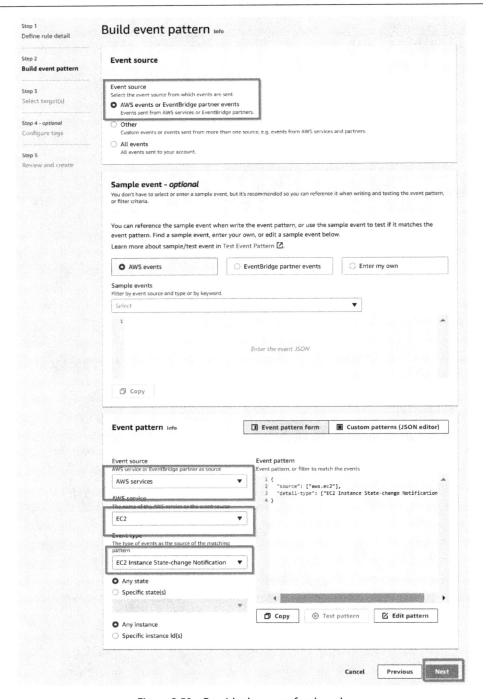

Figure 3.59 – Provide the name for the rule

3. You will set **Event Source** as **AWS events or EventBridge partner events**.

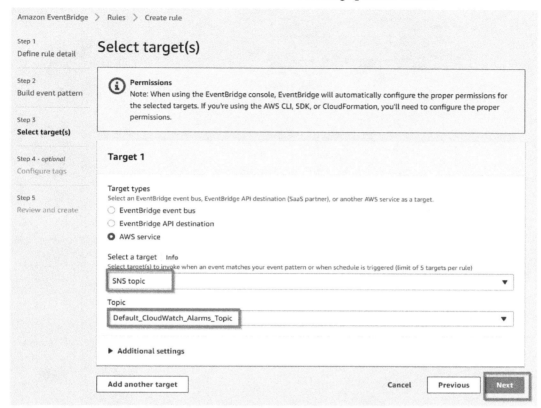

Figure 3.60 – Select Event Source as AWS Events

4. Set **Target 1** as **AWS service**, then choose **SNS Topic** for **Select a target**. Under **Topic**, choose **Default_CloudWatch_Alarms_Topic**, then click **Next**. You will have to create a new SNS topic if not already available.

Figure 3.61 – Select target as SNS topic to send notifications

5. In the **Configure tag(s)** field, do not add anything as they are optional, and click **Next**.

6. In **Review and create**, click on **Create rule**.

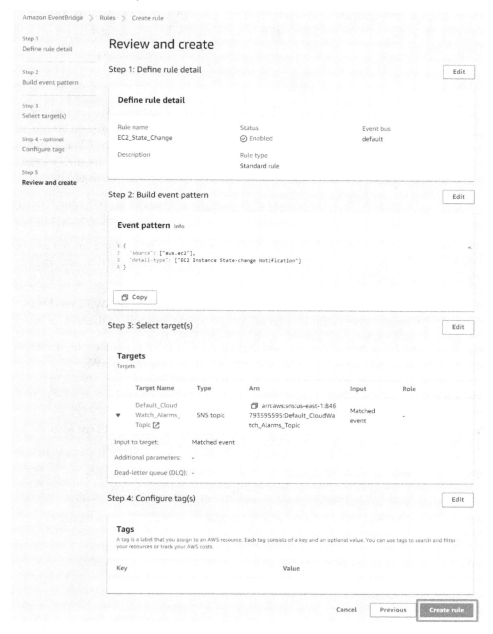

Figure 3.62 – Creating an event rule for the EC2 state changes

You can test whether the rule is working or not by stopping the EC2 instance. You should receive email alerts for state changes of the instance, such as stopping, stopped, and so on.

Amazon EventBridge is a highly durable service that is replicated across **Availability Zones** (**AZs**). Events will be retried for delivery for 24hrs with exponential back-off. Unmatched events will be dropped immediately.

Summary

In this chapter, we learned about the basics of CloudWatch metrics and logs. We learned about the default AWS vended metrics provided by AWS for EC2 instances and have explored on how to gather custom metrics by using CloudWatch agent. We have learned on how to install the CloudWatch agent manually on Windows EC2 instance to gather additional guest OS-level customer metrics and logs and how to automate the CloudWatch agent installation using AWS Systems Manager.

Further, we explored alarms and different types of alarming options available in AWS CloudWatch and how to create alarms for the metrics gathered, and also how to avoid alarm fatigue using composite alarms.

Further, we went ahead and visualized the data generated from CloudWatch using a custom dashboard and utilized widgets to display different types of data, such as AWS vended metrics, custom metrics, logs, and alarms on a single dashboard.

Furthermore, we discussed the importance of events and the requirements of AWS EventBridge. We also discussed the configuration of events in EventBridge and how to subscribe to them as notifications.

In the next chapter, we are going to cover the core concepts of AWS X-Ray and how it can be used to further enhance the observability of your distributed applications running on AWS.

Questions

As we have gone through different concepts related to Amazon CloudWatch, let's see whether we are able to remember what we've learned and answer questions about CloudWatch:

1. What are the different types of namespaces for Amazon CloudWatch metrics?

2. What are the retention settings for metrics and logs in Amazon CloudWatch?

3. What are the different types of alarms you can configure in Amazon CloudWatch?

4. What are CloudWatch custom dashboards used for?

5. What are the components of the Amazon EventBridge?

6. How can you filter events in Amazon EventBridge?

4

Implementing Distributed Tracing Using AWS X-Ray

AWS X-Ray is a versatile and powerful service that allows for detailed analysis and debugging of distributed applications. It can be used to investigate a wide range of applications running on various types of computing on AWS, including EC2 and Serverless. However, its real strength lies in its ability to seamlessly integrate and provide real-time visibility into each service's performance in a microservices architecture. In the microservice architecture, a single request from a user can trigger a complex series of interactions between multiple services. AWS X-Ray helps you understand every aspect of the request-response cycle by tracing each step and allowing you to identify performance bottlenecks and errors, which help you address critical issues that might be impacting the user experience.

In a complex microservices architecture, diagnosing issues can be an incredibly challenging task. Without AWS X-Ray, it can be a time-consuming and error-prone process of manually looking through each service's logs and correlating them in some way. Additionally, the services may be written in different programming languages, use different log formats, and some may log more information than others. All of this can make it difficult to collect all the necessary information and make sense of it. Fortunately, with AWS X-Ray, you can streamline and gather all the necessary information in a single tool and a single place.

AWS X-Ray provides powerful insights into application performance, allowing you to quickly identify the root cause of issues. With X-Ray, you can trace requests as they flow through the various services in your application, from your application code to any AWS services you're using. This end-to-end tracing allows you to gain a deep understanding of how your application is performing and identify the areas that are contributing the most to latency.

By leveraging the data collected by AWS X-Ray, you can build a comprehensive service map that visually depicts the real-time relationships between different services and sources. This service map provides a powerful tool for understanding the complex interactions that occur within your microservices architecture, helping you optimize performance and improve the user experience.

If you see yourself answering *yes* to any of the following statements, we strongly recommend you look at AWS X-Ray:

- Are you building new applications or planning to migrate existing ones to AWS?

- Are you using a microservices architecture?

- Do you want a powerful tool to analyze and debug performance issues and errors within that application?

In this chapter, we will cover the following topics:

- Overview of the AWS X-Ray and concepts of X-Ray

- Overview of CloudWatch ServiceLens map and integration with metrics, logs, and traces

- AWS X-Ray Analytics and troubleshooting application performance issues

- End-to-end instrumentation of a sample application deployed in an EC2 instance

Technical requirements

This chapter is more code-oriented since often, we will need to give up a more black-box approach to modify the code and collect more application context details.

We will use two sample applications to demonstrate the usage of AWS X-Ray in real scenarios, one implemented in JavaScript/Node.js and another in Java. Code experience with those languages is welcome, but we will keep it simple so that no deep understanding is required.

We will rely on infrastructure automation using CloudFormation. Please check the product page and documentation at `https://aws.amazon.com/cloudformation/`.

We will also install and configure software on a Linux server, so some basic bash shell experience is welcome.

Overview of AWS X-Ray

X-Ray is a powerful service and very easy to use, but to understand it better, we can't escape learning its vocabulary. You can find a collection of the main concepts and building blocks, which are well explained, in the AWS documentation (see `https://docs.aws.amazon.com/xray/latest/devguide/xray-concepts.html`). We will provide the same list here for completeness.

X-Ray concepts

Here is a list defining the main concepts related to X-Ray:

- **Segments**: Each application sends information about its unit of work as segments. This segment contains information about the host, the request, the response, start and end times, and any errors or exceptions that happened between the start/end.

- **Subsegments**: Subsegments further break down the work recorded by segments. Subsegments register more granular details about your application code or downstream calls.

- **Service graph**: With segments and subsegments, X-Ray builds a service graph, where each resource is a node of this graph, and the requests are the edges. X-Ray uses this data to generate a visualization called a **service map**. X-Ray combines service segments with the same trace ID in a single service graph.

- **Traces**: With a single trace ID, you can collect all segments and subsegments responsible for a single request. This collection of segments is called a trace.

- **Sampling**: X-Ray records a sample of all traces to ensure you collect relevant tracing while reducing costs. The default sampling rule collects the most pertinent data while minimizing costs, but you can configure the sampling rules to collect more data from sensitive parts of your system.

- **Tracing header**: The first X-Ray-enabled service adds a tracing header to the request, with information such as the trace ID, the parent segment, the sampling decision, and whether the request should be sampled or not.

- **Filter expression**: The amount of trace data generated by a single application can easily overwhelm the operations team when troubleshooting issues. To make life easier when analyzing traces, you can use filter expressions to query for all traces related to the same piece of data.

- **Groups**: You can use filter expressions to define the criteria by which to add traces to designated groups. X-Ray adds any new trace that matches the group's criteria to the group.

- **Annotations** and **metadata**: You can add extra information to segments, such as annotations and metadata. Annotations are key-value pairs, and X-Ray adds them to indexes, which allows the user to use them in filter expressions. Metadata is also key-value pairs, but X-Ray won't add it to indexes, so you should use it when you want to store data together with the trace but do not need to search by it.

- **Instrumentation**: To instrument your application, you need to send trace data for both incoming and outbound requests, as well as other events that occur within your application, along with metadata about the requests. Depending on your specific requirements, there are several instrumentation options available that you can choose from or combine:

- **Auto Instrumentation**: This feature allows you to automatically instrument your application without any code changes. You can use AWS-provided integrations with popular AWS services such as Amazon EC2, Amazon **Elastic Container Service** (**ECS**), AWS Lambda, and AWS Elastic Beanstalk to enable this feature. Once enabled, X-Ray will automatically generate trace data for incoming and outgoing requests to and from these services.

- **Library Instrumentation**: This feature involves adding X-Ray libraries to your application code to enable trace data generation. These libraries are available for a variety of popular programming languages and frameworks, including Java, .NET, Node.js, Ruby, and Python. Once the libraries are added to your code, X-Ray will automatically generate trace data for incoming and outgoing requests.

- **Manual Instrumentation**: This feature involves adding trace generation code directly to your application code. This option provides the most control over the trace data generated and allows you to instrument custom libraries and frameworks. However, it also requires the most effort and may be more error-prone than the other options.

With those concepts in mind, let's go on a guided tour of the X-Ray console.

Navigating the AWS X-Ray console

In this section, I will give you a guided tour of X-Ray. You can just read this section, but I strongly advise you to follow the same steps as me so that you can easily understand how the new tool works:

1. If you are accessing the X-Ray console for the first time, you can search for it and click on the service icon or access it directly via `https://console.aws.amazon.com/xray/home`. You will see a getting started page, as shown in the following screenshot:

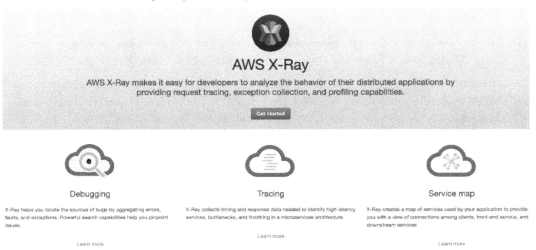

Figure 4.1 – AWS X-Ray start page, first access

Important Note

At the time of writing, the X-Ray product team was about to release a new UI for the service, so you will see a message like the one shown in *Figure 4.2* asking whether you would like to use the new interface. Eventually, the product team will roll out its final version, and all customers will use the new UI, so this message won't exist. To keep this book as up-to-date as possible for longer, we will always opt to use the new UI in these cases. I suggest you do the same; any screenshots from now on will use the new UI.

Let's switch to the new AWS X-Ray console by clicking on **Try out the new console**. This will navigate you to part of the CloudWatch console:

Figure 4.2 – Block message asking whether the user wants to try the new console

2. You can click the links to learn more about the service or the **Getting started** button to skip to the fun part. Click on **Getting Started**.

Step 1 – deploying a sample application

1. After clicking the **Getting Started** button, you will see a page containing instructions on creating a sample application using CloudFormation, the **Infrastructure as Code** offering from AWS. Click on the **Create a sample application with CloudFormation** button:

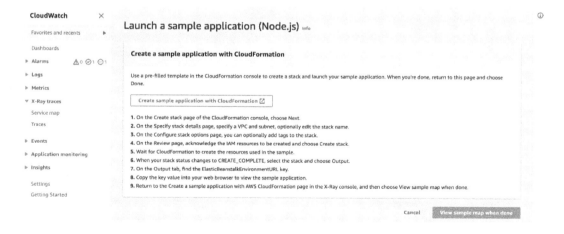

Figure 4.3 – Launching the X-Ray sample application using CloudFormation

2. You will be redirected to the CloudFormation service page when you click the button. You can accept the default values in the first step and click on **Next**:

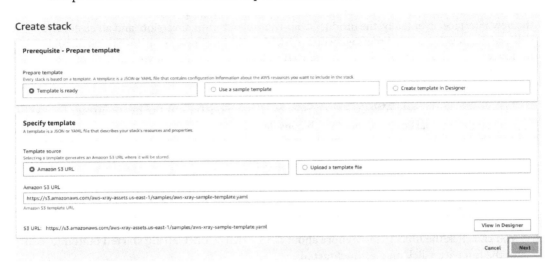

Figure 4.4 – First step in the CloudFormation wizard – launching the X-Ray sample application

3. In the second step, you must select the VPC and subnets where Beanstalk will launch the EC2 instances. If your account has the default VPC, please use it. If you have deleted your default VPC, you can check out how to create it at `https://docs.aws.amazon.com/vpc/latest/userguide/default-vpc.html`. See the following screenshot:

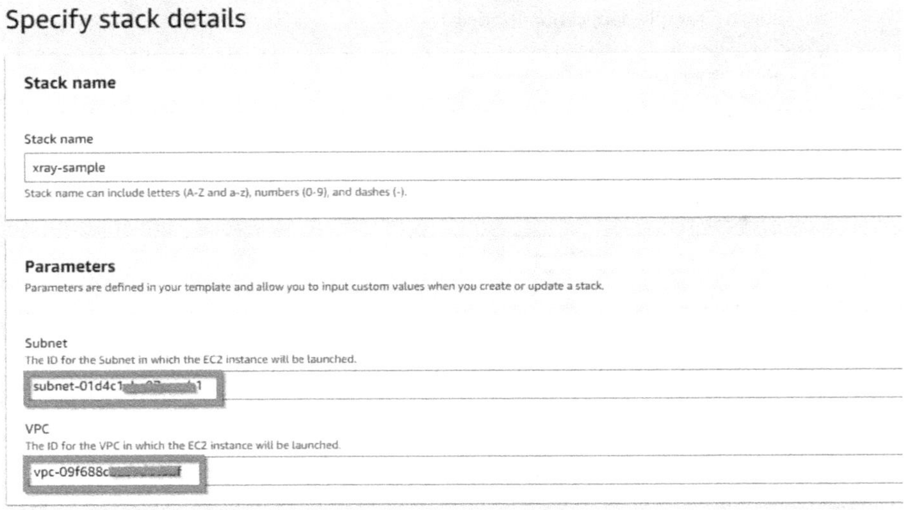

Figure 4.5 – Second step in the CloudFormation wizard – launching the X-Ray sample application

4. You can keep the default values in the third step and click **Next**:

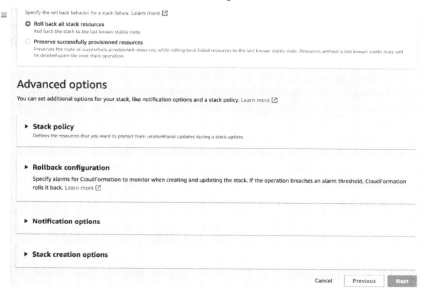

Figure 4.6 – Third step in the CloudFormation wizard – launching the X-Ray sample application

5. Finally, in the fourth step, you can review the configuration, check the box saying you acknowledge CloudFormation will create IAM resources on your behalf, and click **Create stack**:

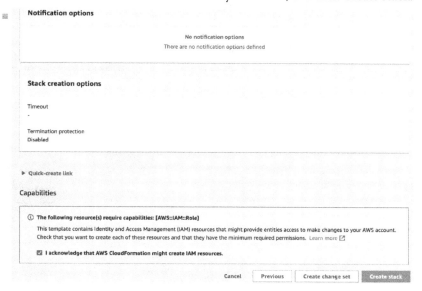

Figure 4.7 – Fourth step in the CloudFormation wizard – launching the X-Ray sample application

6. The process will take a few minutes to finish. You can check the deployment's status by navigating to **CloudFormation | Stacks**, refreshing from time to time using the **Refresh** button, and checking the column status in the **xray-sample** row. Once the deployment finishes, you will see a page similar to the one shown here. This confirms that the application has been successfully deployed:

Figure 4.8 – CloudFormation Stacks with an xray-sample status of CREATE_COMPLETE

Step 2 – navigating the application

1. Now, we need to access the application home page. Click on the **xray-sample** link to see the stack details and, on the **Outputs** tab, copy and paste the value of the **ElasticBeanstalkEnvironmentURL** key into a new browser tab, as shown here:

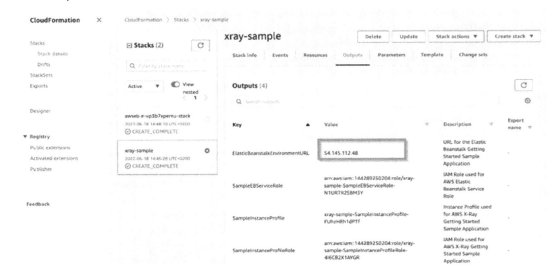

Figure 4.9 – CloudFormation | Stacks | xray-sample details | Outputs tab

2. In this sample application, you can do a signup process manually or automatically generate 10 signup processes every minute. Let's do the latter. Click on the **Start** button:

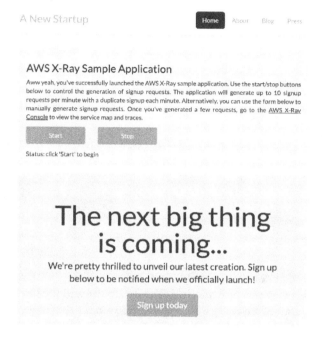

A New Startup   Home   About   Blog   Press

AWS X-Ray Sample Application

Aww yeah, you've successfully launched the AWS X-Ray sample application. Use the start/stop buttons below to control the generation of signup requests. The application will generate up to 10 signup requests per minute with a duplicate signup each minute. Alternatively, you can use the form below to manually generate signup requests. Once you've generated a few requests, go to the AWS X-Ray Console to view the service map and traces.

Start   Stop

Status: click 'Start' to begin

The next big thing is coming...

We're pretty thrilled to unveil our latest creation. Sign up below to be notified when we officially launch!

Sign up today

Figure 4.10 – The A New Startup signup page

The application will show random emails used to emulate the signup process. Wait 5 minutes to make sure enough information has been generated.

Step 3 – navigating the AWS CloudWatch X-Ray user interface

1. We can now open the AWS X-Ray console page to see the service map and traces. Go to https://console.aws.amazon.com/cloudwatch/home#xray:service-map/map?~(query~(filter~())~context~(timeRange~(delta~300000))); you will see a service map similar to the following:

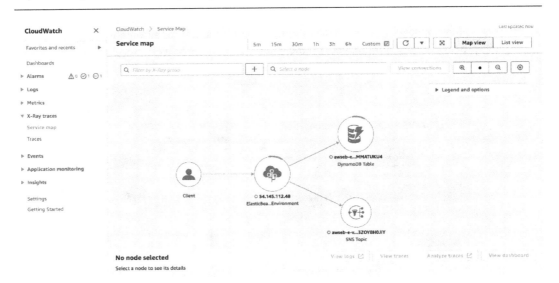

Figure 4.11 – Service map with data from the xray-sample application

Here, we can see the famous *service map*. Each node represents a source/target of requests, while the edges are the requests themselves. The node's size and the line's thickness represent the ratio of the number of requests coming/going from/to the different resources. Nodes with borders in red or orange represent nodes returning Fault (5xx) or Error (4xx) HTTP codes.

This single pane of glass view, which unifies data across different sources, gives you a holistic view of your entire application, from the client request to the application, and finally, the data storage. As you can see, the emulated application also includes some errors to demonstrate how to debug them. We will come back to this in a second.

2. If you click on the application node (the one identified by the application IP), you will see the application's details:

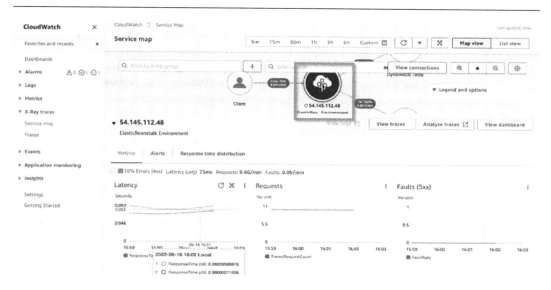

Figure 4.12 – Service map, with the Metrics tab selected

The service map will highlight the service you are checking the details for. You will see **Metrics**, **Alerts**, and **Response time distribution** tabs in the region below the map:

- In the **Metrics** tab, you can see details such as the p50 and p90 latency percentiles, the number of requests per unit of time, and the number of faults served.

- In the **Alerts** tab, we can see any alerts triggered by the application and any X-Ray insights. They are empty for now, but we will return to them in the next section:

Figure 4.13 – Service map with the Alerts tab selected

- The **Response time distribution** tab provides a visual representation of the distribution of response times for each request sent by the client and the distribution across different services in your AWS account. By selecting a particular section of the graph, you can filter out traces that are causing higher latency and focus your analysis on those specific requests. This can help you pinpoint any issues and take the necessary steps to optimize your system's performance:

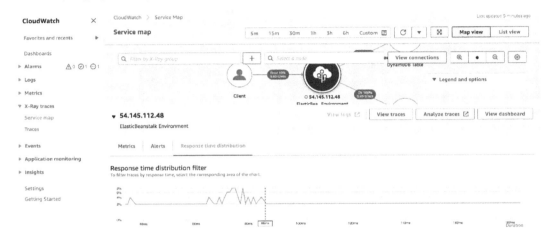

Figure 4.14 – Service map with the Response time distribution tab selected

3. Let's go back to the **Metrics** tabs. We aim to find the root cause of the application errors we can see in the service map. There's a checkbox in the **Metrics** tab labeled **X% Errors (4xx)**. If you click on it, X-Ray will filter and show only the metrics related to traces with application errors. Let's click it and start our investigation:

Figure 4.15 – Service map with the X% Errors (4xx) checkbox selected

4. In the same way, as with **Metrics**, the **View traces** button will change to **View filtered traces**. Initially, clicking on it will redirect you to a list of all application traces, which does not help our investigation. But after you click on the checkbox, the **View filtered traces** button will redirect you to a list of traces related to the application errors. Let's click on it.

5. You will be redirected to **CloudWatch | Traces overview** (see `https://console.aws.amazon.com/xray/home#/traces`). Here, you can build a query to find specific traces, but in our case, it was already pre-populated to retrieve only the traces with errors from the application. We can click on a trace link to drill down into that trace's details:

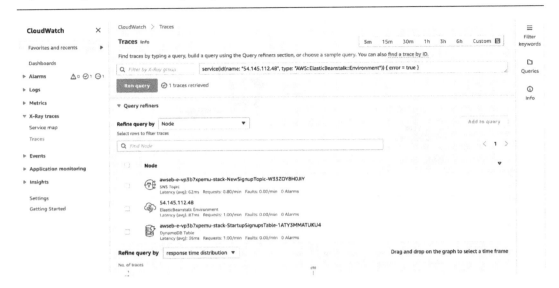

Figure 4.16 – The CloudWatch | Traces page, with a filter to show application errors applied

6. On the trace details page, you will see a **Trace Map**, section where we can see all the resources that took part in this single trace. As we can see, the trace has information about the *Client*, the *Elastic Beanstalk Environment*, and finally, the *DynamoDB Table*. **Trace Summary** contains general information, such as the HTTP method, the response code, duration, and age. The most exciting section is the **Segments Timeline** section:

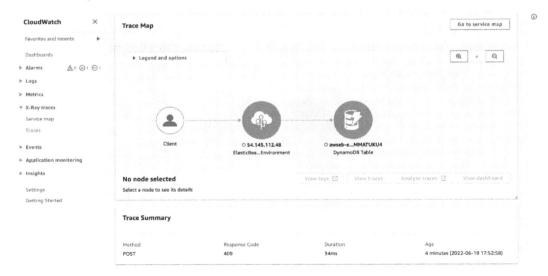

Figure 4.17 – The CloudWatch | Traces page, highlighting the Trace Map section

The **Segments Timeline** section shows the segments and subsegments that are part of this trace, together with temporal data as the start/end times. You can see more details if you click on one of them. We can see that the application segment is returning an error, but the call to the DynamoDB table causes this error. If we click on the **DynamoDB application** segment in the **Segment details** section, we will see what's causing the failure: an exception called `ConditionalCheckFailedException`:

Figure 4.18 – The Segment Timeline section highlighting the DynamoDB application exception

7. If we check the Amazon DynamoDB **Error Handling** page (see `https://docs.aws.amazon.com/amazondynamodb/latest/developerguide/Programming.Errors.html`), we will see an explanation of what the error means:

`ConditionalCheckFailedException`

Message: The conditional request failed.

You specified a condition that is evaluated as false. For example, you might have tried to perform a conditional update on an item, but the actual value of the attribute did not match the expected value in the condition.

Well, what's happening, exactly? Let's look at the application code. You can find the application code at `https://github.com/aws-samples/eb-node-express-sample/tree/xray`, and the section that emulates the end user sign-ups at `https://github.com/aws-samples/eb-node-express-sample/blob/xray/views/index.ejs#L178`. As you can see, every 6 seconds, a new signup is triggered, but 1 in 10 is a duplicated signup that uses a fixed email repeatedly. In the application code (at `https://github.com/aws-samples/eb-node-express-sample/blob/xray/app.js#L81`), we can see that a DynamoDB condition (see `https://docs.aws.amazon.com/amazondynamodb/latest/developerguide/LegacyConditionalParameters.Expected.html`) is used when adding a new item:

```
ddb.putItem({
          'TableName': ddbTable,
          'Item': item,
          'Expected': { email: { Exists: false } }
      },...
```

As soon as you try to execute the signup with an existing email, DynamoDB will return a message stating `The conditional request failed` – bingo!

In this section, you had a little taste of how X-Ray can help you debug your application and connect endpoints and resources in a single pane of glass. It gives you all the tools you need to drill down and find the root cause of an error, even when affecting a small portion of your users, literally helping you find the needle in the haystack.

In the next section, we will discuss another powerful X-Ray feature that allows you to correlate between traces, metrics, and logs, which will give you even more incredible superpowers in your observability journey.

Overview of the CloudWatch ServiceLens map

So far, we have seen how X-Ray traces and service maps can give you a fantastic view of what is going on in your application. If we stop there, that is powerful enough to fulfill your hunger for observability for a long while in many complex cases.

But, as discussed in *Chapter 1, Observability 101*, the exponential growth of your application demands more insights. Traces in X-Ray provide insights into the interaction between different services, but how can we correlate them with all the metrics and logs our application generates? Should we go back and do it manually once more?

That is why we have CloudWatch ServiceLens, which can correlate traces, logs, and metrics in a single place. You can easily find performance issues and bugs affecting your end users with more information.

For the time being, there are some requirements for this magic to happen:

- An updated X-Ray SDK
- Only the SDK for Java supports log correlation
- Only Lambda functions, API Gateway, Java-based applications running on Amazon EC2, and Java-based applications running on Amazon EKS or Kubernetes with Container Insights deployed support log correlation

You can find detailed instructions on setting up CloudWatch ServiceLens for your application in the AWS documentation (see `https://docs.aws.amazon.com/AmazonCloudWatch/latest/monitoring/deploy_servicelens.html`). In the last section of this chapter, we will look at an end-to-end example of how to set it up on a Java application.

Overview of X-Ray Analytics

I saw a post once (and I do not remember exactly where) that resonated with my understanding of observability. I can't remember who said this, and I hope, if someday the author reads these lines, they can forgive my lack of memory and not properly giving them the credit, but the post said the following:

"Observability is not a DevOps problem; it is a data analytics problem." – Unknown

And I cannot agree more. As discussed in *Chapter 1, Observability 101*, if a tool does not provide you with ways to slice and aggregate the information you collect from an application, the tool is not contributing to achieving better observability of your application.

That's where X-Ray Analytics comes in to support you. X-Ray Analytics provides several built-in dashboards, which display aggregate statistics, such as service maps, response time distributions, error rates, and more. You can also create custom dashboards to display the metrics and dimensions that are most relevant to your use case. The X-Ray Analytics traces view is shown in the following figure:

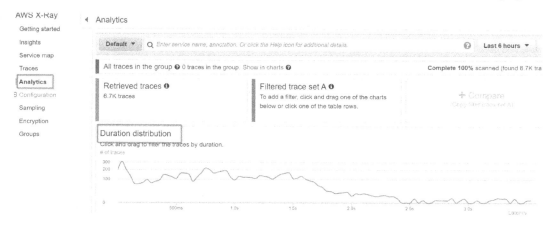

Figure 4.19 – X-Ray Analytics view

Having gained an overview of the concepts of X-Ray and navigated through the various views available in AWS X-Ray, it is time to explore how to instrument an end-to-end sample application using X-Ray for an application that is running on an EC2 instance.

End-to-end instrumentation of a sample application deployed in an EC2 instance

In this section, we will deploy a Java application in an EC2 instance and see the steps necessary to have everything up and running. We usually use the Python language for sample code, but the X-Ray Java SDK, at the time of writing, provides a complete set of features, as we saw in the previous sections.

This section is hands-on, so log in to your AWS account and be prepared for the ride. AWS will charge you for the resources deployed in this section, but for short periods, this is not something that will break the bank. When you finish, remember to clean up the used resources.

This hands-on exercise is spread over three steps:

1. Understanding the application and deploying the applicatio.

2. Testing the sample application to generate X-Ray trace.

3. Exploring the sample application by logging into the EC2 instance

Preparing the environment

Follow these steps to prepare the environment:

1. A CloudFormation template has been provided in this book's GitHub repository for deploying everything necessary so that you have a Linux machine up and running and ready to receive HTTP requests. The following diagram shows the infrastructure you will deploy with it:

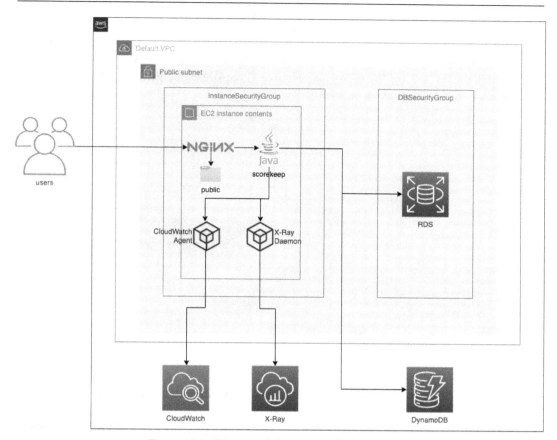

Figure 4.20 – Diagram of the deployed infrastructure

To deploy the application in your AWS Account, log in to your AWS account and click or copy and paste the following URL into your browser window: `https://console.aws.amazon.com/cloudformation/home#/stacks/new?stackName=scorekeep&templateURL=https://insiders-guide-observability-on-aws-book.s3.amazonaws.com/chapter-04/basic-ec2-template.yml`. If you are the type of person who does not trust people deploying things in your account, you can always check the template itself here: `https://insiders-guide-observability-on-aws-book.s3.amazonaws.com/chapter-04/basic-ec2-template.yml`.

2. This action will open the AWS CloudFormation console with all the parameters that are necessary for the template, along with the default values. You can keep the default values and click on the **Next** button on all the subsequent pages. On the last one, named **Review scorekeep**, do not forget to check the box stating **The following resource(s) require capabilities: [AWS::IAM::Role]** and click on the **Create stack** button:

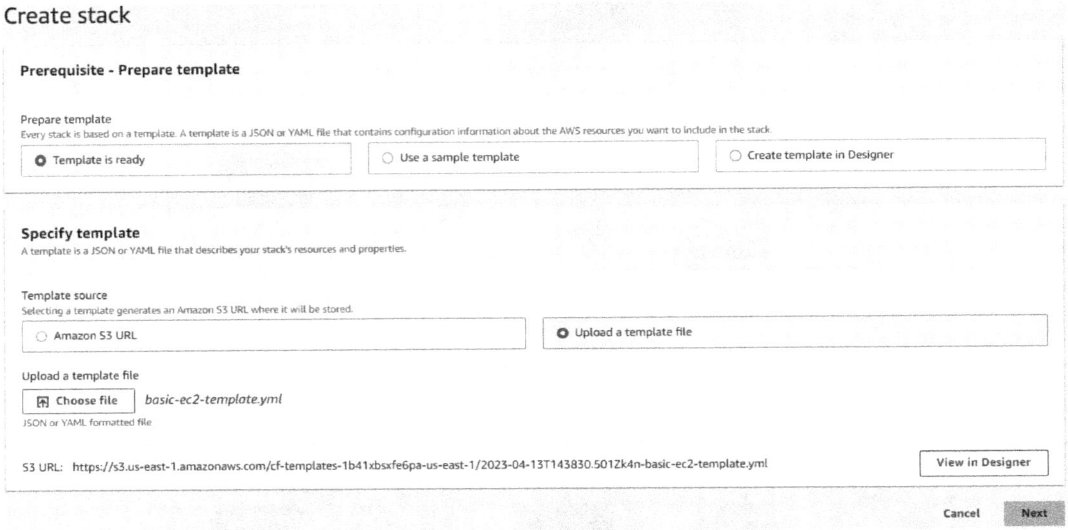

Figure 4.21 – The application stack, loading the template from an external S3 bucket

Let's acknowledge the notification about creating relevant IAM roles for deploying the stack:

Figure 4.22 – Accepting the IAM permission for creating the application stack

3. We will go through some relevant sections of this CloudFormation template in the *Exploring the sample application* section, but if you are curious, you can go to `https://insiders-guide-observability-on-aws-book.s3.amazonaws.com/chapter-04/basic-ec2-template.yml` to see what's inside. We did our best to create a great example with many useful tricks worth checking.

4. Wait until the Stack switches to the **CREATE_COMPLETE** state. After a short timearound 10 to 15 minutes to be more precise), your account will have the following resources deployed for our fun:

   - An EC2 instance
   - A Lambda function
   - A PostgreSQL database
   - Five DynamoDB tables
   - An SNS topic
   - Security groups
   - Roles and permissions

Testing the sample application

Now, it's time to test the sample application:

1. The sample application has been installed and is up and running. You can access the sample application using the URL mentioned for the **PublicDNS** key, which you can find in the **Stacks | Outputs** tab. Click on it:

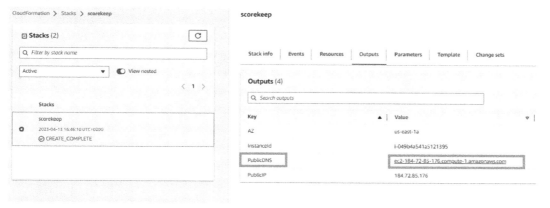

Figure 4.23 – The Stacks | Outputs tab, where you can find the PublicDNS key

2. Once you have clicked on the **PublicDNS** URL, you will see the UI of our small application
    – **Scorekeep**:

Figure 4.24 – The Scorekeep main UI

3. You can enter your name and create a game session (or join an existing one). You can provide a
    game for your session and select the game you want to play (in this MVP, only Tic Tac Toe...):

Figure 4.25 – Creating a new Tic Tac Toe game session

4. Let's play a couple of sessions of the "Tic Tac Toe" game. That is important as it will generate the log and metrics data we will use in the following sections. Or, if you are in a hurry, you can click on the top-right link, **Powered by AWS X-Ray**. You will see two buttons on the following page: **Trace game sessions** and **Trace SQL queries**. If you click on either of them, you will see a simulation of game sessions. You can click on them and leave the browser window/tab open to generate enough traffic. After a while, you can access the X-Ray service map page (go to `https://console.aws.amazon.com/xray/home#/service-map`), and you will see a service map similar to the one shown here. Alternatively, if you opted for a new AWS X-Ray Console, you can navigate to **CloudWatch | X-Ray | Service Map**:

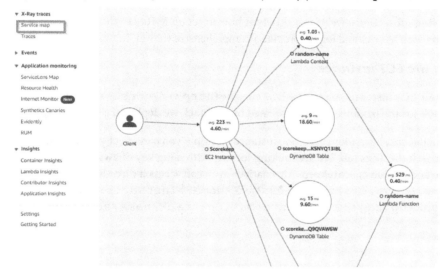

Figure 4.26 – The application service map

Let's continue our exploration by logging into the EC2 instance that's been deployed and exploring the CloudFormation template that's been deployed, along with exploring the important components of the sample application that have been deployed from X-Ray's point of view.

Exploring the sample application running on an EC2 instance

The application we will analyze follows the basic web three-tier architecture. We will use an Nginx server to publish the web UI to the end users and forward any requests starting with the /API path to the Java application port running on the same machine. The Java application will store data using both RDS and DynamoDB databases.

On the same machine, we will run both the CloudWatch agent and the X-Ray daemon, where the application communicates metrics, traces, and logs. The agent and daemon share this data with the CloudWatch and X-Ray APIs.

In the following subsections, we'll look closely at how to set up X-Ray so that it can collect traces and integrate them with metrics and logs to provide a comprehensive view of the application's inner workings.

Accessing the EC2 instance

Let's access our EC2 instance. Access the EC2 console (https://console.aws.amazon.com/ec2) and look for the instance that was created by the stack we deployed a moment ago:

1. If you, like me, have more than one instance running, you can select your instances one by one and check the **Tags** tab. We are looking for one with a tag key of **aws:cloudformation:stack-name** and a value of **scorekeep**. Alternatively, you can access it directly by going to https://console.aws.amazon.com/ec2/v2/home?#Instances:v=3;instanceState=running;tag:aws:cloudformation:stack-name=scorekeep:

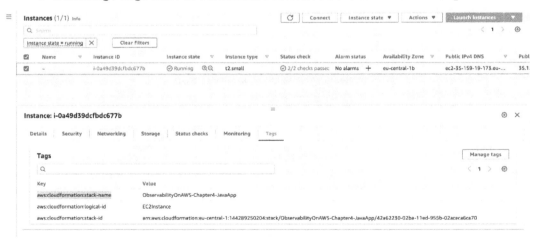

Figure 4.27 – An EC2 instance with a tag of aws:cloudformation:stack-name equal to scorekeep

2. Once we have found our instance, we can connect to it. Click on the **Connect** button at the top of the EC2 console. It will show the **Connect to instance** page. The default tab, **EC2 Instance Connect**, is what we need. You don't need to change any other parameter; click on the **Connect** button. Another tab will open with a prompt connected to our instance, like a charm:

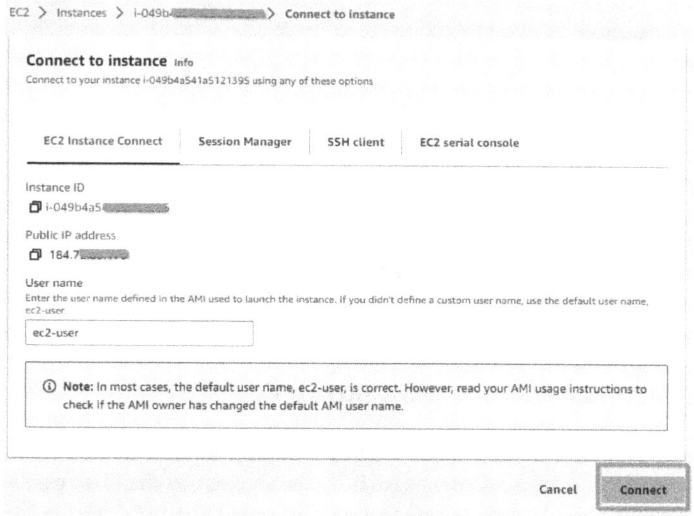

Figure 4.28 – Connecting to the EC2 instance using the EC2 Instance Connect method

You can see the SSH prompt once you connect to the EC2 instance:

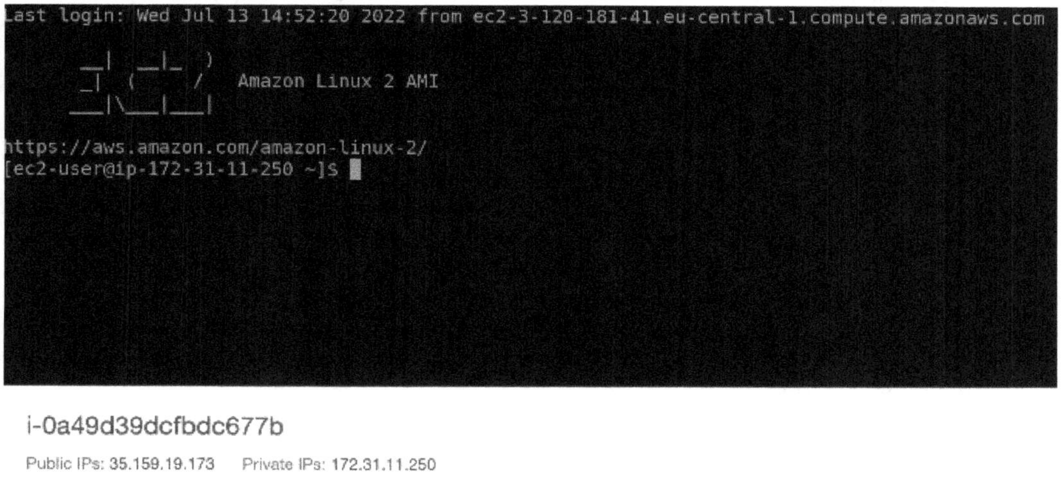

i-0a49d39dcfbdc677b

Public IPs: 35.159.19.173   Private IPs: 172.31.11.250

Figure 4.29 – Logged in to the EC2 instance

3. In this machine, you will find a folder called *java-scorekeep* as this is deployed as part of the CloudFormation template and contains the project source code. In this machine, we have the following processes running:

- CloudWatch agent

- AWS X-Ray daemon

- Nginx

- Scorekeep application

We will discuss each one at a time.

We strongly advise you to keep this X-Ray console window open while reading the following sections so that you can explore and understand the material better. We have both the **nano** (https://www.nano-editor.org/) and the **vim** (https://www.vim.org/) editors installed. The nano editor is easier to use, but vim is an old friend to most system admins, so feel free to pick one.

Installing the AWS X-Ray daemon on Amazon EC2

We will use a feature of the CloudFormation template called AWS::CloudFormation::Init (see https://docs.aws.amazon.com/AWSCloudFormation/latest/UserGuide/aws-resource-init.html). With it, you can provide metadata information to the cfn-init script we will run inside the machine on startup. You can find the X-Ray daemon installation instructions in the AWS documentation (see https://docs.aws.amazon.com/xray/latest/devguide/xray-daemon.html). In our case, you can find the installation instructions in our CloudFormation template between lines 444 and 449 (here's the template file: https://insiders-guide-observability-on-aws-book.s3.amazonaws.com/chapter-04/basic-ec2-template.yml). The following code snippet is for Linux and the rpm package manager since we are using an Amazon Linux 2 image:

```
curl https://s3.us-east-2.amazonaws.com/aws-xray-assets.us-east-2/
xray-daemon/aws-xray-daemon-3.x.rpm -o /home/ec2-user/xray.rpm
yum install -y /home/ec2-user/xray.rpm
rm /home/ec2-user/xray.rpm
```

The daemon will automatically start and restart on every EC2 instance reboot, and the X-Ray daemon will start listening for application connections on port 2000 (see the X-Ray configuration setup at https://docs.aws.amazon.com/xray/latest/devguide/xray-daemon-configuration.html).

Installing the CloudWatch agent on Amazon EC2

In the same way as the X-Ray daemon, we will use `AWS::CloudFormation::Init` to provide instructions on installing and configuring the CloudFormation agent. This process is divided into three steps:

1. Install the CloudWatch agent binaries.
2. Copy the CloudWatch agent configuration file to the right place.
3. Stop/start the agent, passing the new configuration file as a parameter.

If you want more details about the CloudWatch agent, refer to *Chapter 3*, *Gathering Operational Data and Alerting Using Amazon CloudWatch*. In this example, you can see each of these steps in our CloudFormation template, as shown in the *Preparing the environment* section. The binary installation happens in the `packages` section of the CloudFormation Init metadata, as shown in the following code snippet:

```
install_deps:
  packages:
    yum:
      java-1.8.0-amazon-corretto-devel: []
      nginx: []
      amazon-cloudwatch-agent: []
```

You can see the CloudWatch agent configuration file in the CloudFormation template, on lines 302 to 430. It is an extensive file, so I won't reproduce it entirely here, only the initial part:

```
config-amazon-cloudwatch-agent:
  files:
    "/opt/aws/amazon-cloudwatch-agent/etc/amazon-cloudwatch-agent.
json":
      content: |
        {
          "agent": {
            "metrics_collection_interval": 10,
            "logfile": "/opt/aws/amazon-cloudwatch-agent/logs/amazon-
cloudwatch-agent.log"
          },
          "metrics": {
            "namespace": "ScoreKeep",
            "metrics_collected": {
            "cpu": {
            "resources": [
...
```

I have shown a longer portion of the file to show many of the features offered by CloudWatch, among them a detailed memory utilization, processes, and a storage set of metrics. You can learn about the meaning of each of them at `https://docs.aws.amazon.com/AmazonCloudWatch/latest/monitoring/CloudWatch-Agent-Configuration-File-Details.html`.

Application setup

The application setup is simple. In our example, we are using a Java application you can find in the `~/java-scorekeep` folder. Any application must add the required libraries to make X-Ray API calls and publish trace data. In our case, as we are using Gradle (see `https://gradle.org/`) as our build and dependency management tool, we need to add the necessary packages to the build file, which you can find in `~/java-scorekeep/build.gradle`. See the following code snippet:

```
dependencies {
  ...
  compile("com.amazonaws:aws-xray-recorder-sdk-core")
  compile("com.amazonaws:aws-xray-recorder-sdk-aws-sdk")
  compile("com.amazonaws:aws-xray-recorder-sdk-aws-sdk-instrumentor")
  compile("com.amazonaws:aws-xray-recorder-sdk-apache-http")
  compile("com.amazonaws:aws-xray-recorder-sdk-sql-postgres")
  compile("com.amazonaws:aws-xray-recorder-sdk-metrics")
  compile("com.amazonaws:aws-xray-recorder-sdk-slf4j")
  ...
}
dependencyManagement {
  imports {
    mavenBom("com.amazonaws:aws-java-sdk-bom:1.11.761")
    mavenBom("com.amazonaws:aws-xray-recorder-sdk-bom:2.11.0")
  }
}
```

You can find all the X-Ray SDKs for the Java libraries and sub-modules and their meanings at `https://docs.aws.amazon.com/xray/latest/devguide/xray-sdk-java.html`, as well as how to use them for both the Maven and Gradle build managers.

The AWS X-Ray SDK

Once we have the X-Ray **aws-xray-recorder-sdk-aws-sdk-instrumentor** SDK library as one of the application dependencies, any call to any AWS SDK client is instrumented automatically. It doesn't matter if you access Amazon RDS, SQS, SNS, DynamoDB, or any other AWS service; you can remove the library from your dependencies to deactivate the automatic instrumentation. You can also choose to instrument only some AWS SDK calls. In this use case, you can remove the library from your dependencies and add the instrumentation as per the AWS client manually. This can be seen in the following code, where we are only instrumenting **Amazon DynamoDB calls**:

```
public class SessionModel {
    private AmazonDynamoDB client = AmazonDynamoDBClientBuilder.
standard()
  .withRequestHandlers(new TracingHandler())
  .build();
```

Subsegments

With the AWS X-Ray dependencies in place, your application will send segment information from the moment your application receives a request until it's sent the response to the client, which is a great start. But usually, some application flows and/or integrations are more critical and we would like more details. In those cases, we can add subsegments. So, instead of having a big single monolithic segment, we will have separated segments for specific regions of our code, where we can drill down to see more information.

X-Ray annotations and metadata

- You have the option to provide extra information about the execution environment in your segments and subsegments by leveraging annotations and metadata:

- **Annotations**: Annotations are key-value pairs you can use to search for using filter expressions (see https://docs.aws.amazon.com/xray/latest/devguide/xray-sdk-java-segment.html#xray-sdk-java-segment-annotations) (see https://docs.aws.amazon.com/xray/latest/devguide/xray-console-filters.html) or to group segments

- **Metadata**: Metadata is (see https://docs.aws.amazon.com/xray/latest/devguide/xray-sdk-java-segment.html#xray-sdk-java-segment-metadata) pieces of information you want to have stored alongside the segment, but you don't need them indexed for search

Programmatically creating subsegments, annotations, and metadata

In the following figure, you can see a section of the application code that uses subsegments. When you navigate to the `/home/ec2-user/java-scorekeep/src/main/java/scorekeep/GameModel.java` path, as described in the *Accessing the EC2 instance* section, you can understand the addition of sub-segments in AWS X-ray.

The method begins by creating a new subsegment of an X-Ray segment using `AWSXRay.beginSubsegment()`. This allows the method to track performance and record additional information about this `saveGame` operation within the larger request being handled by the application:

```
public void saveGame(Game game) throws SessionNotFoundException {
    // wrap in subsegment
    Subsegment subsegment = AWSXRay.beginSubsegment("## GameModel.saveGame");
    try {
        // check session
        String sessionId = game.getSession();
        if (sessionModel.loadSession(sessionId) == null ) {
            throw new SessionNotFoundException(sessionId);
        }
        Segment segment = AWSXRay.getCurrentSegment();
        subsegment.putMetadata("resources", "game", game);
        segment.putAnnotation("gameid", game.getId());
        mapper.save(game);
    } catch (Exception e) {
        subsegment.addException(e);
        throw e;
    } finally {
        AWSXRay.endSubsegment();
    }
    mapper.save(game);
}
```

Figure 4.30 – Subsegment, annotation, and metadata in AWS X-Ray

Next, the method checks whether the `sessionId` property associated with the game object is valid by using the `sessionModel.loadSession()` method to load the session. If the session cannot be found, a `SessionNotFoundException` error is thrown.

If the session is found, the method adds metadata to the X-Ray subsegment called `subsegment.putMetadata()` and adds an annotation to the X-Ray segment using `segment.putAnnotation()`. These methods allow the developer to attach additional information to the X-Ray segment and subsegment for later analysis and troubleshooting.

Finally, the method attempts to save the game object to the database using the DynamoDB mapper and ends the X-Ray subsegment using `AWSXRay.endSubsegment()`. If any exceptions are thrown during the execution of the method, they are added to the X-Ray subsegment using `subsegment.addException()` before being re-thrown. This allows the X-Ray service to track any errors that occur during the method's execution.

You could search the traces using annotations added by using `gameid` as required, as shown in the following figure:

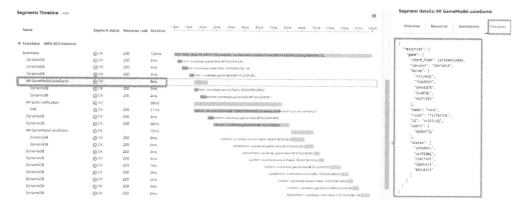

Figure 4.31 – Leveraging an annotation to filter the X-Ray traces

Additionally, you can see the subsegment and metadata information for the traces of **saveGame** in the trace details, as shown in the following figure:

Figure 4.32 – An HTTP POST request highlighting the GameModel.saveGame subsegment and metadata

Instrumenting calls to a PostgreSQL database

You can instrument SQL queries using one of the X-Ray SDK Java JDBC interceptors (see `https://docs.aws.amazon.com/xray/latest/devguide/xray-sdk-java-sqlclients.html`). This action will add information about the start/end time and any errors, but it doesn't add information about the query itself for security reasons. Here, we will look at an example of how to configure the interceptor when using the Spring framework, and the `/home/ec-user/java-scorekeep/src/main/resources/application-pgsql.properties` file.

The `spring.datasource.jdbc-interceptors` property is set to `com.amazonaws.xray.sql.postgres.TracingInterceptor`, which means that the application uses the Tracing Interceptor from the AWS X-Ray JDBC Instrumentation library to trace SQL statements sent to a PostgreSQL database. This allows the X-Ray service to track the performance and other details of the database queries being executed by the application:

```
spring.datasource.continue-on-error=true
spring.jpa.show-sql=false
spring.datasource.jdbc-interceptors=com.amazonaws.xray.sql.postgres.TracingInterceptor
spring.jpa.database-platform=org.hibernate.dialect.PostgreSQL94Dialect

                                              PostgreSQL Tracing Interceptor
```

Figure 4.33 – PostgreSQL database tracing interceptor

The resulting trace looks like this for the PostgreSQL database:

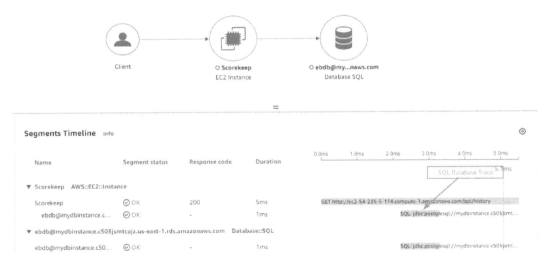

Figure 4.34 – Resulting database trace

Instrumenting AWS Lambda functions

Our sample application uses a single Lambda function called `random-name`. Every time a user starts a session without typing their name, the main application calls this Lambda, generating a random name for the user. The CloudFormation template we are using sets up a function to enable X-Ray tracing. See the following CloudFormation snippet:

```
RandomNameFunction:
  DependsOn: CopyZips
  Type: AWS::Lambda::Function
  Properties:
    FunctionName: random-name
    Handler: index.handler
    Runtime: nodejs14.x
    Role: !GetAtt 'FunctionRole.Arn'
    Code:
      S3Bucket: !Ref 'LambdaZipsBucket'
      S3Key: !Sub 'chapter-04/random-name-lambda.zip'
    Description: Generate random names
    Timeout: 10
    TracingConfig:
      Mode: Active
    Environment:
      Variables:
        TOPIC_ARN: !Ref notificationTopic
```

Figure 4.35 – Tracing Lambda function

When we activate the tracing feature for Lambda functions, we don't need to create new segments to trace the lambda execution; the X-Ray SDK does this for us, adding the `trace-id` property coming from the request and including it in the execution context. We will discuss AWS Lambda instrumentation in *Chapter 7, Observability for Serverless Applications on AWS*.

Instrumenting startup code

So far, we have seen that for any call to our application endpoints, the X-Ray SDK takes care of creating a trace ID and a trace segment, which we can use to collect even more application data. But typically, applications have portions of their code that do not run inside the context of a request, such as initialization code or batch jobs.

In those cases, you need to create segments manually. Let's learn how we can do this in our initialization code as part of the `/home/ec2-user/java-scorekeep/src/main/java/scorekeep/RdsWebConfig` class:

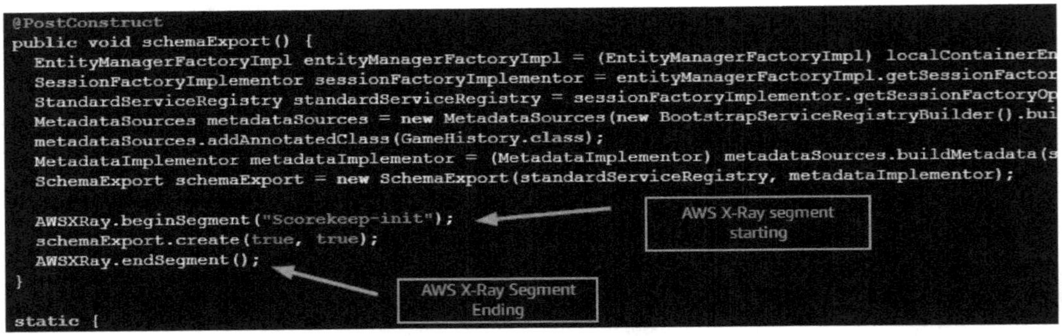

```
@PostConstruct
public void schemaExport() {
    EntityManagerFactoryImpl entityManagerFactoryImpl = (EntityManagerFactoryImpl) localContainerEn
    SessionFactoryImplementor sessionFactoryImplementor = entityManagerFactoryImpl.getSessionFactor
    StandardServiceRegistry standardServiceRegistry = sessionFactoryImplementor.getSessionFactoryOp
    MetadataSources metadataSources = new MetadataSources(new BootstrapServiceRegistryBuilder().bui
    metadataSources.addAnnotatedClass(GameHistory.class);
    MetadataImplementor metadataImplementor = (MetadataImplementor) metadataSources.buildMetadata(s
    SchemaExport schemaExport = new SchemaExport(standardServiceRegistry, metadataImplementor);

    AWSXRay.beginSegment("Scorekeep-init");  ◄────────    AWS X-Ray segment
    schemaExport.create(true, true);                           starting
    AWSXRay.endSegment();
}
                                    AWS X-Ray Segment
static {                                 Ending
```

Figure 4.36 – Tracing startup code

As you can see, this initialization code creates the database schema. To create X-Ray segments and track the duration and any downstream call, we just need to call the `beginSegment()` and `endSegment()` methods.

Using instrumented clients in worker threads

The application uses a worker thread to send a notification to Amazon SNS when a user wins a game. The purpose of this code is to send a notification in a separate thread (asynchronously), using a subsegment to trace the execution of the notification-sending logic. This method is required when you want to propagate the trace ID in a parent-child entity.

The X-Ray Java SDK uses **ThreadLocal** (see `https://docs.oracle.com/javase/7/docs/api/java/lang/ThreadLocal.html`) to store and propagate the `trace-id` property and segment information. This information is attached to the running thread but is lost if a new thread is started. And if you try to use the X-Ray Java SDK without creating a new context, it will throw a `SegmentNotFoundException` error.

To do it right, you need to use the `getTraceEntity` method to get a reference of the parent segment and then reinsert it in the newly created thread. You can see an example in the `/home/ec2-user/java-scorekeep/src/main/java/scorekeep/MoveFactory.java` file:

```
// send notification on game end
Entity segment = recorder.getTraceEntity();
if ( newStateText.startsWith("A") || newStateText.startsWith("B")) {
    Thread comm = new Thread() {
        public void run() {
            segment.run(() -> {
                Subsegment subsegment = AWSXRay.beginSubsegment("## Send notification");
                Sns.sendNotification("Scorekeep game completed", "Winner: " + userId);
                AWSXRay.endSubsegment();
            });
        }
    };
    comm.start();
}
```

Figure 4.37 – Tracing worker subsegments in an application

The result can be seen in the following trace:

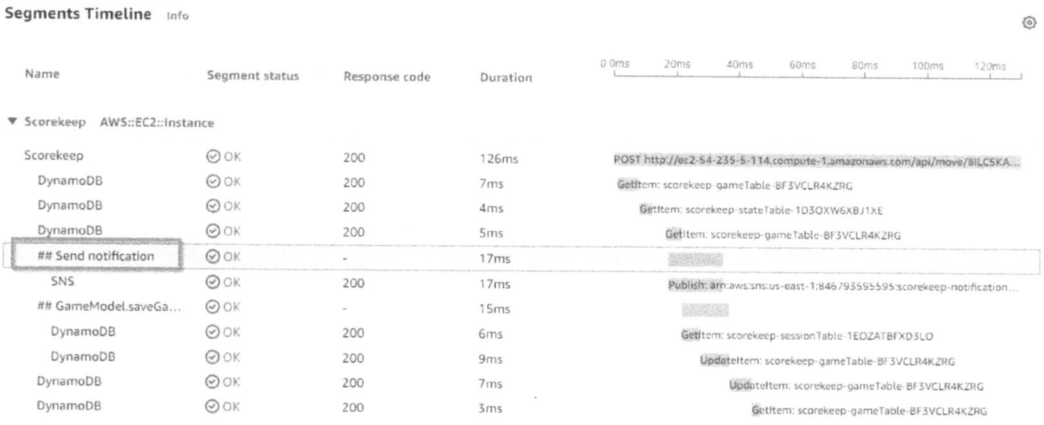

Segments Timeline Info

Name	Segment status	Response code	Duration	
▼ Scorekeep AWS::EC2::Instance				
Scorekeep	⊘ OK	200	126ms	POST http://ec2-54-235-5-114.compute-1.amazonaws.com/api/move/8ILCSKA...
DynamoDB	⊘ OK	200	7ms	GetItem: scorekeep-gameTable-BF3VCLR4KZRG
DynamoDB	⊘ OK	200	4ms	GetItem: scorekeep-stateTable-1D3OXW6XBJ1XE
DynamoDB	⊘ OK	200	5ms	GetItem: scorekeep-gameTable-BF3VCLR4KZRG
## Send notification	⊘ OK	-	17ms	
SNS	⊘ OK	200	17ms	Publish: arn:aws:sns:us-east-1:846793595595:scorekeep-notification...
## GameModel.saveGa...	⊘ OK	-	15ms	
DynamoDB	⊘ OK	200	6ms	GetItem: scorekeep-sessionTable-1EDZATBFXD3LO
DynamoDB	⊘ OK	200	9ms	UpdateItem: scorekeep-gameTable-BF3VCLR4KZRG
DynamoDB	⊘ OK	200	7ms	UpdateItem: scorekeep-gameTable-BF3VCLR4KZRG
DynamoDB	⊘ OK	200	3ms	GetItem: scorekeep-gameTable-BF3VCLR4KZRG

Figure 4.38 – The trace with data from a different thread

In this section, we understood how to instrument a sample application using AWS X-Ray and understood how AWS X-Ray can be utilized to trace the distributed application spanning EC2, DynamoDB, Lambda, Postgres data, and SNS notifications. We also understood enhancements that can be made using subsegments, adding annotations and metadata to the X-Ray traces to provide additional context to the operations team when troubleshooting issues.

Summary

In this chapter, we saw why AWS X-Ray is required in a distributed application and navigated the foundations of AWS X-Ray. We understood various components of the AWS X-ray console, such as the ServiceLens map, AWS X-Ray Analytics, and traces, and how to understand them. Further, we instrumented a Java application running on EC2 using AWS X-Ray and understood how to read the trace data and how annotations and metadata can support troubleshooting operational issues and filter out traces that are relevant to the problem.

At this point, you understand the value of having a distributed tracing tool in your observability toolbelt and how to apply it correctly to collect all the necessary data.

This is the last chapter of *Part 1* of this book, where you gained a good first picture of observability's more fundamental building blocks. *Part 2* will introduce services that remove much of the manual work necessary to gather and show the relevant data and introduce innovative machine learning tools as virtual companions that will assist you in finding the needle in the haystack.

Part 2: Automated and Machine Learning-Powered Observability on AWS

In this part, we will discuss the services available in AWS for automated and machine learning-based experiences for DevOps engineers. We will also discuss the existing AWS services that make the lives of observability practitioners much easier, collecting, aggregating, and inferring the application state automatically with little to no upfront work needed.

This section has the following chapters:

5

Insights into Operational Data with CloudWatch

In the previous chapter, we learned how to implement AWS X-Ray to understand the user journey and how it provides observability for an application running on an EC2 instance. In this chapter, we will see how data in CloudWatch metrics can be presented to solve operational issues and enhance the experience of the operational team. We will then look at instrumenting an application running on an EC2 instance using AWS Application Insights and how that simplifies the job of an operational team due to the auto-instrumenting functionality. Further, we will look into deriving and understanding operational intelligence from CloudWatch Logs using CloudWatch Logs Insights. We will also look at CloudWatch Contributor Insights and its use cases.

In this chapter, we are going to cover the following topics and learn how to configure services practically:

- Deriving operational intelligence from CloudWatch metrics
- Exploring CloudWatch Application Insights
- Exploring CloudWatch Logs Insights
- Exploring CloudWatch Contributor Insights and its use cases

Technical requirements

To be able to carry out the technical tasks in the chapter, you will need to have the following technical prerequisites:

- A working AWS account
- Knowledge of the setup and configuration of an EC2 instance in the AWS account

- An understanding of the deployment of CloudFormation templates
- A fundamental understanding of DynamoDB
- An understanding of resource groups in AWS

Deriving operational intelligence from CloudWatch metrics

In *Chapter 3, Gathering Operational Data and Alerting Using Amazon CloudWatch*, we discussed what CloudWatch metrics are and the different metrics that can be sent to CloudWatch. Also, we briefly explored the CloudWatch metrics explorer in *Chapter 2, Overview of the Observability Landscape on AWS*. Now let's deep dive further into enhancing operational intelligence with CloudWatch metrics!

CloudWatch metrics explorer

CloudWatch metrics explorer is a flexible tag-based tool that filters, aggregates, and visualizes metrics by resource tags. Tag-based monitoring helps you categorize and compare application resources through the use of tags so that you can monitor them efficiently. CloudWatch metric explorer could be used for two different purposes:

- *Monitoring*: With tag-based monitoring, you create a dynamic application infrastructure health dashboard that automatically updates as you deploy new resources. Simply assign tags to your resources and use them to build a comprehensive, real-time view of your infrastructure's health. For example, if you are building an Auto Scaling group of EC2 instances for an application and are looking to provide a dynamic dashboard to the application team whenever a new resource is deployed without manually adding it to the dashboard, you can build a tag-based monitoring dashboard to achieve the functionality.

- *Troubleshooting*: We leverage CloudWatch metrics explorer to troubleshoot by providing sliced and diced metrics and using the resulting visualizations to spot patterns, correlate the results, and reduce the time taken for resolution.

Now let's see how to use the CloudWatch metrics explorer to build a dynamic dashboard for an Auto Scaling web server group:

1. To carry out the exercises in this section, let's build an EC2 Auto Scaling web cluster using AWS CloudFormation using the following template. You can download the CloudFormation template from the following URL and deploy the EC2 Auto Scaling web cluster:

    ```
    https://insiders-guide-observability-on-aws-book.s3.amazonaws.
    com/chapter-05/autoscalingalbcftemplate.json
    ```

    We have built an Auto Scaling EC2 web server with two instances supporting an application behind the load balancer. To effectively plan for capacity, let's say we need to understand the combined utilization of these two instances. Using the CloudWatch metrics explorer, we will merge metrics from both instances to gain a comprehensive understanding of their usage.

2. Navigate to **AWS Console | CloudWatch | Metrics | Explorer**. Metrics explorer provides two different templates – namely, generic templates and service-based templates, or you can select an empty explorer.

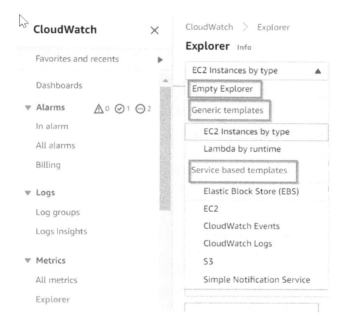

Figure 5.1 – Templates in metrics explorer

3. For this exercise, select **Service based templates** and select **EC2,** and it should select all the metrics related to EC2, as shown in the following screenshot.

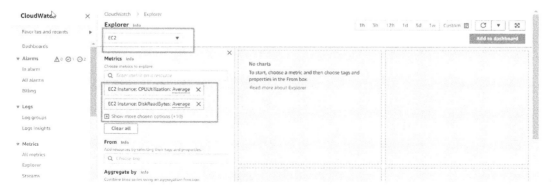

Figure 5.2 – EC2 template

4.   In **From**, select **aws:autoscaling:groupName** and select the created Auto Scaling cluster, as shown in the following screenshot.

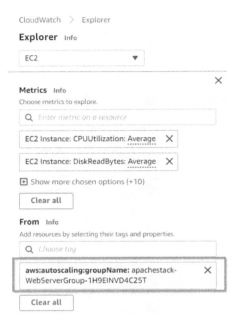

Figure 5.3 – Selecting the Auto Scaling web server group

5.   Look at the **Explorer** page and you will see that EC2 metrics are displayed as line graphs, featuring data from both EC2 instances.

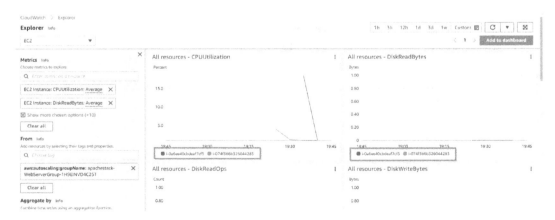

Figure 5.4 – Metrics explorer view

6.   Now, navigate to **Aggregate by** and select **Sum** and **All resources**.

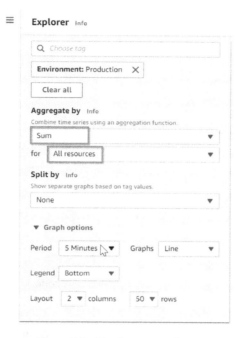

Figure 5.5 – Metric aggregation

This allows the consolidated utilization of data from both EC2 instances and presents a unified view, as shown in the following figure.

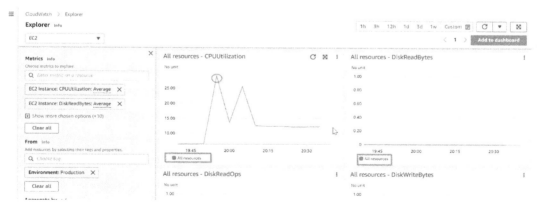

Figure 5.6 – Aggregated view of CloudWatch metrics

7. The view can be further sliced using **Split by** when you want to understand the utilization by availability zone to understand where most of your load is being distributed. This can be helpful and handy if you are troubleshooting an uneven load distribution across **availability zones (AZs)**.

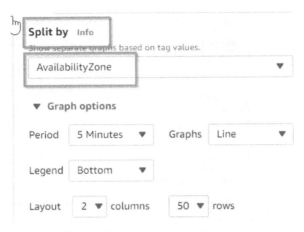

Figure 5.7 – Split by availability zone

8. You can also add this view to the CloudWatch dashboard by clicking on the **Add to Dashboard** button at the top. As the new instances are being built in the Auto Scaling cluster, the new EC2 instances will be added dynamically to the CloudWatch dashboard as we have filtered/grouped them using the Auto Scaling tag (**aws:autoscaling:groupName**).

    Also, you can achieve equal functionality by leveraging the **Explorer** widget on the CloudWatch dashboard.

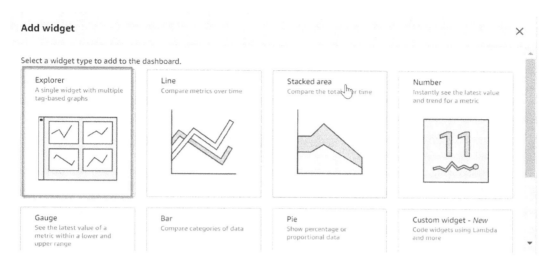

Figure 5.8 – Explorer widget

Through the previous steps, we have gained knowledge of how to use the tag-based metrics explorer to retrieve metrics and tailor the data to meet specific requirements by slicing and dicing the metrics.

Tag-based monitoring and CloudWatch metrics explorer can be leveraged further in the following scenarios:

- **Multi-application environments**: Filter performance by a specific application in a multi-application environment when you are sharing your AWS account

- **Multi-AZ deployments**: Compare performance trends across availability zones (you can use the **Split by** option as discussed earlier)

- **Multi-Region deployments**: Compare performance trends across regions

- **Multi-account deployments**: Compare performance trends or combine metrics running in multiple accounts across multiple regions

- **Multiple sub-components**: Filter performance by a specific sub-component in a complex application (leveraging the **Split by** option)

- **Monitoring deployments**: You can compare performance by instance type and other resource properties that help to isolate issues and correlate with operational events such as upgrades and patching

CloudWatch Metrics Insights

We discussed **CloudWatch Metrics Insights** in *Chapter 2, Overview of the Observability Landscape on AWS*, as a part of **Layer 4**, a newly released AWS service. CloudWatch Metrics Insights provides a fast, flexible, high-performance SQL-based query engine that can query metrics at scale in real time. You can group and aggregate your custom metrics as well as AWS vended metrics in real time based on use cases and business requirements for up to 3 hours (at the time of writing).

These are some of the use cases you can leverage CloudWatch Metrics Insights for:

- Understanding the top 10 highest-consuming compute instances in the account will help you understand over-utilized instances

- The Lambda functions that are running for the longest time will help explain performance issues caused by long-running applications

Metric math expressions

Metric math functionality enables you to query many CloudWatch metrics and use mathematical expressions to derive meaningful insights from them and create new metrics. The following are some intriguing use cases that can be achieved by utilizing metric math functions:

- **Dynamic time threshold**: By utilizing metric math functions, we can create alarms with custom thresholds based on the time and day of the week. This is especially useful for scenarios where traffic patterns vary, such as heavy traffic during weekdays compared to weekends.

- **Daily difference**: When you would like to calculate the difference between the utilization yesterday versus today or between different times of the day.

- **Capacity calculation from metrics**: You can calculate capacity metrics such as **transactions per second (TPS)** and **bytes written to disk per second (BPS)**.

- **Highlighting latency above SLAs**: By utilizing metric math functions, you can identify instances where the latency of certain metrics exceeds the established SLA requirements. An example where this can be useful is to see the load time of a web page from user experience monitoring and determine whether it meets the defined SLA requirements.

- **Generating high-resolution metrics**: You can use metric math functions such as `FILL()` to fill data to generate high-resolution metrics. For example, metrics are published once every 5 minutes and you want to generate a metric every 1 minute. You can use the `FILL()` function to generate a high-resolution metric of one minute each.

- **Filling missing values in nonperiodic traffic**: When a metric is not generated often because of no traffic (for example, website traffic) and you would like to fill in the missing values, you can use the `FILL()` function to generate the missing values.

Let's go ahead and practically look at some of these use cases.

Capacity calculation from metrics

Let's say that you would like to understand the amount of traffic flowing on the load balancer per second and understand any performance bottlenecks. You can leverage the PERIOD() function to achieve that functionality.

In the last section, *Deriving operational intelligence from CloudWatch metrics*, we deployed an application load balancer using a CloudFormation template. We will see how to calculate the overall bytes processed by the Application Load Balancer using the metric math expression:

1. Let's navigate to **CloudWatch | Metrics | All metrics | ApplicationELB | per AppELB Metrics | ProcessedBytes** and navigate to the **Graphed metrics** tab, as shown in the following screenshot, and change **Statistic** to **Sum**.

Figure 5.9 – Bytes processed by the Application Load Balancer

2.   Click on **Add math** and select **PERIOD**.

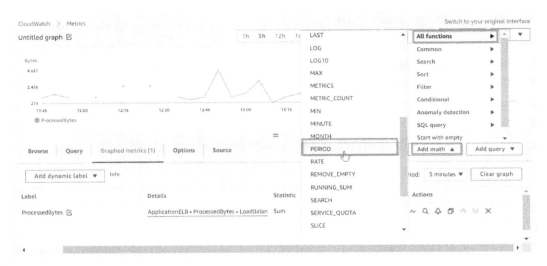

Figure 5.10 – PERIOD() function in metric math

3.   Rename the label from **Expression1** to `ProcessedBytesPerInterval`.

You can observe the traffic flowing on the load balancer per interval, as selected in the **Period** column.

Figure 5.11 – Traffic flowing on the load balancer per interval

4. You can quickly add the widget to the dashboard by clicking on **Actions | Add to dashboard**.

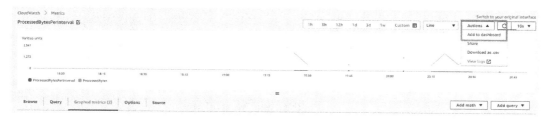

Figure 5.12 – Add the widget to the dashboard

5. You can add to a new dashboard or an existing dashboard by clicking **Create new** or browsing the existing dashboards.

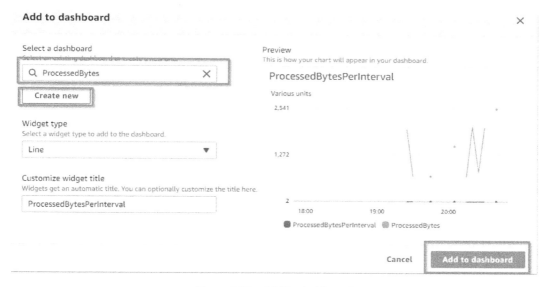

Figure 5.13 – Add to dashboard

Given the metric m1, we can use the PERIOD() function to divide the values of the metric by the period in the **Period** column, which will provide the value per interval for the selected metric.

This PERIOD() function is useful for calculating capacity-related metrics such as **TPS** on **Elastic Block Store (EBS)** volumes, as well as for measuring I/O (bytes written to disk per second, etc).

Horizontal and vertical annotations

The **horizontal annotation** feature provides a quick visual representation of metrics crossing the predefined values, such as SLA limits, offering valuable context for understanding metric values. You can add lines to represent important key values on the y axis. For example, defining maximum and minimum values expected for metrics. This will be especially useful in application operations by incorporating tacit knowledge into dashboards.

If you are visualizing the time series as a part of the x axis, **vertical annotations** will be useful to mark events that happened over time and bring tacit knowledge to dashboards. Let's add annotations to the widget created earlier.

You can edit the widget created in the last exercise and navigate to **Options** and add the horizontal annotation for **Min Limit** and **Max Limit** with values of 500 and 2000. Also, add a vertical annotation called **MajorChange**, as shown.

Figure 5.14 – Annotations on CloudWatch widgets

Then, update the widget. The output will look like the one in the following screenshot:

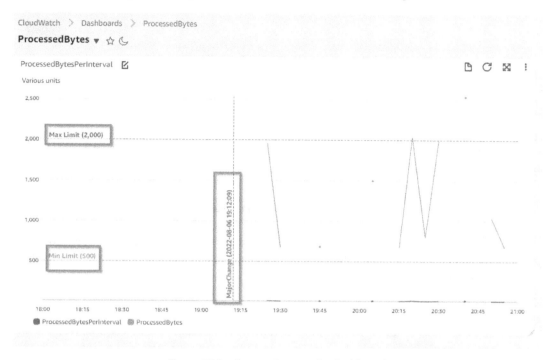

Figure 5.15 – Annotations on the dashboard

By using vertical annotations in this example, the operations team can quickly reference historical data points and determine any changes in the application. By using horizontal annotation, we mapped the expected maximum and minimum threshold. This brings tacit operational knowledge onto the dashboard.

Filling in missing values or generating high cardinality metrics

In the **Calculate capacity from metrics** section, you can see that the **ProcessedBytes** metric is missing values at certain intervals because of no traffic to the website. If you would like to fill in the missing values of the time series using mathematical functions to estimate similar traffic patterns, you can use the FILL command with different parameters:

- FILL(metric, S) – S is the static value to fill. You can see the static value is **1** in the following screenshot. You can edit and add FILL(m1,1) where m1 is the ID of the metric selected.

Figure 5.16 – Traffic flowing through the load balancer per second

- `FILL(metric, REPEAT)` – The `FILL` function is used to fill the missing value with the most recent actual value before the missing value. You can edit and add `FILL(m1, REPEAT)` as shown here.

Figure 5.17 – Traffic flowing through the Load Balancer per second

- FILL(metric, LINEAR) – This is used to fill in missing values and create linear interpolation by rendering a straight line between values at the beginning and the end of the gap. You can edit and add FILL(m1, LINEAR) as shown in the following screenshot.

Figure 5.18 – Traffic flowing through the load balancer per second

Dynamic threshold based on the weekend and weekdays

We can use metric math functions to define a variable alarm threshold based on the weekend and weekdays. The IF () function is powerful, given that you can combine multiple functions to create complex nested expressions. This example shows how to combine IF and TIME_SERIES () and set an alarm based on the time of the day.

ID	Detail	Explanation
Metric	Select metric	Select any metric that you want to calculate dynamic alarms.
Weekday	`TIME_SERIES(1)`	Select the threshold you would like to set for the metric on weekdays. Example 1: **ActiveConnectionCount** on the ALB is set to **1** on weekdays. Example 2: CPU utilization >70% should be represented as `TIME_SERIES(70)`.
Weekend	`TIME_SERIES(2)`	Select the threshold you would like to set for the metric at the weekend. Example 1: **ActiveConnectionCount** on the ALB is set to **2** on weekdays. Example 2: Expecting more traffic over the weekend. CPU utilization >90% should be represented as `TIME_SERIES(90)`.
Dynamic	`IF(((DAY(week) == 7 OR (DAY(week) == 6 AND HOUR(week) > 2)) OR (HOUR(week) < 8 AND DAY(week) == 1)),weekend, weekday)`	This function determines whether a given date and time falls within the weekend or weekday category, using the following criteria: If the date falls on a Sunday, or is after 2 AM on a Saturday, it is considered part of the weekend till 8AM on Monday. Otherwise, the date is categorized as a weekday, and the respective weekday threshold is used.
Alarm	`IF(metric <= dynamic,0,1)`	Set the alarm value to **0** if the metric is below the threshold and **1** if above the threshold. You need to set up notifications when the alarm state has the value of **1**.

You can see from the graph that it does not trigger the weekend threshold alarm when `ActiveConnectionCount=2`, which is equal to or less than the configured threshold of **2**.

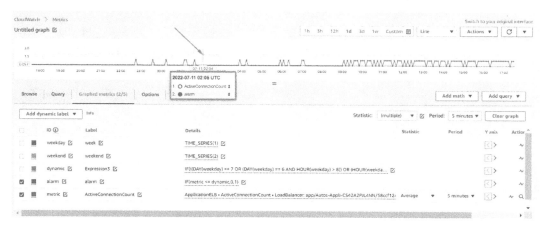

Figure 5.19 – Threshold breach at the weekend

You can see from the graph that the weekend threshold breach happened only when the value was `ActiveConnectionCount=2.5`, which is higher than the configured threshold of **2**.

Figure 5.20 – Threshold breach at the weekend

Whereas on the weekday, the alarm was triggered when the value was `ActiveConnectionCount=2`, which is higher than the configured threshold of **1**.

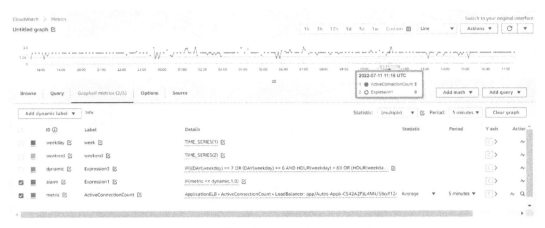

Figure 5.21 – Threshold breach on a weekday

This way, you can configure different thresholds based on the different days of the week and avoid false alarms.

Highlighting latency above SLAs

Another powerful usage of the `IF()` function is to highlight the values that are outside of SLAs without monitoring every data point. This makes it easy to spot breaches of SLAs at a glance. This will be especially helpful for metrics related to latency, availability, and so on. The following screenshot shows a view of the target response time view of the load balancer outside the defined SLA.

Figure 5.22 – Traffic flowing through the Load Balancer breaching the SLA

You can create visualization for target response time breaches in a load balancer using the metric math, as follows.

ID	Labels	Details	Explanation
sla	expression1	TIME_SERIES(0.004)	Define the SLA value of the metric in the time series.
n2	TargetResponseTime	Metric to define the SLA	Select the metric for which you would like to visualize SLA breaches.
e2	belowsla	IF(n2<sla,n2)	Provides time series metrics that are below the SLA.
e1	abovesla	IF(n2>sla,n2)	Highlights the time series that are above the SLA.

This provides a view to visualize breaches above the threshold value to understand issues quickly.

CloudWatch search expressions

Search expressions in metric math help you to search and group metrics across multiple CloudWatch metric namespaces. Search expressions will help you query and quickly add multiple related metrics to a graph. They also enable you to create dynamic graphs that automatically add metrics to their display, even if those metrics don't exist when you first create the graph. This will be especially useful when you are transferring CloudWatch dashboards/queries to a new AWS account.

If you would like to understand the overall size of the S3 buckets in the AWS account, we can use the metric math SEARCH() and SUM() functions to achieve the required functionality. Here is the search query that will provide a sum of the size of all the buckets in an AWS account:

```
SUM(SEARCH('{AWS/S3,BucketName,StorageType}
MetricName="BucketSizeBytes"', 'Sum', 300))
```

You can see the sum of S3 buckets shown in *Figure 5.23*.

Figure 5.23 – Sum of the data in all S3 buckets in the account

In this section, we have looked at how to leverage metric math to derive intelligence from CloudWatch metrics using different functions. Let's go ahead and understand how to use the anomaly detection feature in CloudWatch metrics.

CloudWatch anomaly detection

We discussed configuring an alarm for anomalous behavior in *Chapter 3*, *Gathering Operational Data and Alerting Using Amazon CloudWatch*, but let's understand how to configure or adjust the anomalous settings and understand the parameters in anomaly detection. In the last section, we discussed setting up dynamic thresholds based on the day/time of the week, but this is still a manual methodology to determine thresholds. If you are looking to automate this and leverage machine learning, consider using the **CloudWatch anomaly detection tool**.

CloudWatch anomaly detection applies machine-learning intelligence to automate, accelerate, and improve the detection of abnormal system and application behavior. CloudWatch anomaly detection uses predictive modeling to compare expected behavior against actual metric behavior, providing developers, systems engineers, and operators with real-time, targeted insights into abnormal system and application changes. The anomaly detection model is generated by algorithms by continuously analyzing the metrics of systems and applications and determining normal baselines and identifying anomalies with minimal user intervention. The model generates a range of expected values that represent normal metric behavior.

Some of the use cases for applying anomaly detection are as follows:

- **Detect Unexpected Volume Changes**: Capture unexpected volume drops or increases in CloudWatch metrics, such as abnormal decreases in load balancer requests. Though these changes may be within the bounds of static alarm thresholds, they can also be indicative of impending issues, such as resource exhaustion or DDoS attacks.

- **Monitor Dynamic Applications**: Monitor applications that exhibit cyclical or seasonal behavior with dynamic alarms that auto-adjust alert thresholds. Examples include increased CloudFront requests during peak hours or organically changing trends, such as gradual decreases in database writes as a system is deprecated.

- **Identify Deployment Side Effects**: When code deployments or resource changes are made, you can apply anomaly detection to monitor unexpected behavioral changes post-deployment, such as utilization spikes. This can be useful to monitor whether deployments cause unintentional side effects and determine whether rollbacks are necessary.

- **Proactive Troubleshooting**: Identify abnormal metric behavior before critical thresholds are breached to remediate potential issues. CloudWatch anomaly detection predicts behavior up to 2 hours into the future and enables you to preemptively identify resource exhaustion and add capacity in anticipation of future demand.

- **Monitor Business Metrics**: You can publish custom metrics that measure business KPIs, such as website order rate, video streaming latency, or API call rate to identify abnormal customer and business-impacting behavior.

Let's see how to configure CloudWatch anomaly detection and also delete existing models from CloudWatch:

1. In the AWS Management Console, on the **Services** menu, click **CloudWatch**.
2. In the left navigation menu, click on **Metrics**.
3. Click on **ApplicationELB | per AppELB metrics | TargetResponseTime** and navigate to the **Graphed metrics** tab. Your screen should look like the following.

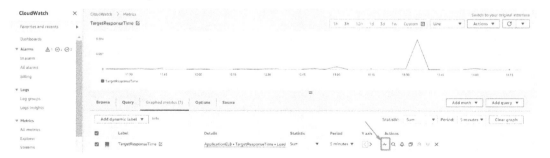

Figure 5.24 – Enable anomaly detection

4. Click on the **Anomaly detection** icon, as shown in the preceding screenshot.

5. **Anomaly detection** (**AD**) will be enabled immediately. A model is created based on the metric data from the past 2 weeks. AD will also be enabled even if there is no data available for the two-week period.

Figure 5.25 – Generated anomaly detection band

6. Notice the expression in the **ANOMALY_DETECTION_BAND(m1,2)** graph. This indicates that the metric m1 anomaly detection has been enabled with a standard deviation of 2. You can adjust the standard deviation number to increase the deviation scope for the metric data point if required. You can simply edit the expression as follows and click **Apply**.

Figure 5.26 – Editing anomaly detection band

7. At times, there may be certain data that you wish to exclude from the training model, such as periods during which traffic deviates from the norm, such as a promotional campaign. You could do that by clicking on **ANOMALY_DETECTION_BAND(m1,3)** and clicking on **Edit anomaly detection model** as shown.

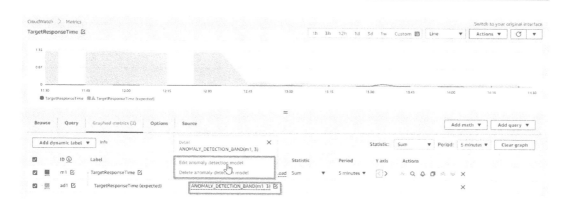

Figure 5.27 – Editing anomaly detection model

8. Select the exclusion time from the model training and click on the **Apply** button.

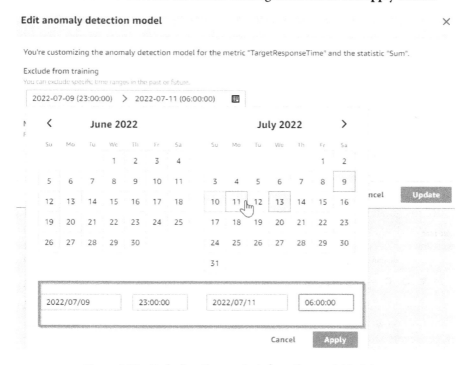

Figure 5.28 – Excluding time periods from the model training

9. Select the time zone of the period to exclude or add additional timings to exclude from the AD model training and click **Update**.

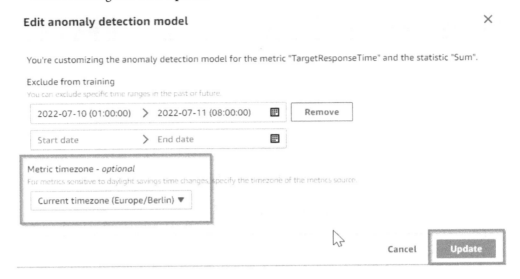

Figure 5.29 – Adding a time zone

10. To delete the model, you can simply click **Delete anomaly detection model** and then click **Delete**.

You can also create alarms, as discussed in *Chapter 3, Gathering Operational Data and Alerting Using Amazon CloudWatch*, for anomaly detection based on the model created.

This marks the end of the exploration of metrics and the various operational enhancements that can apply to metric data generated from various sources. The functionalities discussed in this section can also be extended to containers and serverless application metrics generated using CloudWatch. We will see how to gather metrics, logs, and traces from containerized applications and serverless applications in *Chapter 6, Observability for Containerized Applications on AWS*, and *Chapter 7, Observability for Serverless Applications on AWS*.

We have seen in earlier chapters that metrics, logs, events, alarms, and dashboards form the foundational components of observability in CloudWatch. In *Chapter 3, Gathering Operational Data and Alerting Using Amazon CloudWatch*, we learned how to instrument an EC2 instance manually using the CloudWatch agent and using SSM. In *Chapter 4, Implementing Distributed Tracing Using AWS X-Ray*, we saw how to instrument an application running on EC2 instance using CloudFormation templates.

One challenge with the approaches discussed is where to start, what to set up, and how to set it up. That is where **CloudWatch Application Insights** comes into play. In the next section, we will look at how CloudWatch Application Insights will be useful in instrumenting custom applications, along with the discovery process. Let's understand what CloudWatch Application Insights is and how it will ease the operational burden.

Exploring CloudWatch Application Insights

CloudWatch Application Insights streamlines the monitoring of enterprise applications with its intuitive and automated setup process. This reduces the time and effort required to configure monitoring, as it automatically sets up metrics, telemetry, logs, and alarms. The key advantage of using Application Insights is that it automatically discovers and configures application-specific monitoring, making it easy to effectively monitor your applications. Furthermore, it leverages ML analysis to perform in-depth problem analysis based on the gathered data, allowing you to quickly identify correlations between issues and relevant events.

It provides a dashboard for detected problems and provides insights and observations. It has predefined, customizable rules for alerting. It also helps you create an **AWS System Manager (SSM)** OpsItem to take remediation action with SSM runbooks.

Supported data sources on the Windows operating system include the following:

- **Platform Metrics**: Built-in platform metrics, advanced custom metrics, CloudWatch metrics
- **Application Data**: Microsoft IIS Logs, Microsoft SQL Server error log, custom .NET applications, .NET Core, Prometheus Java metrics
- **Windows Performance Counters**: .NET CLR, W3SVC_W3WP, Interop CLR, SQL server metrics, IIS counters
- **Windows Event Logs**: System, security, and application
- Extensibility for custom log patterns and custom application tiers

CloudWatch Application Insights supports gathering metrics, logs, and application-specific metrics using the console. Currently, X-Ray tracing is not supported.

You can use two different methods to discover the application using Application Insights:

- **Resource group based**: You can create a new resource group for an application using AWS resource group and tags as a component of your CloudFormation template. As part of the exercise in this section, we will use this method to create a new resource group.
- **Account based**: By implementing the account-based method, you can create application monitoring for all the resources in the AWS account. This is ideal when you are looking to instrument all future applications deployed in your AWS account.

In the previous chapter, we deployed a Java application on an EC2 instance and set up observability as a part of CloudFormation. We will set up observability for the same application using Application Insights and get additional insights related to the application besides the path of the user journey provided by X-Ray.

You can download the changed template for this exercise. To do so, log into your AWS account and click or copy and paste into your browser window the following Quickstart URL:

```
https://console.aws.amazon.com/cloudformation/home#/stacks/
new?stackName=scorekeepappinsight&templateURL=https://insiders-
guide-observability-on-aws-book.s3.amazonaws.com/chapter-05/basic-
ec2-template_rg_nomonitoring.yml
```

If you would like to download the template and verify it before deploying, you can download it from the following URL:

```
https://insiders-guide-observability-on-aws-book.s3.amazonaws.com/
chapter-05/basic-ec2-template_rg_nomonitoring.yml
```

This action will open the AWS CloudFormation console with all the parameters with default values. You can keep the default values and click on the **Next** button on all the subsequent pages. On the last one, named **Review scorekeepappinsight**, do not forget to check the box with the text **The following resource(s) require capabilities: [AWS::IAM::Role]** and click on the **Create Stack** button.

What we have changed in the CloudFormation template is the following:

- The addition of YAML to create a resource group

- Modification of the IAM role to allow management using AWS Systems Manager (as Application Insights leverages AWS SSM to install and configure the CloudWatch agent)

- The removal of the installation and configuration of the CloudWatch agent on the EC2 instance

Let's deploy the application using CloudFormation and provide the stack name to start with: scorekeep. Once it's successfully deployed, let's discover the application using CloudWatch Application Insights:

1. Navigate to **Application Insights** | **Add an application** | **Resource group based application** | **Confirm**.

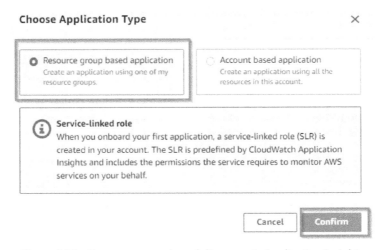

Figure 5.30 – Resource group based discovery in Application Insights

2. From the dropdown, select **my_resource_group** and select **Automatic monitoring of new resources**, **Monitor EventBridge events**, and **Integrate with AWS Systems Manager OpsCenter** and click **Next**.

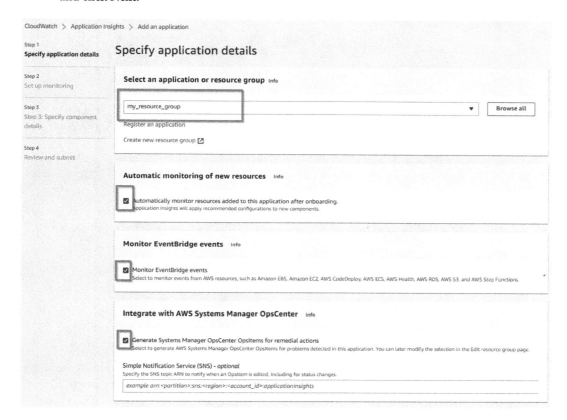

Figure 5.31 – Discovery of resources using resource groups

3. CloudWatch Application Insights discovers all the resources that are deployed as a part of the CloudFormation template. You can add the components installed on EC2 from the dropdown as shown in the following screenshot, which will add additional applications to CloudWatch monitoring.

Figure 5.32 – Application components discovered by Application Insights

4. In **Specify component details**, you can also add additional logs to be gathered from the Java application leveraging the CloudWatch agent. Add the log file `/tmp/scorekeep.log`. We can also change it at a later stage to accommodate additional run operations if necessary.

Figure 5.33 – Adding logs to CloudWatch monitoring

5.   Select **Next** and **Submit** to start the onboarding of the application.

Figure 5.34 – Submit application monitoring

6.   You can verify that the resource group has been added successfully using Application Insights.

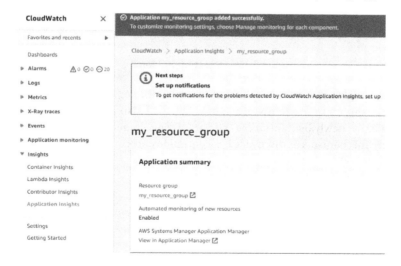

Figure 5.35 – Successfully added application

You can see that it is additionally asking to set up X-Ray, which is already done as a part of the CloudFormation template and application instrumentation, as discussed in *Chapter 4, Implementing Distributed Tracing Using AWS X-Ray*.

7. Once you navigate to Application Insights, you will see monitored assets, telemetry, and a summary along with the detected problems as a unified dashboard.

Figure 5.36 – Detected problems

Application Insights uses CloudFormation to deploy the required resources to monitor EC2 instances, as shown in the following figure. This will also address the challenge of agent installation in the newly built instances as a part of the resource group when you are leveraging Auto Scaling mechanisms as discovery is a continuous process.

Figure 5.37 – CloudFormation template created by Application Insights

8. When you navigate to any problem summary generated by Application Insights, it will provide you with an insights dashboard for the issue and also request feedback. In the following screenshot, I have navigated to the CPU issue, where the RDS database instance has high latency and the proposed action is to scale up the RDS instance.

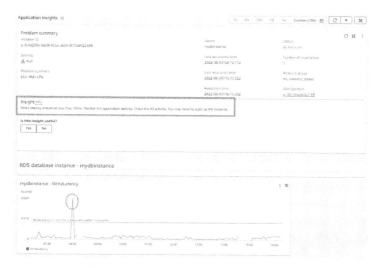

Figure 5.38 – Problem navigation in Application Insights

9. As we also created an OpsItem in AWS Systems Manager, it has provided a summary of the related resources and provided navigation of CloudTrail (audit events), CloudFormation (resources), and also CloudWatch alarms in a single view.

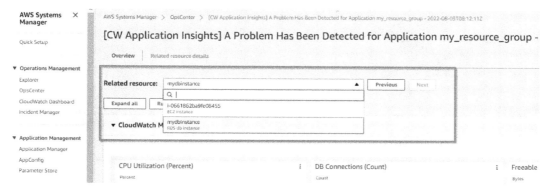

Figure 5.39 – Related resources view in AWS Systems Manager OpsCenter

Application Insights provides a way to setup collecting infrastucutre, application related metrices and logs and setup alarms and anomaly detection automatically along with installation of CloudWatch agent in in few clicks compared to instrumenting the application manually. We can also implement Application Insights using CloudFormation by using the resource type *AWS::ApplicationInsights:Application*: `https://docs.aws.amazon.com/AWSCloudFormation/latest/UserGuide/aws-resource-applicationinsights-application.html`.

Now let's explore how to derive operational intelligence from CloudWatch Logs using CloudWatch Logs Insights.

Exploring CloudWatch Logs Insights

We discussed CloudWatch Logs Insights as a part of *layer 4* in *Chapter 2, Overview of Observability Landscape on AWS*, and learned about the functionalities provided by Logs Insights. Logs Insights is a feature of CloudWatch Logs that allows querying log groups without exporting them to an external tool.

CloudWatch Logs Insights features a purpose-built query language with the key benefits of fast execution, query auto-completion, and log field discovery. It provides the ability to save queries and organize them into folders. It has built-in support for service logs, including discovered fields and sample queries, and makes log analysis effortless.

Logs Insights discovers fields automatically and works with both AWS logs and custom logs from custom applications that are in JSON format. Logs Insights generates field names automatically. If the custom logs are not in JSON format, we can still query them in Logs Insights and split them into fields using the parse command using the regex model. Logs Insights supports querying multiple CloudWatch Logs groups – up to 20 (as of this writing) – and visualizing them. We also have support for adding the log details as a part of the CloudWatch dashboard.

By default, there are five discovered system fields – namely, `[@message]`, `[@timestamp]`, `[@ingestiontime]`, `[@logStream]`, and `[@log]`. Field names that start with the @ character are automatically generated by Logs Insights.

Figure 5.40 – Default fields in CloudWatch Logs Insights

As discussed, it is simple yet powerful as there are only seven commands in Logs Insights – namely, `display`, `fields`, `filter`, `stats`, `sort`, `limit`, and `parse`. It has built-in support for functions and operators such as arithmetic operators (addition, subtraction, multiplication, etc.), Boolean operators (`and`, `or`, `not`), comparison operators (`=`, `<`, `>`, `>=`, `<=`, `=!`), numeric operators (`greatest`, `least`, `ceil`, etc.), **Datetime** functions, general functions (`ispresent`, `coalesce`), string functions (`replace`, `strlen`, etc.), and stats aggregation functions (`avg`, `count`, etc.). Each query can include one or more query commands separated by a Unix-style format.

Let's look at examples of how to leverage CloudWatch Logs Insights. We have generated a few logs as a part of application logs for AWS Lambda with the log group name `/aws/lambda/random-name`. Let's explore the logs in the log group using Logs Insights:

1. In the AWS Management Console, navigate to **CloudWatch | Logs | Logs Insights**, as shown in the following screenshot.

Figure 5.41 – Logs Insights view

2. From the dropdown, select the **random-name** log group or **IISLogs** (if you have executed exercises in *Chapter 3, Gathering Operational Data and Alerting Using Amazon CloudWatch*) and click **Run query**, which will execute the default query from the log. If you don't see any results, you can adjust the period in the right corner as required to query the logs:

```
fields @timestamp, @message
| sort @timestamp desc
| limit 20
```

Executing the default query, as in the preceding block, will display the **timestamp** and **message** fields, sort them by timestamp in descending order, and limit the number of results to the top 20, as shown in the figure.

Figure 5.42 – Default query execution

3. When you click on the **Fields** button on the right side, you will see the additional discovered fields from the logs, and you can also check the percentage field availability against each column.

Figure 5.43 – Discovered fields in Logs Insights query

4. If you would like to know the overall billed duration for a specific period for a Lambda function(s), you can select the log groups accordingly and execute the following query to get the total billed duration for a specific period:

```
fields @timestamp, @message
| sum(@billedDuration) as TotalDuration
```

You can see in the following screenshot the total duration of the Lambda function executed over the selected time period:

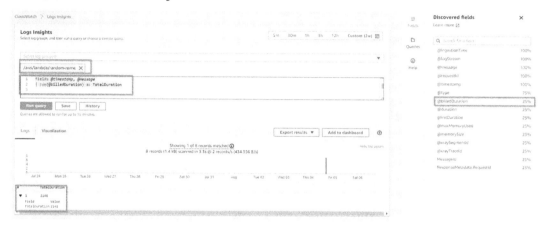

Figure 5.44 – Custom query to calculate the total duration of the Lambda function

Let's explore what **CloudWatch Contributor Insights** is and the use cases for it.

Exploring CloudWatch Contributor Insights and its use cases

We briefly discussed **CloudWatch Contributor Insights** (**CCI**) in *Chapter 3, Gathering Operational Data and Alerting Using Amazon CloudWatch*, as a part of *Layer 4* and discussed its functionalities. CloudWatch Contributor Insights, as the name suggests, will allow you to know the top contributors to a specific log group(s) or natively supported AWS services such as DynamoDB. CloudWatch Contributor Insights supports logs in the format of JSON or **CLF** (short for **Common Log Format**) (https://en.wikipedia.org/wiki/Common_Log_Format).

A *contributor* can be any field within a log entry that can be aggregated. The main configuration object is known as a rule, which can be applied to one or multiple log groups, up to 20 at a time. The log entries in the selected log groups are evaluated against the rule, and any matching logs are referred to as events. Rules can be either custom or pre-defined. CCI rules help you evaluate patterns in structured events in CloudWatch Logs from services such as AWS CloudTrail, Amazon VPC Flow Logs, and so on, and also for any custom logs sent from your applications or on-premises servers. You can see metrics generated by CloudWatch Contributor Insights about the top-N contributors, the total number of unique contributors, and their usage, for example, the top 10 source IPs that queried Route53. The source IP is a contributor in this case.

Built-in rules currently supported on DynamoDB are as follows:

- Most Accessed Items (Primary Key)
- Most Throttled Items (Primary Key)
- Most Accessed Items (Primary Key + Secondary Key)
- Most Throttled Items (Primary Key + Secondary Key)

By enabling CCI, you are not impacting the performance of DynamoDB as it is asynchronous in nature. As a part of use cases, you can identify the following:

- Hotkeys in a DynamoDB table
- DynamoDB table access patterns over time
- The top 25 most frequent keys

The output of the CCI can be displayed as a part of CloudWatch dashboards. Let's look at an example and enable CCI for DynamoDB.

In the *Exploring CloudWatch Application Insights* section, we deployed an application with DynamoDB as the database. Let's enable CCI for DynamoDB and see how it works:

1. Navigate to **DynamoDB | Tables | scorekeep-stateTable-1random | Monitor | Enable CloudWatch Contributor Insights**.

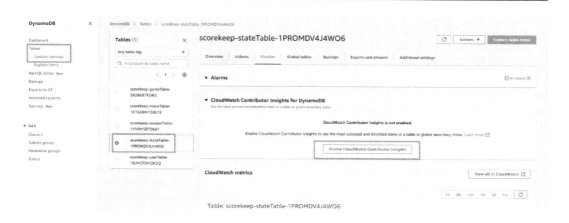

Figure 5.45 – Enable CloudWatch Contributor Insights

2.   Select **Enable** for both the primary key and secondary key.

Manage CloudWatch Contributor Insights settings   ✕

Use CloudWatch Contributor Insights for DynamoDB to see the most accessed and throttled items in a table or global secondary index. Additional costs might apply. Learn more ☑

Name	Resource type	Partition key	Sort key	Enable
scorekeep-stateTable-1PROMDV4J4WO6	Table	id	-	⬤
game-index	Index	game	-	⬤

⚠ Users who have the appropriate CloudWatch permissions will be able to view primary keys protected by fine grained access control (FGAC) in CloudWatch Contributor Insights graphs. If the primary key contains FGAC-protected data that you don't want published to CloudWatch, you should not enable CloudWatch Contributor Insights for DynamoDB for this table.

CloudWatch Contributor Insights for DynamoDB graphs display the partition key and (if applicable) sort key of frequently accessed items and frequently throttled items in plaintext. If you require the use of AWS Key Management Service (KMS) to encrypt this table's partition key and sort key data with an AWS managed key or customer managed key, you should not enable CloudWatch Contributor Insights for DynamoDB for this table.

Cancel   Save changes

Figure 5.46 – Enable CloudWatch Contributor Insights settings on a DynamoDB table

3. After a few application plays of the sample score-keep application, you can verify the most accessed keys by navigating to **CloudWatch | Insights | Contributor Insights.** Select the **scorekeep-statetable** rule to verify the consumed throughput units

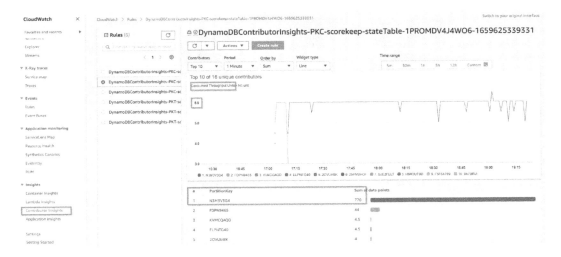

Figure 5.47 – Most accessed keys and consumed throughput units

You can see that the top key in terms of consuming DynamoDB capacity is highlighted in the preceding screenshot. We learned how to leverage CloudWatch Contributor Insights for DynamoDB in this section. Further, you can also leverage CloudWatch Contributor Insights for your custom logs.

Summary

In this chapter, we learned how to derive operational intelligence from CloudWatch metrics using the metrics explorer and query metrics using CloudWatch Metrics Insights. We looked at how to derive operational intelligence using metric match expressions and search expressions, and also how to use anomaly detection in metrics. Further, we explored leveraging CloudWatch Application Insights to observe an application running on an EC2 instance and gather metrics and log files, and how observability onboarding can be simplified during Day 2 operations with ease. We learned how to derive operational metrics from CloudWatch Logs and how to leverage CloudWatch Logs Insights to understand log files in a better way. Finally, we discussed CloudWatch Contributor Insights and how that can be used to understand the top contributors from DynamoDB tables and the supported log formats for CCI.

The skills learned in this chapter will help you reduce alarm fatigue, knowing how to leverage features in CloudWatch to enhance the operational experience.

In the next chapter, we will learn how to set up observability for applications running as container workloads on AWS. We will also explore how **AWS AppMesh** can help in extending the observability of the network layer in *Chapter 6, Observability for Containerized Applications on AWS.*

Questions

1. If you would like to add tacit knowledge of the operational team to the CloudWatch dashboard, what features could be leveraged?

2. What is the internal mechanism used by Application Insights to roll out monitoring of EC2 instances?

3. What are the different commands available in CloudWatch Logs Insights to query log files?

4. What AWS services are supported by CloudWatch Contributor Insights?

6

Observability for Containerized Applications on AWS

Malcolm Purcell McLean revolutionized the international transport business by introducing modern containers to reduce the labor required to load ships and trucks with goods of different sizes, reducing costs, improving reliability, and shortening transit time.

In the same way, container technology is revolutionizing how businesses ship production code to data centers and the public cloud. It is becoming the standard way of transporting code to run elsewhere, allowing the standardization of tools and pipelines, realizing the dream "compile once, run everywhere" promise for any technological stack.

A recent Gartner report says that we will have an increase from 20% in 2019 to 70% in 2023 of global organizations running more than two containerized applications in production (see `https://www.gartner.com/en/newsroom/press-releases/2020-06-25-gartner-forecasts-strong-revenue-growth-for-global-co`). With all this growth, sooner or later, if not now, you as a practitioner will face the challenge of a single application or a distributed application running on containers.

It is better to be prepared and understand all the tools AWS has to offer to help you in your observability journey for containers.

In this chapter, we will go through some manual and automated tools designed to reduce the effort to create and collect metrics yourself and what else to do to improve your observability powers on the two primary container services on AWS: Amazon EKS and Amazon ECS. We will look at AWS App Mesh, which extends your observability reach to the network layer, which is extremely important for distributed applications. And finally, we will check how to use the newly acquired knowledge in a scenario to solve performance bottlenecks in containerized applications. We will go through the following topics:

- Introduction to CloudWatch Container Insights
- Implementing observability for a distributed application running on Amazon EKS

- Implementing observability for a distributed application running on Amazon ECS
- End-to-end visibility of containerized applications using AWS App Mesh
- Understanding and troubleshooting performance bottlenecks in containers

Technical requirements

In your AWS account, it is advisable to experiment with the commands and procedures shown in this chapter.

Most commands are Bash shell commands, so access to a Unix-like environment is necessary, and some experience with Bash and command-line interfaces would be helpful.

Introduction to CloudWatch Container Insights

Talking to customers and practitioners, we see a common set of requirements for any observability solution:

- **Understand distributed architecture**: Bird's eye view of the application to locate the source of the problem
- **Zoom into individual components**: Drill down into specific components or workloads to learn more
- **Derive insights from data collected**: Connect the dots across metrics and logs collected

Launched in 2019, the AWS observability team specially tailored **CloudWatch Container Insights** to containers on AWS or any other system that runs containers.

CloudWatch Container Insights mainly performs the following functions:

- Collects and aggregates metrics and logs
- Provides reliable, secure metrics and log collections
- Enables automated dashboard creation and analysis
- Provides observability experience across metrics, logs, and traces
- Provides ad hoc analysis

It's the right tool if you are beginning your journey or want to scale. With CloudWatch Containers Insight, we don't need to deploy agents to collect metrics (if you are using **Fargate**), a time series database to store them, or even a visualization tool to graph them. All is done for you using the **Amazon Elastic Kubernetes Service (Amazon EKS)**, **Amazon Elastic Container Service (Amazon ECS)**, or Kubernetes platforms on Amazon EC2.

When you open **CloudWatch Containers Insights**, you see a list of all your resources, such as clusters, namespaces, services, and pods. See the following screenshot:

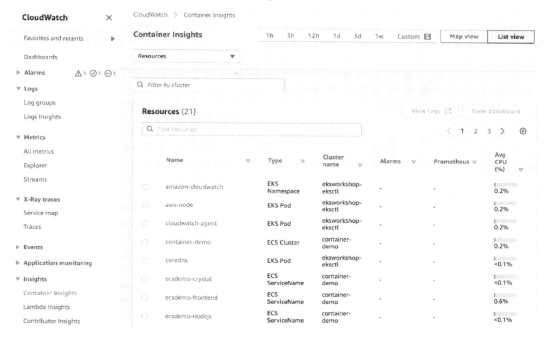

Figure 6.1 – A list view of container resources

You can click the **Map view** button to see a graphical representation of the same resources, showing the relationship between them. You can select one of the existing heatmaps, CPU-based or memory-based, to graphically represent the resource usage against its limits. If you hover the mouse over one of the resources, you can see more details, such as memory utilization, CPU, and network traffic. The graphical representation is easier to read for a few resources, but as soon you reach a bigger scale, you will likely use the list view and filters to find a specific resource. **List view** and **map view** give you a birds-eye view of your workloads to locate an application issue. See the following screenshot:

Figure 6.2 – Map view of container resources

You can use the map or list view and click on any resource to see more details. As soon as you click on one resource, you will see a dashboard with relevant data for clusters, nodes, services, namespaces, and pods.

Let's open the **EKS Nodes** view:

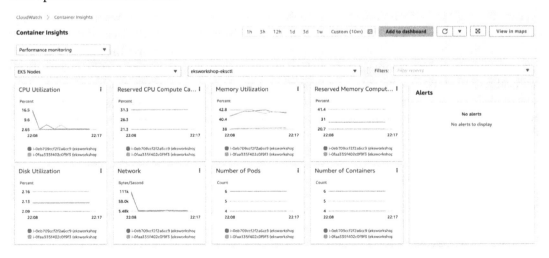

Figure 6.3 – Container Insights EKS Nodes view

It will show all cluster nodes and the usual metrics, such as CPU, memory, disk usage, and so on. You can hover the mouse over a metric to filter all dashboards on a single node. See the following screenshot:

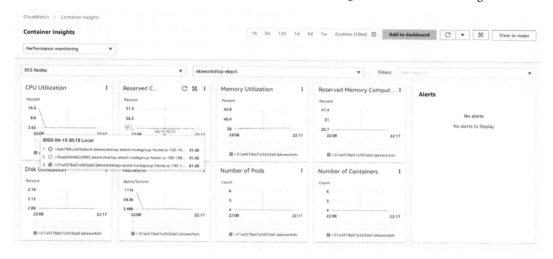

Figure 6.4 – Container Insights EKS Nodes, filtering on a single node

You can click on the Container insight metrics, open metrics groups created by Container Insights, and create alarms based on them, as we saw in the previous chapter, in the *Exploring CloudWatch Logs Insights* section.

Figure 6.5 – Logs Insights, showing one of the automatically created log groups

If you have AWS X-Ray installed in your environment, you can click on a pod and drill down on the details of the traces related to the pod. See the following screenshot:

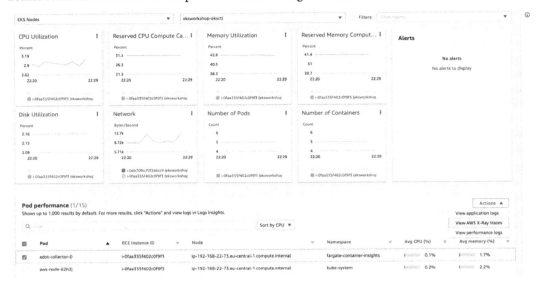

Figure 6.6 – Container Insights, showing the option to see the AWS X-Ray traces

And finally, you can access the performance logs. You can access performance-related metadata and query it using CloudWatch Analytics:

Figure 6.7 – Performance logs showing some of the discovered fields

In this section, we saw how Container Insights fulfills all the requirements of a containerized observability tool, such as the following:

- A bird's eye of your workload
- How you can drill down to see the details of a particular component
- How to derive data using CloudWatch Analytics

In the next subsection, will set up a sandbox environment and see how to use Container Insights for Amazon EKS.

Set up a Cloud9 development workspace

To have a standard **integrated development environment** (IDE) with the required set of tools, let's create an AWS Cloud9 environment:

1. Please click on the following link to create the required set of resources in your AWS account based on the AWS CloudFormation template:

   ```
   https://console.aws.amazon.com/cloudformation/home#/
   stacks/quickcreate?templateURL=https://insiders-guide-
   observability-on-aws-book.s3.amazonaws.com/common/cloud9.
   yaml&stackName=InsidersGuideCloud9Chapter6
   ```

2. You can keep the default values and click on the checkbox asking for extra capabilities. Click on **Create stack**, and in a few minutes, you can find the environment URL in the CloudFormation **Outputs** tab, as in the following screenshot:

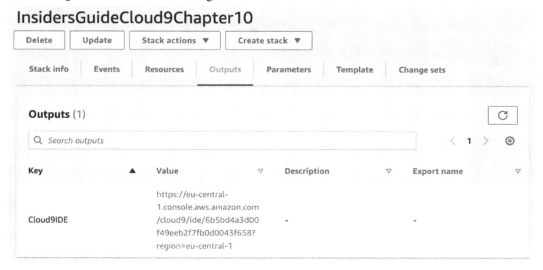

Figure 6.8 – CloudFormation Outputs tab, showing the AWS Cloud9 URL

3. The setup of this environment requires some extra time after the CloudFormation execution to install extra tools. So, after the CloudFormation execution, wait for around 15 minutes just to be sure. You can check the setup progress by accessing **AWS Systems Manager | Run command**, as in the following screenshot:

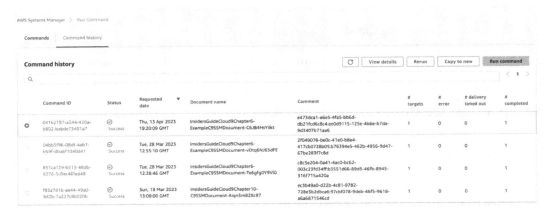

Figure 6.9 – Systems Manager | Run command – installing tools on the Cloud9 environment

4. If you click on it, you will find a newly configured environment, as in the following screenshot:

Figure 6.10 – AWS Cloud9 welcome page

Now, let's set up an Amazon EKS cluster as our sandbox environment.

Set up an Amazon EKS cluster

We will now configure a sandbox environment to use for the exercises in the rest of this chapter:

1. On the Cloud9 environment, run the following command to download the required Bash script:

    ```
    wget https://insiders-guide-observability-on-aws-book.
    s3.amazonaws.com/common/create-eks-ec2-eksctl.sh
    ```

    After the download, don't forget to check the content of this script (always a good practice).
    Execute the script as follows:

    ```
    bash create-eks-ec2-eksctl.sh
    ```

    The preceding command will start the creation of a new EKS cluster. You can see the cluster
    status in the command output. The process of creating a new cluster may take a few minutes.

2. After creating the cluster, let's check the communication between the Cloud9 environment and
    the new cluster. Run the following command:

    ```
    kubectl get svc
    ```

    You should see an output like this:

    ```
    ec2-user:~/environment $ kubectl get svc
    NAME           TYPE        CLUSTER-IP    EXTERNAL-IP
        PORT(S)    AGE
    kubernetes     ClusterIP   10.100.0.1    <none>
            443/TCP    7m52s
    ```

This command retrieves all services deployed in the default namespace. It shows you can communicate
with the Kubernetes API, and that you have the required permissions to execute commands.

Set up an Amazon ECS cluster

We will set up an Amazon ECS cluster as well to run some of our sample applications. For that, please
run the following command:

```
aws ecs create-cluster --cluster-name olly-on-aws
```

In this section, we had a brief overview of Amazon CloudWatch Container Insights and its features,
and we set up our environment to experiment with it. In the next section, we will see how to apply
CloudWatch Container Insights to one important AWS container service: Amazon EKS.

Implementing observability for a distributed application running on Amazon EKS

In this section, we will see how to set up your Amazon EKS cluster to use native CloudWatch features and Container Insights. We will postpone alternative methods using Amazon OpenSearch or Prometheus and Grafana to later chapters.

We will see two different methods to set up metrics collection on Amazon EKS. Amazon EKS lets you host your pods using EC2 worker nodes or AWS Fargate. When using EC2 worker nodes, we can collect metrics directly from the kubelet agent. But, when using AWS serverless compute for Fargate containers, no pod has access to the kubelet, so we need a different approach.

Let's see the two different methods in the next subsections.

Container Insights metrics on your EKS EC2 or customer-managed Kubernetes clusters

Whenever you want a system or workload to call AWS APIs on your behalf, you need to give it the correct permissions, which is similar to publishing metrics and logs to CloudWatch. Let's attach the required policy to your worker node's role.

Assuming you created the cluster using the script given in the previous section, you can attach `CloudWatchAgentServerPolicy` to your worker node's role by executing the following command:

```
$ curl -sSL https://insiders-guide-observability-on-aws-book.
s3.amazonaws.com/chapter-06/attach_role_cw_agent.sh | bash
```

> Important
>
> This method is the easiest way to set up the correct permissions for your worker nodes. It works regardless of using Amazon EKS or deploying Kubernetes on Amazon EC2. But it gives permissions to all pods running into the worker nodes to write data to CloudWatch. It may be okay for some workloads, but it doesn't follow the least privilege security principle. Another option is to bind IAM roles to service accounts, which allows you to give permissions to only the required pods. If you prefer this way, see the documentation at `https://docs.aws.amazon.com/AmazonCloudWatch/latest/monitoring/Container-Insights-prerequisites.html`.

We are ready to install the necessary software to collect metrics and logs from your cluster. The product team did a fantastic job creating Docker containers with the required agents and plugins to make them integrate seamlessly into the AWS ecosystem. We will install two `DaemonSet` resources, one for the CloudWatch agent and another for Fluent Bit. **Fluent Bit** (`https://fluentbit.io/`) is a fast, lightweight processor, logging forwarder, and a **Cloud Native Computing Foundation (CNCF)** graduate project. Instead of recreating the wheel, the product team used an existing standard to send the cluster logs to CloudWatch.

Even though we have the necessary Docker containers in public repositories (see `https://gallery.ecr.aws/cloudwatch-agent/cloudwatch-agent` and `https://gallery.ecr.aws/aws-observability/aws-for-fluent-bit`), we do have the task of writing the Kubernetes manifests to deploy the `DaemonSet` resources and all the surrounding objects such as `Namespace`, `ServiceAccount`, `ClusterRole`, `ClusterRoleBinding`, and `ConfigMaps`. The product team raises the bar once again and provides a Quick Start setup with all the necessary resources (see `https://docs.aws.amazon.com/AmazonCloudWatch/latest/monitoring/Container-Insights-setup-EKS-quickstart.html`).

To install all the necessary resources, you need to run the following command:

```
$ curl -sSL https://insiders-guide-observability-on-aws-book.
s3.amazonaws.com/chapter-06/install_cw_agent_fluentbit_k8s.sh | bash
```

After installing the `DaemonSet` resources, you should see the cluster metrics and logs in the **Container Insights** console, as shown in the following screenshot:

Figure 6.11 – Container Insights, showing a single EKS cluster metrics

Container Insights metrics on EKS Fargate

Let's see a different method to enable Container Insights on an EKS cluster, but now using Fargate. We will use the **AWS Distro for OpenTelemetry (ADOT)**. We have covered ADOT in detail in *Chapter 9, Collecting Metrics and Traces Using OpenTelemetry*. However, we will briefly discuss this here for the sake of continuity on EKS observability.

ADOT is an AWS-supported OpenTelemetry distribution. **OpenTelemetry** (see `https://opentelemetry.io/`) is a collection of tools, SDKs, and APIs. It can be used to generate, collect, instrument, and export telemetry data (for example, logs, traces, and metrics) in order to analyze the behavior and performance of your software.

It is not a fork but a packaging of libraries and plugins to make it easy to integrate with the AWS ecosystem. Because of the way ADOT collects metrics (using the Kubernetes API), ADOT is well suited to collect metrics when using Fargate.

The following script will install the necessary permissions, and as we have done in the EKS EC2 case, it will install the ADOT agent, but now as Kubernetes StatefulSets. For that, execute the following command:

```
$ curl -sSL https://insiders-guide-observability-on-aws-book.
s3.amazonaws.com/chapter-06/enable_container_insights_eks_fargate.sh |
bash
```

The preceding procedure is another Quick Start recipe provided by the product team. As before, this command will create not only the ADOT DaemonSet but also the `Namespace`, `ServiceAccount`, `ClusterRole`, `ClusterRoleBinding`, and `ConfigMap` objects required to make it work.

After the installation, you can find your cluster metrics in Container Insights.

In this section, we saw how to publish metrics to Amazon CloudWatch Container Insights when we use Amazon EKS and Amazon EC2 for worker nodes, or Amazon EKS and the serverless container compute Fargate. Next, we will see how to publish metrics when we are using the AWS-native container orchestration service, Amazon ECS.

Implementing observability for a distributed application running on Amazon ECS

Amazon ECS is similar to Amazon EKS, but as an AWS native container orchestration engine, it is more integrated into the AWS ecosystem and allows some simplifications.

We will see four procedures to collect metrics and logs from ECS clusters:

1. Collect cluster and service-level metrics.
2. Collect instance-level metrics using the CloudWatch agent.

3. Collect instance-level metrics using ADOT.

4. Collect logs and send them to CloudWatch Logs using FireLens.

Container Insights on Amazon ECS for the cluster- and service-level metrics

You can activate Container Insights on Amazon ECS for the cluster- and service-level metrics on the account and cluster levels.

When you activate Container Insights on an account level, every new Amazon ECS cluster created after that will have Container Insights data collection activated by default.

You can activate Container Insights on the account level using the AWS Management Console UI or the CLI.

To activate it using the AWS Management Console, go to `https://console.aws.amazon.com/ecs/`. In the menu on the left of the page, select **Account Settings**, click on the **Update** button in the top-right corner, and then at the bottom of the page, click on the checkbox under **CloudWatch Container Insights**, and then click on **Save changes**. See the following screenshot:

Figure 6.12 – Activate Container Insights on account level

To activate Container Insights on the account level using the CLI, execute the following command:

```
$ aws ecs put-account-setting --name "containerInsights" --value
"enabled"
```

Now, you can activate Container Insights per cluster if you want to save some costs on non-critical/low-margin environments or non-production environments. For new ECS clusters, you can also do it using the AWS Console Management UI or the CLI.

To create a new ECS cluster using the AWS Management Console with Container Insights enabled, you can follow the usual procedure to create a new cluster at `https://console.aws.amazon.com/ecs/`, making sure you click on the **Enable Container Insights** checkbox. See the following screenshot:

Figure 6.13 – Activate Container Insights on a new cluster using the graphical console

If you are creating a new cluster using the CLI, you can do it with Container Insights enabled using the following command:

```
$ aws ecs create-cluster --cluster-name myCICluster --settings
"name=containerInsights,value=enabled"
```

Activating Container Insights on an existing cluster is also easy. Assuming you have a variable named `clustername` with the cluster name, you need to execute the following command:

```
$ aws ecs update-cluster-settings --cluster ${clustername}  --settings
name=containerInsights,value=enabled --region ${AWS_REGION}
```

Container Insights on Amazon ECS for instance-level metrics using ADOT

As an alternative to the CloudWatch agent, you can use ADOT to collect instance- and application-level metrics. As already mentioned, we have covered ADOT in detail in *Chapter 9*, *Collecting Metrics and Traces Using OpenTelemetry*. However, we will briefly discuss this here for the sake of continuity on ECS observability.

Please execute the following script to see a deployment example:

```
$ curl -sSL https://insiders-guide-observability-on-aws-book.
s3.amazonaws.com/chapter-06/enable_container_insights_ecs_adot.sh |
bash
```

The preceding script will deploy an Amazon ECS service comprising the AWS ADOT collector and two sample applications. It will set up all required permissions, and once done, you can check to see the deployed workload accessing the Amazon ECS console and `o11y-on-aws` cluster (check `https://console.aws.amazon.com/ecs/v2/clusters/o11y-on-aws/services`), as in the following screenshot:

Figure 6.14 – Amazon ECS console, o11y-on-aws cluster, highlighting the o11y-on-aws-adot service

You can then check the Amazon Cloudwatch Container Insights details for the Amazon ECS `o11y-on-aws` task, as in the following screenshot:

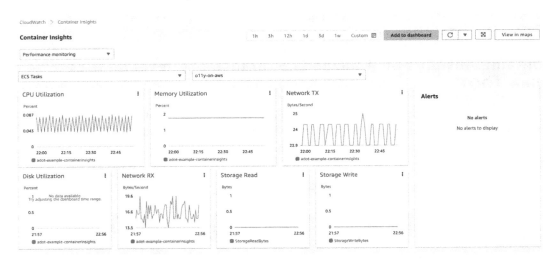

Figure 6.15 – Amazon CloudWatch Container Insights, ECS Tasks, o11y-on-aws

Collect logs and send them to CloudWatch Logs using FireLens

You can use FireLens to route logs to CloudWatch Logs for storage. Run the following script to deploy an Amazon ECS task with an example of how to do it:

```
$ curl -sSL https://insiders-guide-observability-on-aws-book.
s3.amazonaws.com/chapter-06/enable_firelens_ecs.sh | bash
```

The preceding script will deploy an Amazon ECS service that uses FireLens to forward the applications log to Amazon CloudWatch Logs. After the execution of the script, you will find a new task accessing the Amazon ECS console and the `o11y-on-aws` cluster (check `https://console.aws.amazon.com/ecs/v2/clusters/o11y-on-aws/services`), as in the following screenshot:

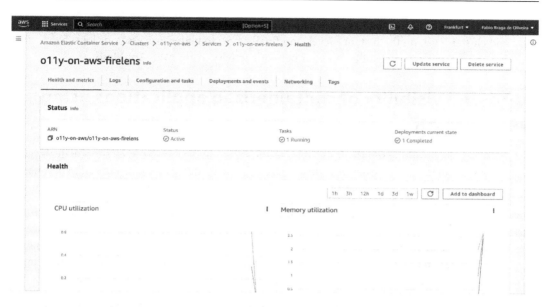

Figure 6.16 – Amazon ECS console, o11y-on-aws cluster, highlighting the o11y-on-aws-firelens service

From there, you can visit the Amazon CloudWatch logs console. There's a new log group created by the FireLens agent, named `firelens-blog`. Check the following screenshot:

Figure 6.17 – Amazon CloudWatch, log group named firelens-blog

In this section we saw different methods to collect observability signals from Amazon ECS, on the service and cluster level, on the instance level, and finally, how to better process your logs. Next, we will see how to achieve better visibility on your network activity using AWS AppMesh.

End-to-end visibility of containerized applications using AWS App Mesh

AWS App Mesh is a networking service that enables seamless communication between services deployed across different computing infrastructures. This service helps standardize the communication between your services, thus providing end-to-end visibility and ensuring high availability for your applications.

Modern applications are typically built using multiple services deployed across various computing infrastructures, including Amazon EC2, Amazon EKS, and AWS Fargate. As the number of services grows, it becomes challenging to identify the source of errors, reroute traffic after failures, and safely deploy code changes. In the past, addressing these challenges required you to incorporate monitoring and control logic directly into your code and redeploy your service every time there were changes.

AWS App Mesh simplifies service deployment by providing consistent visibility and network traffic controls for services deployed across different computing infrastructures. With App Mesh, there is no need to modify your application code to alter the way monitoring data is collected or traffic is routed between services. Instead, App Mesh configures each service to export monitoring data and ensures that communication control logic is standardized across your application. This enables you to quickly identify the source of errors and reroute network traffic automatically when there are failures or code changes.

You can use App Mesh to run your application at scale with various computing infrastructures, including AWS Fargate, Amazon EC2, Amazon ECS, Amazon EKS, and Kubernetes running on AWS. App Mesh leverages the open source Envoy proxy, which makes it compatible with a wide range of AWS partners and open source tools.

You can access the AWS App Mesh documentation to learn about its components (see `https://docs.aws.amazon.com/app-mesh/latest/userguide/what-is-app-mesh.html#app_mesh_components`) or the AWS App Mesh Workshop to learn about its benefits (see `https://www.appmeshworkshop.com/introduction/appmesh_benefits/`).

Add monitoring and logging capabilities

One of the main benefits of implementing App Mesh is the ability to have visibility around your microservices and their communications.

AWS App Mesh allows you to integrate the logs generated by the Envoy proxies running in your infrastructure with Amazon CloudWatch. Let's see how to do that.

The first step is to activate Container Insights in your cluster by following the instructions in the *Implement observability for a distributed application running on Amazon EKS* section if you are using an Amazon EKS cluster, or the instructions in the *Implement observability for a distributed application running on Amazon ECS* section if you are using an Amazon ECS cluster. Enable CloudWatch for the container and the Envoy sidecar following the instructions at `https://docs.aws.amazon.com/AmazonECS/latest/userguide/using_awslogs.html`.

After a few minutes, you should see the logging details provided by the envoy sidecar, as follows:

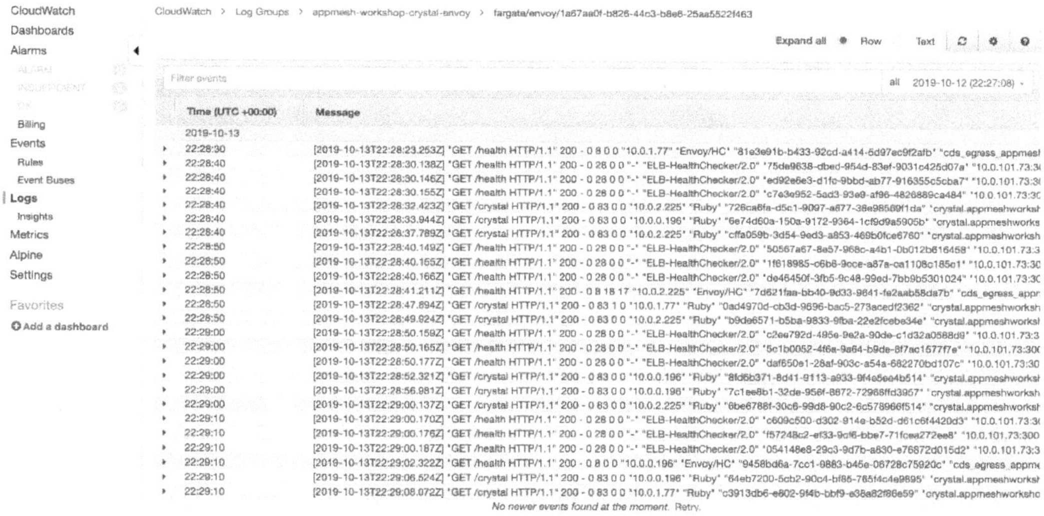

Figure 6.18 – Networking details provided by App Mesh Envoy sidecar

Add end-to-end tracing capabilities

The integration of AWS X-Ray and AWS App Mesh allows for the effective management of Envoy proxies utilized by microservices. Envoy, which is available in App Mesh, can be customized to transmit trace data to the X-Ray daemon, which operates in a container located in the same pod or task. To activate the integration of X-Ray with App Mesh Envoy proxies, see the following details.

An envoy task definition for ECS should have the following properties:

```
{
    "name": "envoy",
    "image": "840364872350.dkr.ecr.us-west-2.amazonaws.com/aws-appmesh-envoy:v1.15.1.0-prod",
    "essential": true,
    "environment": [
      {
```

```
            "name": "APPMESH_VIRTUAL_NODE_NAME",
            "value": "mesh/myMesh/virtualNode/myNode"
        },
        {
            "name": "ENABLE_ENVOY_XRAY_TRACING",
            "value": "1"
        }
    ],
    "healthCheck": {
        "command": [
          "CMD-SHELL",
          "curl -s http://localhost:9901/server_info | cut -d' ' -f3
| grep -q live"
          ],
          "startPeriod": 10,
          "interval": 5,
          "timeout": 2,
          "retries": 3
    }
}
```

In the preceding example, the App Mesh envoy image is set together with the ENABLE_ENVOY_ XRAY_TRACING environment variable to 1. That is all that is necessary to activate the X-Ray integration. Now, for EKS, it is slightly different. Assuming you use Helm as the tool to deploy the App Mesh Envoy, the deployment command looks like this:

```
$ helm upgrade -i appmesh-controller eks/appmesh-controller \
--namespace appmesh-system \
--set region=${AWS_REGION} \
--set serviceAccount.create=false \
--set serviceAccount.name=appmesh-controller \
--set tracing.enabled=true \
--set tracing.provider=x-ray
```

In this section, we saw how to add network observability capabilities using AWS App Mesh. The next will cover how to use what we have seen so far in a sample application.

Understanding and troubleshooting performance bottlenecks in containers

Now, let us practice what we have learned in this chapter and troubleshoot container performance bottlenecks. Let's get started.

Workspace

In this section, we will use the same Cloud9 workspace we created in the *Set up a Cloud9 development workspace* section. It already contains all the necessary tools.

Build the environments

In the Cloud9 workspace, to ensure the service-linked roles exist for load balancers and ECS, run the following commands:

```
aws iam get-role --role-name "AWSServiceRoleForElasticLoadBalancing"
|| aws iam create-service-linked-role --aws-service-name
"elasticloadbalancing.amazonaws.com"

aws iam get-role --role-name "AWSServiceRoleForECS" || aws iam create-
service-linked-role --aws-service-name "ecs.amazonaws.com"
```

Please download the script that will set up a sample application in our AWS environment with the following command:

```
wget https://insiders-guide-observability-on-aws-book.s3.amazonaws.
com/chapter-06/deploy_demo_application.sh
```

Now execute the script. Feel free to check it beforehand to understand all steps taken:

```
bash deploy_demo_application.sh
```

This script will take many minutes to execute, so go and grab a coffee while it goes through the deployment process of all the different microservices. Now we have the infrastructure and sample application deployed. Next, let's set up our container service to publish observability signals.

Set up Container Insights

With the sample application deployed, let's start to publish observability signals to Amazon CloudWatch Container Insights.

Execute the following command to enable Container Insights for our cluster:

```
aws ecs update-cluster-settings --cluster $(aws ecs list-clusters
--query "clusterArns[*]" --output text | sed 's/\s\+/\n/g' | grep
olly-on-aws-) --settings name=containerInsights,value=enabled --region
${AWS_REGION}
```

Explore Container Insights

After we have deployed the sample application and set the publication of observability signals using Container Insights, we are ready to monitor the application. Let's check what we have available to do so.

Check that the logs are streaming into CloudWatch Logs. Navigate to CloudWatch Logs (`https://console.aws.amazon.com/cloudwatch/home#logs:`) and search for a log group identified by `/aws/ecs/containerinsights/cluster-name/performance`.

Now, navigate to the Amazon CloudWatch **Container Insights** console (`https://console.aws.amazon.com/cloudwatch/home#container-insights:infrastructure`). From the first drop-down box, select **Performance monitoring**, and in the two new drop-down boxes that will appear below, select **ECS Clusters** and **o11y-on-aws** (from the second drop-down box), as shown in the following screenshot:

Figure 6.19 – Select ECS Clusters in Container Insights

You should see a dashboard automatically created with the key metrics of your cluster, as in the preceding figure.

If you return to the resource list, you can select the `olly-on-aws` cluster and click on the **View logs** button. You will see the CloudWatch Logs Insights, where you can select a log group such as `/aws/ecs/containerinsights/olly-on-aws/performance`, and run queries against more detailed data points. Check the following screenshot:

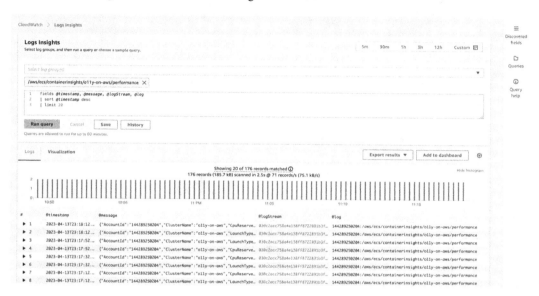

Figure 6.20 – Log Insights showing performance logs

Set up load tests

We now have monitoring enabled for our cluster. Let us push the limits of our system to see how the metrics may change. To perform our load test, we will use the tool **Siege** (`https://github.com/JoeDog/siege`). To install Siege on Cloud9, execute the following command:

```
$ sudo yum -y install siege
```

Perform a load test

Run the following command in a terminal window:

```
$ curl -sSL https://insiders-guide-observability-on-aws-book.
s3.amazonaws.com/chapter-06/run_load_test.sh | bash
```

This command will execute the Siege tool and it will drive 200 concurrent connections to the ECS application. You should see an output like the following:

Figure 6.21 – Siege tool output

You can leave the tool running for 15-20 seconds and then you can kill the process with *Ctrl + C*.

Load testing metrics

Go back to the Container Insights metrics and select **Performance monitoring | ECS Service | o11y-on-aws** and the time range of **5 min**, as in the following screenshot:

Figure 6.22 – Select Container Insights and a time range of 5 minutes

Or you can access the same dashboard directly from the following link: `https://console.aws.amazon.com/cloudwatch/home#container-insights:performance/ECS:Service?~(query~(controls~(CW*3a*3aECS.cluster~(~'o11y-on-aws)))~context~(timeRange~(delta~300000)))`.

You can see that the metrics are starting to show up on the graph widgets in the following screenshots. Notice the CPU utilization increasing as Siege increases the load on the application.

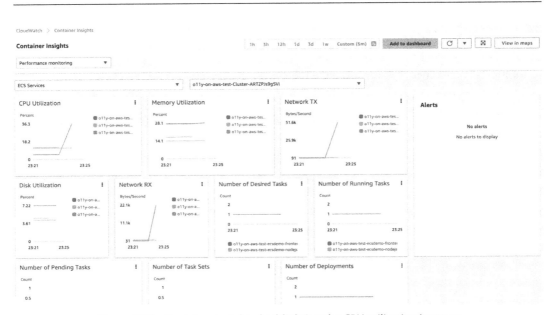

Figure 6.23 – Container Insights, highlighting the CPU utilization increase

Accessing CloudWatch Logs Insights

Let us explore the CPU utilization increase from another angle, using CloudWatch Logs Insights:

1. Navigate to CloudWatch Logs Insights (`https://console.aws.amazon.com/cloudwatch/home#logs-insights:`) and select the `/aws/ecs/containerinsights/cluster-name/performance` log group, as shown here:

Figure 6.24 – Logs Insights, cluster group selected

2. Copy and paste the following filter command:

```
stats avg(MemoryUtilized) as Avg_Memory, avg(CpuUtilized) as
Avg_CPU by bin(5m)
 | filter Type="Task"
```

3. Select the **Visualization** tab, then the **Bar** chart. You will see a screen like the following:

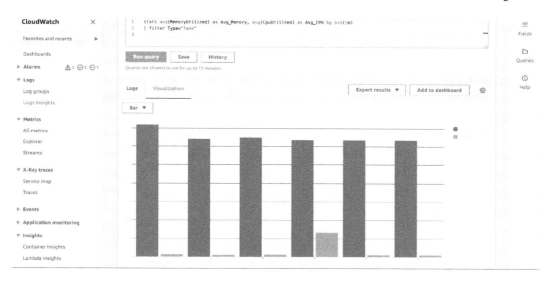

Figure 6.25 – Metrics statistics using Amazon CloudWatch Logs Insights

As you can see, getting CloudWatch Container Insights to work and setting alarms for CPU and other metrics is pretty easy. With CloudWatch Container Insights, we remove the need to manage and update your monitoring infrastructure and allow you to use native AWS solutions for which you don't have to manage the platform.

Summary

In this chapter, we saw how Container Insights helps you automate the collection, aggregation, and visualization of key container metrics. We saw how to activate it on both Amazon ECS and Amazon EKS, two major container orchestration services on AWS. We also saw how to integrate App Mesh to add network spice to our observability recipe. In the end, we saw how to use the acquired new skills to identify and isolate a CPU peak using the tools.

As we highlighted in the introduction, all those skills will be invaluable as more and more organizations move their workloads to containers.

As you move forward modernizing your workloads, the next natural step is to use Lambda functions. In the next chapter, we will explore how to speed up the process of collecting, aggregating, and visualizing metrics for Lambda functions.

Observability for Serverless Applications on AWS

To summarize the chapters so far, in *Chapter 3, Gathering Operational Data and Alerting Using Amazon CloudWatch, Chapter 4, Implementing Distributed Tracing Using AWS X-Ray*, and *Chapter 5, Insights into Operational Data with CloudWatch*, we focused on observing applications running on EC2 using CloudWatch. In *Chapter 6, Observability for Containerized Applications on AWS*, we delved into observability for applications running on containers. In this chapter, we will explore the observability of serverless applications running on AWS, specifically for those running on AWS Lambda.

Amazon Web Services (**AWS**) introduced AWS Lambda in 2015 as a solution for developers to create software without the overhead of managing operating systems and scalability. Observability for Lambda will be important, as the functions are event-driven and loosely coupled, making it challenging to understand the interactions between them and troubleshoot issues. The stateless nature of the Lambda functions and the requirement to load some state information from other services only add to the complexity and make it complex to understand issues affecting performance. Additionally, the cost of running Lambda functions is directly proportionate to the duration of their execution, making observability even more important.

In this chapter, we will navigate through the out-of-the-box metrics and logs that come with AWS Lambda function deployment. We will discuss the Lambda extensions and see how they will support enhancing observability for Lambda functions. Additionally, we will go through Lambda Insights and the benefits of leveraging Lambda Insights.

By the end of this chapter, you will gain a comprehensive understanding of application-related observability through the use of **Powertools for logging** for Lambda functions. We will equip you to effectively instrument a Node.js application running on serverless components. Most importantly, this chapter will help you understand and troubleshoot performance bottlenecks in Lambda compute.

We will cover the following main topics in the chapter:

- Deploying a basic serverless application running on AWS Lambda
- Understanding CloudWatch Lambda Insights
- End-to-end tracing of Node.js application running on a serverless component
- Troubleshooting performance issues using X-Ray groups

Technical requirements

To carry out the technical tasks in the chapter, you should have the following:

- A working AWS account
- A fundamental understanding of AWS DynamoDB and API Gateway
- Knowledge of AWS Lambda and basic setup of AWS Lambda
- An understanding of fundamental Node.js functionality
- An understanding of fundamental building blocks of observability

Deploying a basic serverless application running on AWS Lambda

To gain practical knowledge of the various built-in metrics and logs available in Lambda functions, we will deploy a basic serverless application on AWS Lambda in this section. This will serve as a hands-on opportunity to learn how to trace the end-to-end performance of an application running on serverless components using AWS observability services.

Let's deploy a basic serverless application using AWS services, namely Amazon API Gateway, AWS Lambda, and DynamoDB, following the architecture outlined in this diagram:

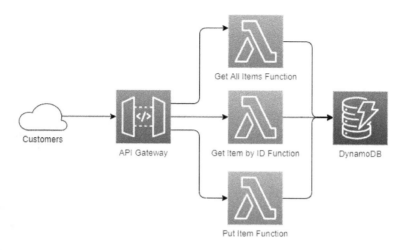

Figure 7.1 – Sample app architecture

You should click on the following CloudFormation YAML file to deploy the sample application. Once the application is deployed in your AWS account, it will only contain the necessary IAM roles for accessing and storing data in CloudWatch and other relevant services. It is not yet instrumented for end-to-end observability. Let's deploy the sample application by clicking the following link:

```
https://console.aws.amazon.com/cloudformation/home#/stacks/
new?stackName=serverless-app&templateURL=https://insiders-guide-
observability-on-aws-book.s3.amazonaws.com/chapter-07/init/template.
yaml
```

You can find the API Gateway URL of the deployed application in the **Outputs** tab of the CloudFormation template, as shown in *Figure 7.2*. The Lambda functions and API Gateway deployed as a part of the CloudFormation are set to capture only the default out-of-the-box metrics. Further deployed Lambda functions are enabled to log to **CloudWatch log groups**:

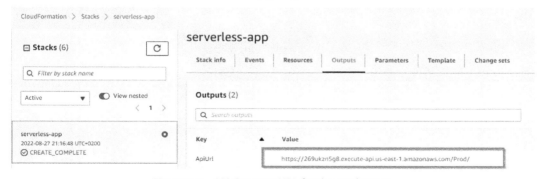

Figure 7.2 – API Gateway URL for the application

A Postman configuration has been made available at the following URL, allowing non-developers to easily insert data into the application. You can access this URL to find it: `https://documenter.getpostman.com/view/3349468/VUqoRe3n`

Figure 7.3 – Inserting data using the Postman app

You can insert a few items into the application using the Postman application by replacing `<REPLACE-ME>` with the value of `APIurl`, as shown in *Figure 7.3*. You are inserting the API Gateway information for the application deployed, which is obtained from the **Outputs** tab of the CloudFormation template deployed, as shown in *Figure 7.2*.

Figure 7.4 – Post the items or items

You can examine the items you have posted by navigating to DynamoDB or by using the **Get All Items** option in the Postman app.

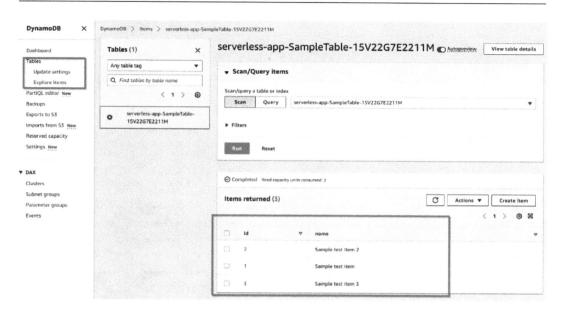

Figure 7.5 – Sample Items from DynamoDB

This shows that the serverless application is successfully deployed and you are ready to use the AWS observability tools to instrument the application. In the next section, we will look into the built-in metrics and default Lambda logging that comes out of the box.

Built-in metrics

Let's look at the default out-of-the-box metrics and logs available when you deploy an application on AWS Lambda. Lambda comes with 15 default metrics (please visit https://docs.aws.amazon.com/lambda/latest/dg/monitoring-metrics.html for more information) divided into three categories: **Invocation metrics**, **Performance metrics**, and **Concurrency metrics**. Some of the important Lambda metrics are **Invocations**, **Errors**, **Throttles**, **Duration**, and **ConcurrentExecutions**. You can create a widget with the important metrics as shown here:

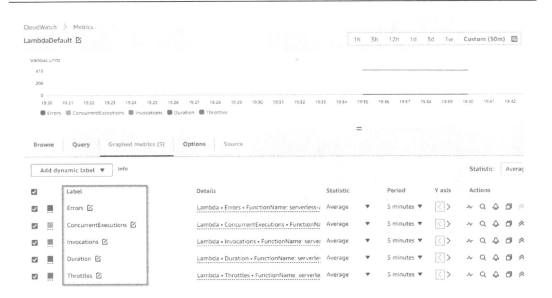

Figure 7.6 – View of important built-in Lambda metrics

To fully understand the performance of Lambda functions, it is crucial to track the *duration of the execution* and *percentile of requests completed within a period* to analyze the performance impact. Analyzing the P90 metric (the P90 metric provides you with the average duration of execution of less than or equal to 90% of requests executed for a given period), as seen in *Figure 7.7*, reveals that 50% of requests are processed within 92.5 ms (milliseconds), 90% within 556 ms, and 95% within 565 ms. These statistics provide valuable insights and can guide you toward optimizing the function's configuration for improved performance. For more information on utilizing the statistics and displaying them in numeric widgets, as shown in *Figure 7.7*, please refer to *Chapter 5, Insights into Operational Data with CloudWatch*.

Figure 7.7 – Percentile metrics for the duration

Now that we have explored the default metrics generated by AWS Lambda in CloudWatch, let's explore the default logging being provided in AWS Lambda when the out-of-the-box logging feature is enabled.

Lambda logging

When you deploy a Lambda function, logging is enabled by default. It provide logs for every execution with information about billed duration, memory use, and memory configured, as shown in *Figure 7.8*.

To verify the logs generated due to Lambda invocation, you can navigate to **CloudWatch | Logs | Log groups** and verify the log group name starting with `/aws/lambda/serverless-app-getAllItemsFunction-%%%%`. The last few characters are randomly generated. You can select the log stream to view the logs generated by the Lambda function.

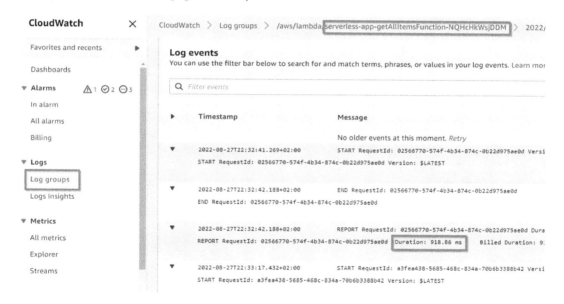

Figure 7.8 – Default Lambda logs

CloudWatch Logs Insights from Lambda logs

You could derive operational intelligence from the Lambda logs using **CloudWatch Logs Insights**. For instructions on how to use Logs Insights, please refer to *Chapter 5, Insights into Operational Data with CloudWatch*. Here is a sample query to estimate the number of cold starts, the average duration of the Lambda function execution, and memory usage grouped by a 5-minute time period:

```
filter @type = "REPORT"
| stats
  count(@type) as countInvocations ,
```

```
count(@initDuration) as countColdStarts ,(count(@initDuration)/
count(@type))*100 as percentageColdStarts,
 max(@initDuration) as maxColdStartTime,
 avg(@duration) as averageDuration,
 max(@duration) as maxDuration,
 min(@duration) as minDuration,
 avg(@maxMemoryUsed) as averageMemoryUsed,
 max(@memorySize) as memoryAllocated,  (avg(@maxMemoryUsed)/max(@
memorySize))*100 as percentageMemoryUsed
 by bin(5m) as timeFrame
```

The output of this code, when executed, will provide a view as follows:

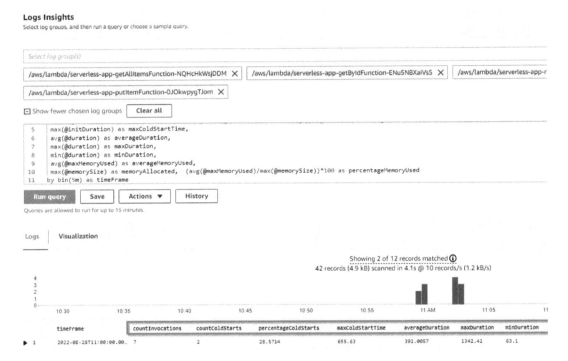

Figure 7.9 – Intelligence from CloudWatch Logs Insights

To see the important metrics in a unified dashboard, let's create a CloudWatch dashboard for this custom application. You can refer to *Chapter 4, Implementing Distributed Tracing Using AWS X-Ray*, to create a dashboard for the serverless app, as in the following, for metrics generated from the API Gateway, Lambda functions, and DynamoDB.

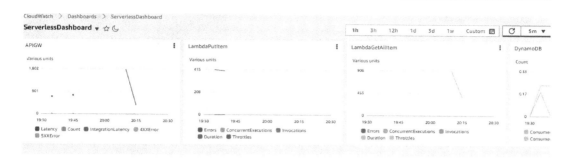

Figure 7.10 – Serverless app dashboard

API Gateway metrics and logs

AWS API Gateway provides several default metrics that include `Latency`, `Count`, `Integration Latency`, `4XX Error`, and `5XX Error`. To assess the impact of including API Gateway in the overall solution, it's important to observe the difference between the *Latency* metric and the *Integration Latency* metric. This difference represents the extra overhead added to the application performance because of including API Gateway in the architecture.

API Gateway allows you to log two types of logs: **API Gateway execution logs** and **API Gateway access logs**.

API Gateway execution logs can have three different logging levels: namely *errors only*, *error and info logs*, and *full request and response logs*. You can set logging globally for the entire API Gateway or in different stages of API Gateway.

API Gateway access logs provide a comprehensive view of who is accessing the API including information such as `IP`, `HttpMethod`, `User`, `Protocol`, and `Time`. These logs provide valuable insights into the origin of API Gateway invocations.

> **Please note**
>
> We have not enabled logging in the CloudFormation template for this exercise, and it should be done manually, as shown in *Figure 7.11*.

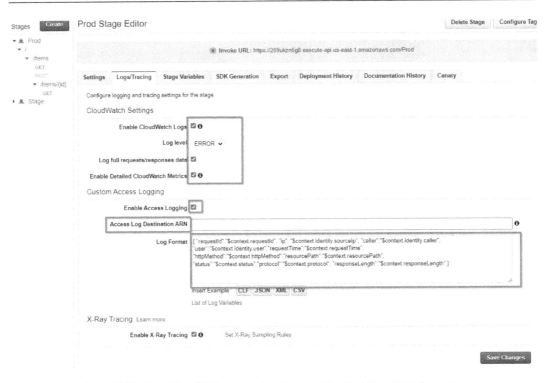

Figure 7.11 – Enabling API Gateway logging and detailed CloudWatch metrics

Relying solely on the built-in metrics for API Gateway and Lambda function(s) presents difficulties in understanding the performance issues related to the Lambda functions – for example, network performance and cold starts. Although we can estimate the count of cold starts from logs, we can achieve a more comprehensive understanding of performance through **Lambda Insights**.

CloudWatch Lambda Insights

CloudWatch Lambda Insights is a powerful extension for understanding the performance of Lambda functions. It provides valuable insights into a range of issues that can affect the performance of Lambda such as memory leaks, identifying high-cost functions, identifying performance impact caused by new versions of Lambda functions, and also understanding latency drivers in the Lambda functions.

The traditional method of gathering process-level metrics and logs in an EC2 environment involves using the CloudWatch agent. In the container setup, we use either a sidecar or a daemon where the compute power is continuously available. However, in the case of Lambda functions, this method may not be optimal. Using the same type of model for Lambda, which is a short-lived compute that is billed for the time it is running, can result in increased costs. This is because the agent would need to run continuously in the background, even between invocations, resulting in a waste of resources and an increase in cost, as illustrated in *Figure 7.12*.

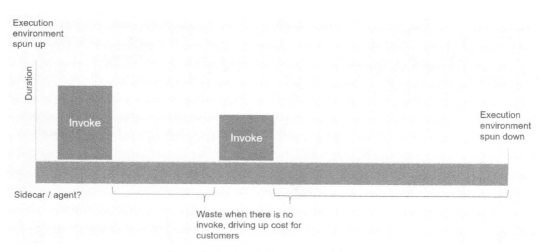

Figure 7.12 – Lambda invocation life cycle

To overcome this challenge, AWS released Lambda extensions in May 2021 to augment Lambda functions and provide easy-to-plug-in tools to integrate deeply with Lambda with no complex installation, configuration, or operational overhead. Extensions extend the invocation of the Lambda life cycle and only run when there is some functionality to carry. Lambda extensions can run after a function is invoked to send telemetry data about the invocation, or they can run before the runtime starts or can do clean-up tasks before the execution environment is spun down. The additional overhead time, as depicted in *Figure 7.13*, is minimal compared to running the agent for the full duration. This is because of the efficient execution of Lambda extensions, resulting in a smaller impact on the overall runtime.

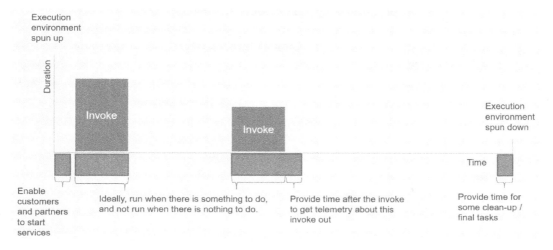

Figure 7.13 – With Lambda extensions

Lambda extensions serve several purposes, and one of the use cases is observability and logging. An example of a Lambda extension available for this purpose is **CloudWatch Lambda Insights**, which provides the ability to monitor, troubleshoot, and optimize the performance of your Lambda functions. CloudWatch Lambda Insights allows you to capture diagnostic information before, during, and after function invocation, requiring no code changes.

You can enable CloudWatch Lambda Insights from the AWS console. You can navigate to the Lambda function and then click on **Configuration** | **Monitoring and operations tools** | **Edit configuration** and then enable **Enhanced monitoring**, as shown in *Figure 7.14*:

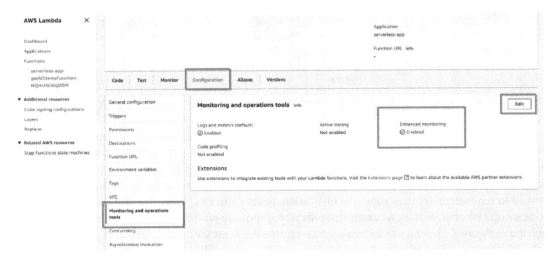

Figure 7.14 – Lambda enhanced monitoring

When you enable enhanced monitoring, the Lambda Insights extension will add to the Lambda function as a layer, as shown here:

Figure 7.15 – Lambda Insights extension

Rather than enabling the same from the AWS console, let's enable CloudWatch Lambda Insights for all the functions in the application deployed using CloudFormation.

You can download the CloudFormation template from the following URL and update the CloudFormation stack to enable the Lambda extension for all three Lambda functions.

Please find the `template.yaml` file here with Lambda Insights enabled: https://insiders-guide-observability-on-aws-book.s3.amazonaws.com/chapter-07/enableinsights/template.yaml.

What we have changed in the CloudFormation template is the addition of the following YAML code, which adds a new Lambda layer with the Lambda Insights extension with the latest version in the `template.yaml` file:

```
Layers:
  - !Sub "arn:aws:lambda:${AWS::Region}:580247275435:layer:
LambdaInsightsExtension:21"
```

Enabling Lambda Insights for a Lambda function provides additional metrics related to execution and also provides logs about the Lambda execution. A dashboard view is available out of the box for visualizing performance monitoring for a single function or a performance monitoring view for multiple functions.

Let's navigate to the Lambda Insights dashboard from the CloudWatch console and understand the metrics and insights available from the Lambda Insights dashboard.

Single-function view

Single-function view provides detailed performance metrics and logs for a single Lambda function in your AWS account.

Let's navigate to the single-function view. Navigate to **CloudWatch | Insights | Lambda Insights**. Select the Lambda function of the serverless application deployed in the *Deploying a basic serverless application running on AWS Lambda* exercise, namely `"serverless-app-getAllItemsFunction-xyz"`. It provides CPU, memory, and network usage of the Lambda function. Additionally, it provides a detailed breakdown of the last 1,000 invocations, including initialization duration, overall duration, memory used, CPU time, and network I/O. You can also view the application logs generated for your function by simply navigating to the second tab labeled **Application logs**, as highlighted in the following figure:

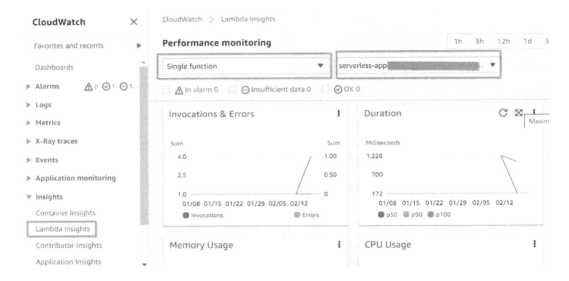

Figure 7.16 – Lambda Insights dashboard view

Multifunction view

The multifunction view offers a comprehensive overview of all Lambda functions in your current AWS account and region. In the **Function summary** view, you can examine the aggregate information for each function, including **Invocations**, **Cold starts**, **Total cost**, **Max. memory**, **CPU time**, and **Network IO** for all Lambda functions. Important metrics to focus on in this are the cost (**Total cost**) and usage metrics (**Max. memory** and **Cold starts**).

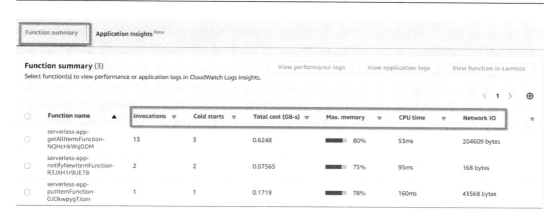

Figure 7.17 – Multifunction view dashboard

While the default out-of-the-box metrics, CloudWatch Lambda Insights metrics, and Lambda logs generated by Lambda Insights provide valuable insight into the performance of an individual Lambda function, they lack context regarding the overall performance of the larger application. The performance metrics generated by each component of the application, such as API Gateway, Lambda, and DynamoDB, offer only a limited understanding of their individual performance, without considering their interconnectedness and impact on the overall application.

To get a complete picture of how a single function affects the overall system, you need to look at its performance as part of the whole application. This requires enabling tracing for the entire end-to-end application. Enabling tracing for end-to-end applications provides us with the ability to examine correlations and pinpoint potential root causes for a specific request. By doing so, we can gain a deeper understanding of the performance of our application and take action to improve it.

End-to-end tracing of the Node.js application

In the last section, we saw how to enable enhanced metrics for gathering additional metrics for Lambda functions. Let's enable X-Ray active tracing. Active tracing is a feature of AWS X-Ray that provides visibility into the performance of your serverless applications. When enabled, active tracing captures detailed information about the flow of a request as it travels through the different components of your application, such as AWS Lambda functions, API Gateway, and DynamoDB. To gain a deeper understanding of AWS X-Ray and explore the various console-level options available in CloudWatch X-Ray, please refer to *Chapter 4, Implementing Distributed Tracing Using AWS X-Ray.*

Let's enable active tracing for both the API Gateway and the Lambda function using the CloudFormation template in *step 3*. I describe changes made in the CloudFormation template in *steps 1* and *2* as follows:

1. To enable active tracing for the Lambda function(s), I have added a `Mode:Active` line to enable active tracing for the Lambda function:

    ```
    Function:
        Runtime: nodejs16.x
        Timeout: 100
        Layers:
            - !Sub "arn:aws:lambda:${AWS::Region}:580247275435:layer:
    LambdaInsightsExtension:21"
        TracingConfig:
            Mode: Active
    ```

2. To enable active tracing for API Gateway, I have added a `TracingEnabled: true` line in the CloudFormation template:

    ```
    Api:
        TracingEnabled: true   # <----- ADD FOR API Tracing
    ```

3. To implement the changes, you can download the completed CloudFormation template and update your CloudFormation stack using the template:

    ```
    https://insiders-guide-observability-on-aws-book.s3.amazonaws.
    com/chapter-07/tracingenabled/template.yaml
    ```

4. After successfully deploying the updated template, you can confirm that tracing is enabled for your Lambda function(s) by navigating to the Lambda function's configuration page. Navigate to the Lambda function, click on the **Configuration** tab, and then verify that **Active tracing** is **Enabled**, as highlighted in *Figure 7.18*.

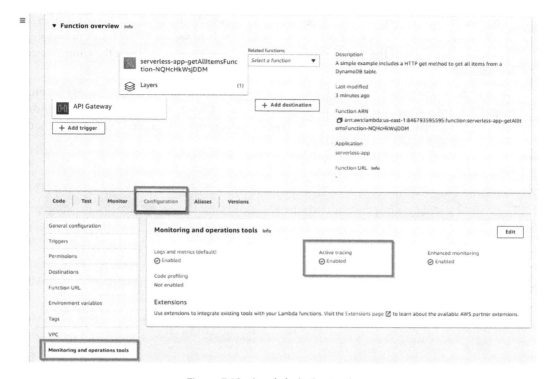

Figure 7.18 – Lambda Active tracing

5. To ensure tracing is enabled for the API Gateway *Prod* stage, navigate to **API Gateway | Stages | Prod** and verify that the **Enable X-Ray Tracing** option is selected, as shown in *Figure 7.19*.

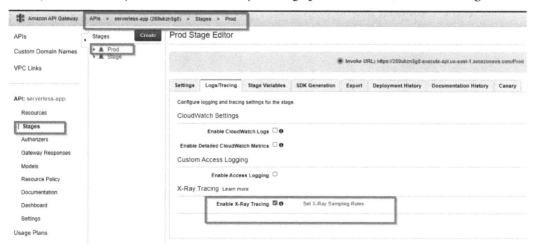

Figure 7.19 – API Gateway tracing

To test the end-to-end tracing of the application, you can invoke the `GetAllItems` function or insert the items using the `Put` function. You should be able to see the X-Ray traces showcasing the end-to-end view of the user transaction, as shown in *Figure 7.20* from the CloudWatch service map.

To view the X-Ray trace in the service map, navigate to **CloudWatch | X-Ray traces | Service map**. Here, you can see the flow of the requests between **Client**, **API Gateway**, and **Lambda**, including additional properties, such as each node's latency, requests/sec, and 5xx errors, without the need for any code-level instrumentation.

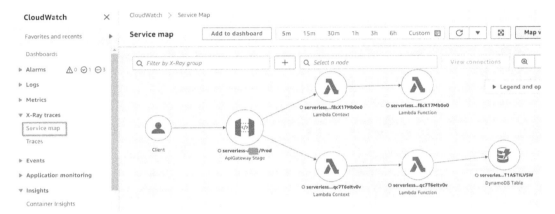

Figure 7.20 – Service map X-Ray

6. By navigating to the **Trace details** section and viewing **Segments Timeline**, as shown in *Figure 7.21*, you can gain insight into the amount of time spent at each stage and identify whether the major delay is at the invocation of the Lambda function.

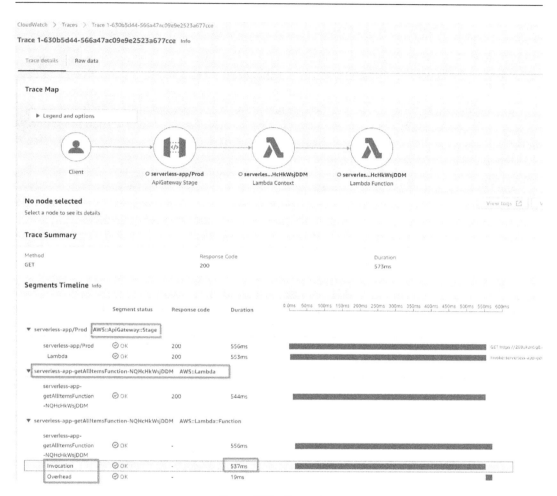

Figure 7.21 – Trace view

The **Trace** view provides a high-level overview of the user journey, but it may not give enough detail on where the user is encountering an issue. To get this information, we can use segments in AWS X-Ray and trace database calls with X-Ray. Leveraging Lambda Powertools can simplify the implementation process and add these details without sacrificing the maintainability or readability of the application code.

Exploring Lambda Powertools

Lambda Powertools is a suite of utilities and libraries that will help in adopting best practices for tracing, structured logging, and so on in Lambda functions that are built around the AWS SDKs. Lambda Powertools will help in implementing observability best practices by keeping the Lambda functions lean and allowing you to focus on the business logic without writing complex code for logging and tracing requirements. Lambda Powertools currently supports Lambda functions written in Python, Java, Node.js, and .NET Lambda runtimes. There are three main core utilities, namely **Logger**, **Metrics**, and **Tracer** to support the three pillars of observability:

- The **Logger** utility offers a simple and standardized way to format your logs as a structured JSON. It allows you to pass in the string or more complex objects and will take care of serializing the output as a structured JSON. Common use cases include logging the Lambda event payload and capturing Lambda cold starts. The Logger also supports the use of your own custom log formatted by "bring your own formatter" to meet your organization's specific standards.

- The **Tracer** utility of Lambda Powertools helps you effortlessly monitor serverless functions by sending traces to AWS X-Ray. This tool provides a clear view into function calls, interactions with other AWS services, or even external HTTP requests, allowing you to identify and resolve performance bottlenecks. The Tracer utility is a streamlined and user-friendly interface to the AWS X-Ray SDK, making it easier to implement observability best practices in your serverless architecture. With the ability to add annotations to traces, you can better categorize and analyze your traces, such as grouping by Lambda cold start information or transactions such as buying, selling, putting, or getting. The Tracer utility is also smart enough to automatically disable tracing when it is not running in the Lambda environment.

- The **Metrics** utility in Lambda Powertools makes it effortless to collect custom metrics from your application by utilizing **Embedded Metric Format** (**EMF**). As we briefly discussed in *Chapter 2, Overview of the Observability Landscape on AWS*, EMF is a feature in CloudWatch that allows you to submit custom metrics to custom namespaces in AWS asynchronously. CloudWatch EMF provides a scalable and reliable solution for collecting custom metrics. Lambda Powertools for the Metrics utility leverages this EMF functionality, making it easy to store important business and application metrics in a CloudWatch custom namespace.

Now that you understand Lambda Powertools and its utilities, the next step is to explore how to effectively integrates these tools into your Lambda functions. Lambda Powertools offers three different ways to instrument your code, offering flexibility and convenience to suit your specific needs:

- **Middy**: This is a middleware engine specifically designed for AWS Lambda functions that offers a quick and straightforward way to incorporate Lambda Powertools into your code with the fewest lines of code.

- **Method Decorator:** This method decorator approach is the best method for Lambda functions written in Typescript as classes and has only limited JavaScript support.

- **Manual:** The manual approach provides more control and is more verbose, but it also offers the greatest level of control over how Lambda Powertools is integrated into your Lambda function.

In the example presented in the section, we will use the manual approach to generate logs, metrics, and traces for the Lambda function.

Now, let's extend the observability for the application referred to in the *Deploying a basic serverless application running on Lambda* section using Lambda Powertools.

Before incorporating Lambda Powertools into our Lambda function, let's look into current logging in the CloudWatch logs. As we examine the logs, we'll notice they are lacking in detail and are not structured in any particular format. This presents a challenge when trying to gain insight into the inner workings of the Lambda function, especially when retrieving items from DynamoDB.

Figure 7.22 – Lambda logs in CloudWatch logs

To make it easy for you to carry out the exercise on Lambda Powertools, an updated CloudFormation template is provided with all the changes we are discussing in this section. You could use this CloudFormation template and create a new application named `serverless-app2`. As the application is in Node.js, I have imported the npm libraries of Lambda Powertools and included them in the CloudFormation deployment:

```
https://console.aws.amazon.com/cloudformation/home#/stacks/
new?stackName=serverless-app2&templateURL=https://insiders-guide-
observability-on-aws-book.s3.amazonaws.com/chapter-07/final/template.
yaml
```

Let's explore the changes made and why the changes are made from the viewpoint of Lambda Powertools.

I have included Lambda Powertools only in the `get-all-items.js` Lambda function in this exercise. So, please execute the Postman configuration for inserting the records and retrieving the details, as described in the *Deploying a basic serverless application running on AWS Lambda* section. We will examine each modification made to the `GetAllItems` Lambda function step by step in order to expand observability for metrics, logs, and traces. We will see how it enhances our ability to troubleshoot applications and add business context to our observability, resulting in an improved overall experience.

Lambda Powertools for enhanced logging

The `GetAllItems` Lambda function has been added with Lambda Powertools to capture information using the structure JSON format and retrieve additional information about the context of the Lambda function, such as cold starts, runtime, and so on. Let's look at the additional code added for structure in the Lambda function:

1. The code first imports the `Logger` and `injectLambdaContext` modules from the `@aws-lambda-powertools/logger` library:

    ```
    //Logging using Lambda powertools with Lambda context support.

    //Inclusion of Logger PowerLambda Tools
    const { Logger, injectLambdaContext } = require('@aws-lambda-
    powertools/logger');
    ```

2. A new `Logger` instance is created where the `serviceName` property is set to `get-all-items`. This is used to identify the source of the logs in the Amazon CloudWatch logs:

    ```
    //Servicename of the lambda function shown in the CloudWatch
    Logs
    const logger = new Logger({serviceName: 'get-all-items'});
    ```

3. The `injectLambdaContext` function is then used to log the Lambda context, which contains information about the function's execution environment and its interaction with the AWS infrastructure. You can see that the added Lambda context provides information about cold starts, function information, and service name. We have also logged output received from `get-all-items` into the CloudWatch log:

    ```
    //Logging Lambda Context and the output of items as a JSON using
    logger standard format
            logger.addContext(context);

    //Logging Items retrieved as a JSON in CloudWatch Logs.
            logger.info('Items in list:', { items });
    ```

4. You can see the full Lambda context along with the number of items retrieved and the details in the CloudWatch logs, as shown in *Figure 7.23*.

Figure 7.23 – Enhanced Lambda logging with Lambda Powertools

Lambda Powertools – custom metrics

We have further enhanced our application by retrieving and adding custom business metrics using the Lambda Powertools metrics functionality. The Lambda function retrieved the item count metric namely *Count of retrieved items* as a custom metric and added it into CloudWatch metics using **Embedded Metric Format** (**EMF**). Let's understand the changes made in the get-all-items.js file:

1. The code first imports the Metrics, MetricUnits, and logMetrics modules from the @aws-lambda-powertools/metrics library:

```
//Inclusion of Metrics from Lambda PowerTools
const { Metrics, MetricUnits, logMetrics } = require('@
aws-lambda-powertools/metrics');
```

2. Next, a new `Metrics` instance is created, where the `namespace` property is set to `getitems` and the `serviceName` property is set to `get-all-items`. These properties identify the source of the metrics in the Amazon CloudWatch custom metrics:

```
//Custom Metric namespace and service name in CloudWatch Custom
Metrics
const metrics = new Metrics({ namespace: 'getitems',
serviceName: 'get-all-items' });
```

3. The code then adds a custom metric called `'itemcount'` to the `Metrics` instance, using the `metrics.addMetric` method. The metric is set to the count of items in the `items` array and has a unit of `MetricUnits.Count`. Finally, the code calls the `metrics.publishStoredMetrics` method to publish the custom metrics to CloudWatch:

```
// Adding the total retrieved items as a metric in the Custom
namespace called "getitems"
        metrics.addMetric('itemcount', MetricUnits.Count, items.
Count);
        metrics.publishStoredMetrics();
```

4. You can verify from the CloudWatch logs that the metric count for the total items retrieved is published to the CloudWatch metric custom namespace called `getitems`:

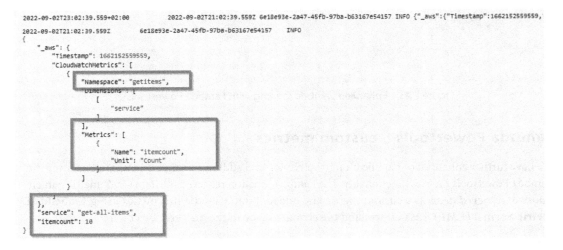

Figure 7.24 – Custom metric data in CloudWatch logs

5. Navigate to **All metrics | Custom namespaces | getitems**.

The `getitems` namespace is published with the details of the *count of items* retrieved from DynamoDB from the structure JSON output from the Lambda log:

Figure 7.25 – The getitemsmetric namespace

6. If you graph the metric over the period, you can understand how the number of items increased in DynamoDB over some time:

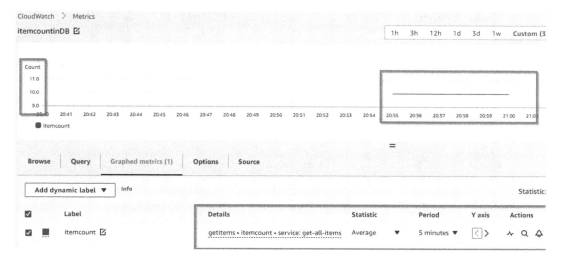

Figure 7.26 – Items in DynamoDB over a period

Lambda Powertools – tracing

Now, let's look into the traceability enhancement made using Lambda Powertools by adding annotations for the cold start of the Lambda function and adding segments to capture the DynamoDB calls using the tracer functionality:

1. The code first imports the `Tracer` and `captureLambdaHandler` modules from the `@aws-lambda-powertools/tracer` library:

    ```
    //Inclusion of tracer from Lambda Powertools
    const { Tracer, captureLambdaHandler } = require('@aws-lambda-
    powertools/tracer');
    ```

2. Next, a new `Tracer` instance is created, where the `serviceName` property is set to `get-all-items`. This is used to identify the source of the traces in the AWS X-Ray service:

    ```
    //X-Ray traces will be added with the servicename "get-all-
    items"
    const tracer = new Tracer({ serviceName: 'get-all-items' });
    ```

3. The code then imports the AWS DynamoDB SDK and creates an instance of the `DocumentClient` class. The `tracer.captureAWSClient` method is used to capture and trace all calls to the DynamoDB service through this client:

    ```
    //Inclusion of dynamodb sdk to support tracing
    const { DocumentClient} = require('aws-sdk/clients/dynamodb');

    //Trace the dynamoDB calls
    const docClient = tracer.captureAWSClient(new DocumentClient());
    ```

    You can look at the X-Ray trace to see that the DynamoDB information is captured along with the details about the time spent on the DynamoDB querying, as shown here:

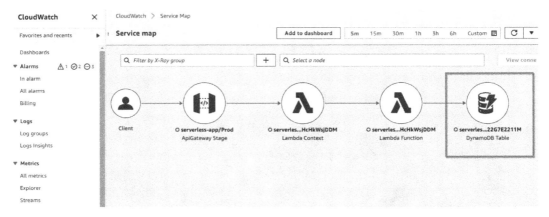

Figure 7.27 – AWS X-Ray tracing along with DynamoDB info

4. We have also added annotations to the Lambda function about the cold start with the service name. We have added the following code to annotate the Lambda function in the X-Ray tracing, and also added metadata about the event payload to the X-Ray tracing:

```
tracer.putAnnotation('awsRequestId', context.awsRequestId);
tracer.putMetadata('eventPayload', event);
```

In *Figure 7.28*, you can see the annotations about the Lambda cold start, which could help you in filtering the traces using the search functionality.

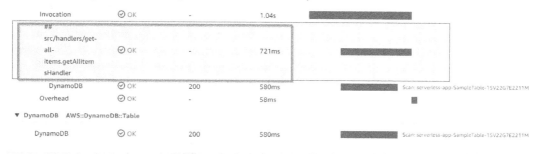

Figure 7.28 – Lambda annotations in AWS X-Ray

5. Close the segment using the `handlerSegment.close()` function:

```
finally {
    // Close subsegment (the AWS Lambda one is closed
automatically)
    handlerSegment.close();

    // Set back the facade segment as active again
    tracer.setSegment(segment);
}
```

We learned how to utilize the AWS Lambda Powertools library to improve the observability of a Lambda function by adding X-Ray tracing and DynamoDB information.

To summarize what we have learned in the overall end-to-end tracing of the Node.js application, with structured logging, we can log important information, such as the Lambda context and the output of the function, in a consistent and well-organized manner. This makes it easier to search and analyze the logs in Amazon CloudWatch Logs.

By adding custom business metrics, we can track key performance indicators and other relevant metrics related to the function's behavior and impact on the overall business. This data can be stored and visualized in Amazon CloudWatch custom metrics, allowing us to monitor the performance of the function over time and make informed decisions about the function's development and operation.

In addition, by using X-Ray tracing, we can gain visibility into the performance and behavior of the function as it interacts with other AWS services, such as DynamoDB. By capturing and tracing all calls to the DynamoDB service through the `DocumentClient`, we can see a complete picture of the function's behavior, from the time it is invoked to the time it returns a response. This information can be analyzed in the X-Ray service, providing us with valuable insights into the performance and behavior of the function and its interactions with other services.

Overall, leveraging Lambda Powertools to enhance the structured logging, custom metrics, and X-Ray tracing of a Lambda function can greatly improve the overall observability of the function, making it easier to troubleshoot issues, understand its impact on the business, and optimize its performance.

In the next section, we will understand how to troubleshoot performance issues and focus on important traces using AWS X-Ray groups.

Troubleshooting performance issues using X-Ray groups

It would be practically difficult to analyze all the X-Ray traces generated by a complex system and look at each trace to understand the issues. That's where X-Ray groups will be helpful. X-Ray groups will help simplify the process by focusing on the filtered traces based on rule-based criteria when there is a breach in a specific parameter. For example, if you would like to focus on the traces where the response time is greater than 3 seconds, you can create an X-Ray group with the criteria of `responseTime > 3`. This way, you can quickly isolate and analyze only the traces that indicate a problem, making it easier to identify and resolve issues. Let's create an X-Ray group and understand only the problematic traces from the generated traces:

1. You can see from the following figure that there are three traces with different response times. You can filter and focus only on the traces with a response time greater than (>) 3 seconds.

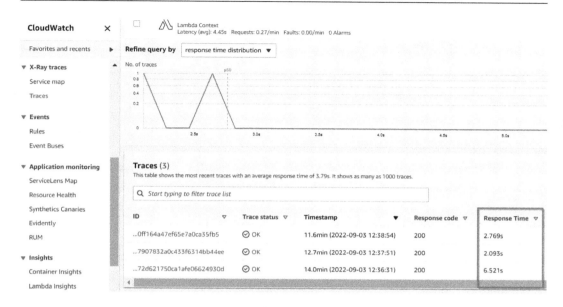

Figure 7.29 – X-Ray traces

2. Now, let's create an X-Ray group by navigating to **CloudWatch** | **Settings** | **Groups** | **View settings**:

Figure 7.30 – X-Ray groups

3. Click **Create group**, name the group as `LatencyGreaterthan3s`, and set **Filter expression** to `responseTime > 3`.

Figure 7.31 – X-Ray group creation

4. Navigate to **Traces | Filter by X-Ray group | LatencyGreaterthan3s**.

Figure 7.32 – Filter traces based on X-Ray group

5. You should see that only one trace is visible instead of three, as the other traces are filtered out due to the X-Ray group.

Troubleshooting performance issues using X-Ray groups   227

Figure 7.33 – Visible traces in the AWS X-Ray console

6. You can also repeat the same process to filter the traces based on annotations. In this example, we have annotated `ColdStart` in Lambda functions. You can create the filter using annotation as follows:

```
annotation.ColdStart = true
```

You could write a filter expression to focus on only the traces when the AWS Lambda is having a cold start and observe the pattern of those traces exclusively for any other issues, as shown in *Figure 7.34*.

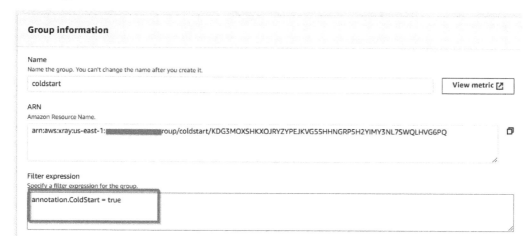

Figure 7.34 – X-Ray group based on annotations

X-Ray groups will be fundamentally helpful to filter out traces based on metrics and annotations and focus on only the problem traces to get a clearer understanding of the issues at hand. This will help you focus on traces that may have a potential impact on your application rather than focusing on every trace.

Summary

In this chapter, we looked at how to instrument a serverless application from end to end. We began by examining the default metrics and logs generated by the Lambda function and then expanded our monitoring capabilities using Lambda Insights and looked at how Lambda Insights works.

Furthermore, we discussed the importance of Lambda Powertools and walked through an example, and discussed the step-by-step process of changes made to include them in your Node.js function in each area of metrics, logs, and traces and how they can enhance the operational experience.

As a part of the custom metrics, we have discussed how to set up important business metrics as part of Lambda Powertools metrics. Finally, we talked about the benefits of using X-Ray groups to filter and focus on the traces that are of most interest, making it easier to troubleshoot and resolve issues.

The overall Lambda observability could be summarized in a diagram, as follows:

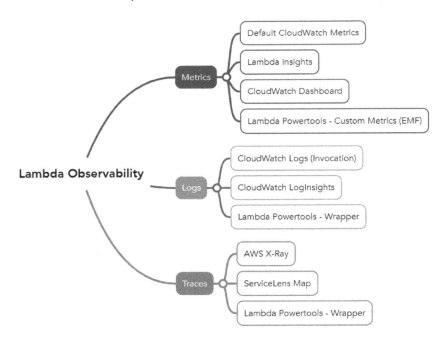

Figure 7.35 – Lambda observability summary

In the next chapter, we will look at open source observability options available on AWS and how to instrument applications using Open Telemetry.

Questions

1. What are the drawbacks of only utilizing default metrics and tracing in Lambda?

2. What are Lambda extensions?

3. What are Lambda Insights and the advantages of using Lambda insights?

4. What languages does Lambda Powertools support?

5. What are the utilities in Lambda Powertools?

8

End User Experience Monitoring on AWS

Welcome to the exciting world of end user experience monitoring on AWS! In *Chapters 3, Gathering Operational Data Using Amazon CloudWatch*, through *Chapter 7, Observability for Serverless Application on AWS*, we discussed how you can observe the applications running on AWS across different workloads including EC2, containers, and serverless compute systems such as Lambda, based on the building blocks of observability of gathering metrics, logs, and traces. While this provided us with a solid foundation to tackle issues we know about, it's equally important to consider issues we might not know about yet. We build and run applications to serve the users, and it is always good to understand how our applications are performing from the user's perspective, not just from the server's standpoint. This is where end user experience monitoring comes into play, and AWS offers a wide range of services to help you understand the end user experience.

In this chapter, we will explore different AWS services that allow you to measure the end user experience and learn how you can leverage them proactively and observe your applications. By doing so, you can move from being reactive to being proactive in your observability approach, ensuring that you are always ahead of the game for keeping your users happy and satisfied. Let's dive into the fascinating world of end user experience monitoring on AWS.

In this chapter, we are going to understand the following:

- Fundamentals of synthetic monitoring
- Overview of AWS CloudWatch Synthetics canaries
- Implementing synthetic monitoring for a sample website using CloudWatch Synthetics canaries
- Overview of **Real User Monitoring (RUM)** on AWS
- Implementing and tracking end user experience with RUM

Technical requirements

To follow along with this chapter, you need to have the following:

- A working AWS account

- An understanding of S3 and the setup of an S3 static website

- Fundamental knowledge of CloudFormation and setup of CloudFormation templates

End user experience monitoring

End user experience monitoring is also known as **digital experience monitoring** (DEM). As defined by *Gartner*, DEM is an *"availability and performance monitoring discipline that supports the optimization of the operational experience and behavior of a digital agent, human or machine as it interacts with enterprise applications and services."* It offers insights into how the application performs from the user's point of view, allowing you to understand the impact on users of the application. DEM provides the outside-in view of the application performance on how a user is experiencing the application when accessed. The outside-in view begins with establishing what looks good from the end user's point of view. Examples include web page response times, client-side JavaScript errors, visual stability, interactivity, API latencies, and so on.

If you look at the observability stack referred to in *Chapter 2, Overview of the Observability Landscape on AWS*, DEM comes as a part of layer 3 and consists of three different services from AWS, namely **Amazon CloudWatch Synthetics**, **Amazon CloudWatch RUM**, and **Amazon CloudWatch Evidently**. From now on, we will refer to them as **CloudWatch Synthetics**, **CloudWatch RUM**, and **CloudWatch Evidently** for simplicity. Let's understand what they do briefly before going into the details of each service:

- **CloudWatch Synthetics** helps you ensure that web applications are available and performing as expected. The modular and lightweight canary scripts in CloudWatch Synthetics help you monitor your web applications 24x7 and quickly identify and help address any issues that arise.

- **CloudWatch RUM** empowers you with valuable insights into the real-time experience of your end users. By monitoring actual user interactions, CloudWatch RUM provides you with a comprehensive understanding of your application's performance and usability based on actual user data.

- **CloudWatch Evidently** allows you to launch new application features and validate your web application decisions by running online experiments. With this feature, you can safely test new ideas and configurations to determine their impact on application performance and user experience and measure business outcomes. This will help you make data-driven decisions for any new application releases.

Let's go into the details of what CloudWatch Synthetics and CloudWatch RUM are and explore their unique features and functionalities in this chapter.

CloudWatch Synthetics

CloudWatch Synthetics allows you to proactively monitor the website and API endpoints every minute, 24x7, using modular canary scripts. This service helps you receive instant alerts when your application does not behave as expected, allowing you to quickly identify and address issues before they impact your users. Synthetic monitoring is essential for organizations to gain insights into their application performance from the perspective of their users, and identify intermittent issues that may go unnoticed with traditional monitoring tools. CloudWatch Synthetics will act like a user and perform a health check on a defined schedule, even when there is no user traffic, helping you to catch issues proactively and prevent user-facing problems.

CloudWatch Synthetics helps you gain a comprehensive understanding of the availability and performance of your web applications and ensures you meet your **Service-Level Agreement (SLA)** requirements.

Now, let's explore how CloudWatch Synthetics operates and gain a deeper understanding of its underlying mechanisms. By learning about its workings, you can effectively leverage the tool to monitor your web applications and ensure the best possible user experience.

How CloudWatch Synthetics works

Let's take a closer look at how CloudWatch Synthetics works. *Figure 8.1* provides a summary of the overall operations of the tool:

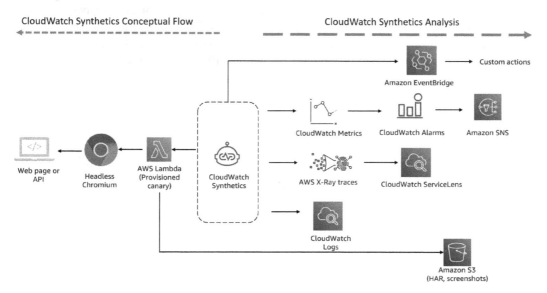

Figure 8.1 – How CloudWatch Synthetic works

In *Figure 8.1*, the left-hand side shows the conceptual flow of CloudWatch Synthetics. CloudWatch Synthetics leverages a Lambda function that executes browser-based testing using a headless Chromium browser, which is a part of the managed fleet. This means that AWS automatically manages Lambda functionality, so you don't have to worry about managing the underlying infrastructure.

Headless Chromium (`https://developer.chrome.com/blog/headless-chrome/`) works by polling the website or API, rendering the results (including JavaScript), and following a predefined path inside an application based on your scripts. By using this process, CloudWatch Synthetics can simulate user interactions with your application and identity issues that may affect user experience.

Synthetic canaries check the availability and latency of endpoints. They monitor website URLs, REST API, and website content. Synthetic canaries use Puppeteer, Node.js, or Python Selenium scripts. They create Lambda functions in your AWS account that use these languages as a framework. CloudWatch Synthetics completely manages these Lambda functions and requires no intervention from the users.

CloudWatch Synthetics testing can also take place inside AWS VPCs for applications that are not accessible from the internet, such as intranet applications or internal application APIs.

Now, let's turn our attention to the right side of *Figure 8.1*. The output of CloudWatch Synthetics consists primarily of **metrics, events, logs, and traces** (**MELT**), which are stored in CloudWatch primary services:

- **CloudWatch Metrics**: CloudWatch Synthetics provides a range of metrics including **2xx**, **4xx**, **5xx, duration, failed request count**, and **success percent**. These metrics are available as a custom CloudWatch metric namespace, which can be used to create custom dashboards, set alarms, and detect anomalies. You can browse the metrics in two ways: by navigating to the **Synthetics canaries** section of CloudWatch and selecting your canary to see its **Monitoring** tab, as shown in *Figure 8.2*, or by viewing **All metrics** in the CloudWatch console.

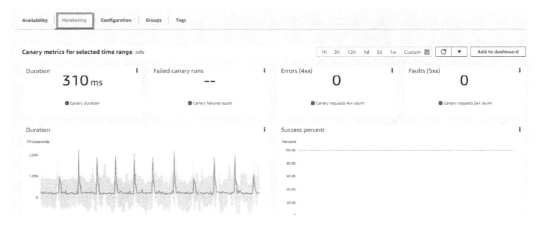

Figure 8.2 – CloudWatch Synthetics metrics

- **X-Ray Traces**: AWS X-Ray can record the details of a trace, offering a complete perspective of the synthetic transaction and enabling the results to be viewed through Service Lens. As of this writing, X-Ray tracing is only available for Node.js-based Puppeteer scripts. You can visualize the traces generated by CloudWatch Synthetics, as shown in *Figure 8.3*:

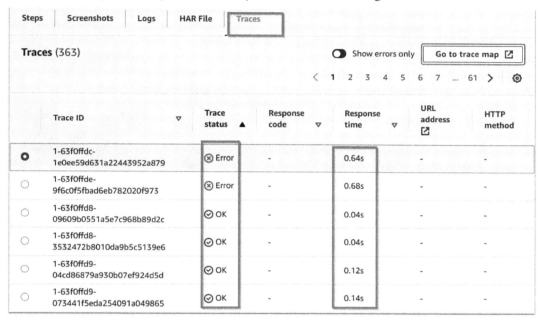

Figure 8.3 – CloudWatch Synthetics Traces

- **CloudWatch Logs**: Logs generated during the execution of the synthetic canary are stored in CloudWatch logs, offering a comprehensive view of activities performed. I show a view of the CloudWatch logs for a synthetic canary in *Figure 8.4*:

Figure 8.4 – CloudWatch Synthetics Logs

- **CloudWatch Events**: Through CloudWatch Events, you can understand the status of the Synthetics canaries, including changes in status, failure, and success. You can also configure notifications for these based on the EventBridge rules, as shown in *Figure 8.5*:

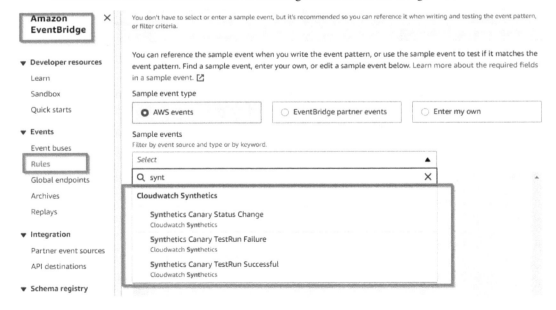

Figure 8.5 – CloudWatch Synthetics Events

- **Data Retention Settings**: The output data from the execution of Synthetic canaries is saved in an S3 bucket, and we can adjust the retention settings for this data to manage storage costs. You can see the configuration settings for this in *Figure 8.6*:

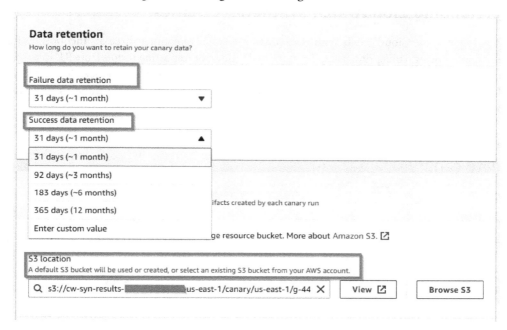

Figure 8.6 – Data retention settings

Having gained an understanding of how CloudWatch Synthetics operates, let's explore the various use cases of this service.

Use cases of CloudWatch Synthetics monitoring

Use cases of CloudWatch Synthetics monitoring include the following:

- **Availability and latency monitoring**: To monitor URL availability and understand the latency in loading the website by the users

- **Easy web testing**: Quick testing of the website experience without any user traffic

- **SLAs/SLOs made easy**: To measure the availability of the website against the SLA

- **Visual regression monitoring**: To understand issues in the visual components of the website

- **Proactive alerting**: To alert for any issues proactively before affecting the end users

- **Anomaly detection**: To understand any anomalies in the website performance before it affects the end users

Now let's try to understand the configuration supported in CloudWatch Synthetics and understand how to configure for some of the options.

Understanding CloudWatch Synthetics canaries

Upon logging in to the CloudWatch Synthetics console, you will encounter three distinct options for creating canaries: using a blueprint, using the inline code editor, or importing scripts from S3. However, I have grouped these options into five categories based on my practical experience. Specifically, I consider two of the blueprints, namely **Canary Recorder** and **GUI workflow builder**, as a method to create a synthetic canary. *Figure 8.7* shows the classification:

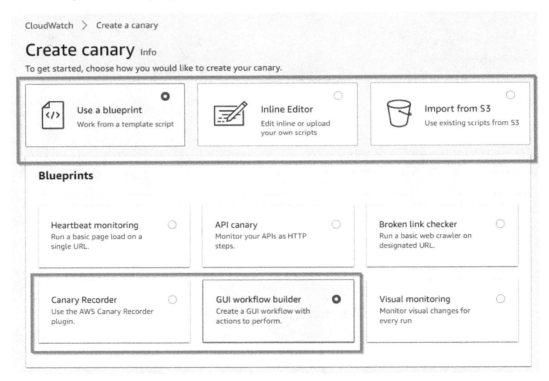

Figure 8.7 – Synthetic canaries methods

Let's delve into each of these five methods and gain a comprehensive understanding of them. We can use the following five methods for creating canaries in CloudWatch Synthetics, depending on their feasibility and requirements:

- **Blueprints**: Predefined blueprints are available to generate canary scripts automatically based on configured options for specific use cases.

- **Canary Recorder**: The AWS Canary Recorder is a Chrome browser extension that allows you to record clicks and actions on a website and auto-generate the script based on the actions performed.

- **GUI Workflow Builder**: The GUI workflow builder is ideal for creating secure monitoring of user workflows. For instance, if you want canaries to mimic usernames and passwords but do not want to store them in the script, this tool allows you to pass them as variables and leverage AWS Secrets Manager to store the credentials.

- **Inline Code Editor**: The inline code editor is suitable for creating custom scripts or importing existing scripts from your own local system.

- **Import Scripts from S3**: You can also upload your existing Puppeteer or Selenium scripts into S3 storage and import them into the Synthetic canaries.

Now let's look into the other default blueprints available in CloudWatch Synthetics after our recategorization. There are four default blueprints, namely **Heartbeat monitoring, API canary, Broken link checker**, and **Visual monitoring**, as shown in *Figure 8.8*:

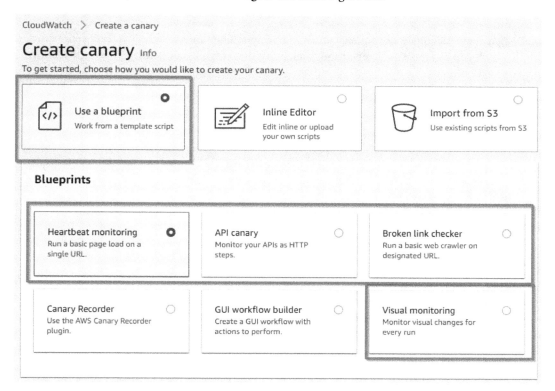

Figure 8.8 – Blueprints

Let's go into the details of each blueprint highlighted in *Figure 8.8*:

- **Heartbeat monitoring**: Heartbeat monitoring allows you to check the availability of a website by measuring its basic page-load of a single URL. It's useful for measuring the **Service Level Objective (SLO)** of a website's availability.

- **API canary monitoring**: This blueprint is designed to test REST APIs and their availability. By invoking the methods in the API and verifying the response time and output, you can ensure that APIs are functioning properly. This will be quite useful for testing the availability of APIs, as many modern applications rely heavily on APIs.

- **Broken link checker**: When updating a website with multiple URL references, it's common for some of the referenced URLs to become unavailable. This blueprint provides a solution to this issue by checking all the URLs on the listed website based on the blueprint.

- **Visual monitoring**: This blueprint is useful for understanding the visual differences in your web interfaces and comparing any changes to the baseline. It provides a way to monitor the user experience and quickly identify any issues with the design or layout of the website.

CloudWatch Synthetics offers a comprehensive monitoring solution for proactively tracking endpoints. With this tool, you can measure latency and availability, detect GUI anomalies, and quickly identify any visual regressions.

Now let's look at configuring CloudWatch Synthetics canaries using some methods.

Configuring CloudWatch Synthetics canaries

Let's take a scenario where you are looking to configure the availability monitoring for a public-facing website such as aws.amazon.com. You can use the heartbeat blueprint to configure and measure the availability of the website.

Heartbeat monitoring

Let's create a heartbeat canary blueprint on the URL aws.amazon.com and monitor the availability of the URL using the Synthetics canaries:

1. Navigate to **AWS Console | CloudWatch | Application monitoring | Synthetics Canaries | Create canary**:

Figure 8.9 – Creating Synthetics canaries

2.   Select **Use a blueprint | Heartbeat monitoring**:

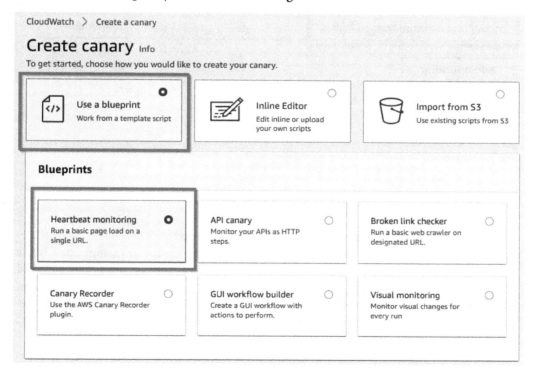

Figure 8.10 – Selecting Heartbeat monitoring

3. Provide `awswebsite` as the name and `https://aws.amazon.com` as the application or endpoint URL. In one synthetic canary, you can add up to five different endpoints. If you would like to take a screenshot of the website during each canary run, you can select the **Take screenshots** checkbox:

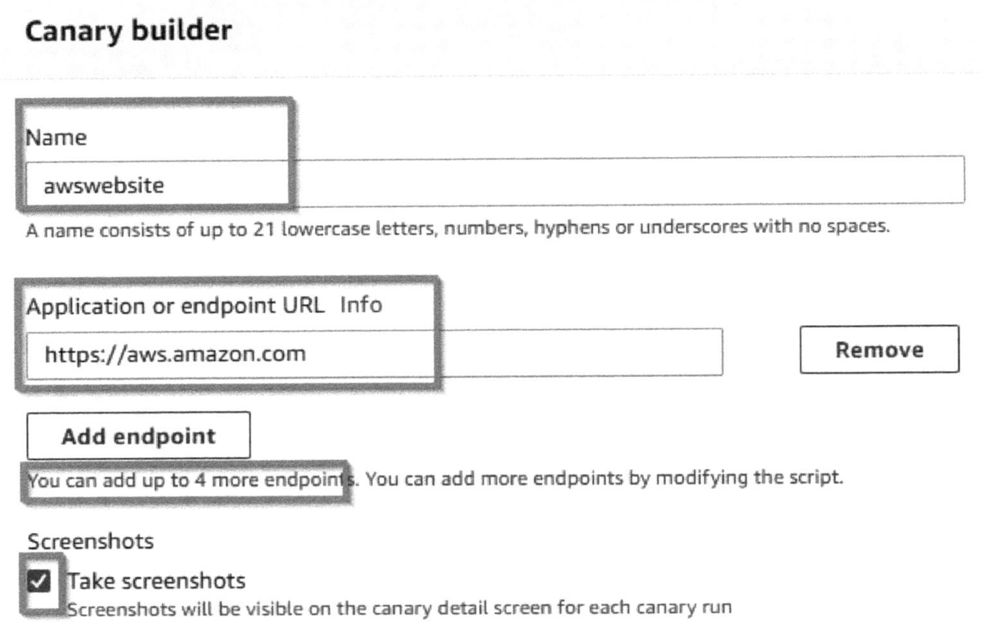

Canary builder

Name

awswebsite

A name consists of up to 21 lowercase letters, numbers, hyphens or underscores with no spaces.

Application or endpoint URL   Info

https://aws.amazon.com   Remove

Add endpoint

You can add up to 4 more endpoints. You can add more endpoints by modifying the script.

Screenshots

☑ Take screenshots
Screenshots will be visible on the canary detail screen for each canary run

Figure 8.11 – Input to canary endpoint

4. When you navigate to **Script editor**, you will notice that the script is generated automatically and populated with the parameters based on your selected options. For example, the URL is included in the script, and the **Screenshots** option is enabled. You can select your preferred **Runtime version** from the drop-down list, with options for either Puppeteer or Selenium Python scripts, depending on the specific capabilities available with your organization. This streamlined process makes it easy to customize your script and run it efficiently, without needing to manually write the code from scratch:

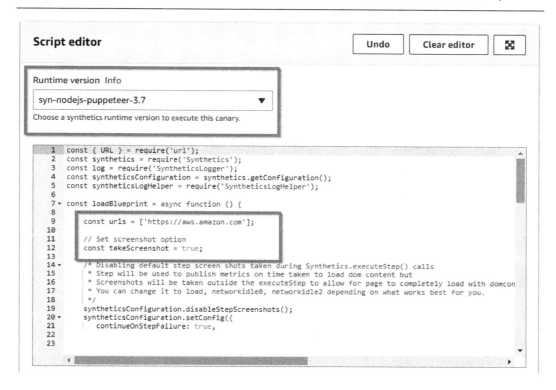

Figure 8.12 – Script editor, auto-populate

5. You can easily schedule your canaries to run at specific intervals, either continuously or just once. You can choose a preset interval or customize your schedule using a CRON expression. For this exercise, we set the interval to every 5 minutes. This flexibility in scheduling allows you to tailor your canary runs to your specific needs and also control costs.

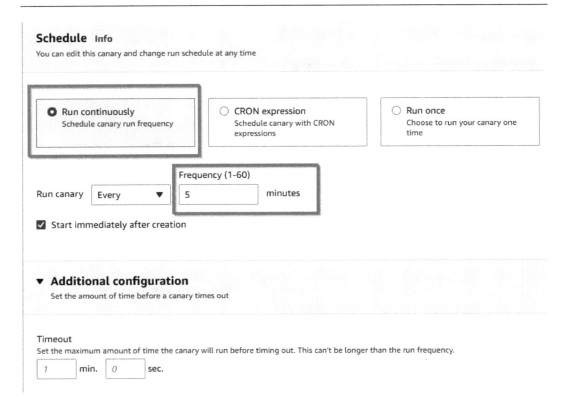

Figure 8.13 – Scheduler for Synthetics canaries

6. Select the **Failure data retention** and **Success data retention** intervals. Understand that this retention setting affects the S3 storage cost. **S3 location** is auto-populated if it is the first time, otherwise, select a bucket to store the canary artifacts such as screenshots, HAR (short for **HTTP Archive**) information, and so on:

Data retention

How long do you want to retain your canary data?

Failure data retention

31 days (~1 month) ▼

Success data retention

31 days (~1 month) ▼

Data Storage

Select an S3 folder where you would like to store the artifacts created by each canary run

Canary run data is stored in an Amazon S3 storage resource bucket. More about Amazon S3. ↗

S3 location

A default S3 bucket will be used or created, or select an existing S3 bucket from your AWS account.

🔍 s3://cw-syn-results-███████████-us-east-1/canary/us-east-1/aws ✕ | View ↗ | | **Browse S3** |

▶ **Additional configuration** Info

Encrypt the canary artifacts using SSE-S3 or AWS KMS.

Figure 8.14 – Storage settings

Synthetics canaries will create a new AWS IAM role by default to execute the conceptual flow:

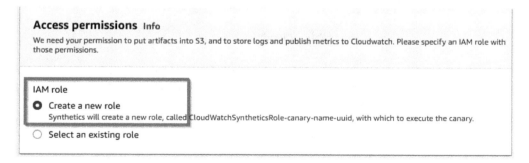

Access permissions Info

We need your permission to put artifacts into S3, and to store logs and publish metrics to Cloudwatch. Please specify an IAM role with those permissions.

IAM role

⦿ Create a new role

Synthetics will create a new role, called CloudWatchSyntheticsRole-canary-name-uuid, with which to execute the canary.

◯ Select an existing role

Figure 8.15 – Access permissions

7. You can create CloudWatch alarms by clicking on the **Add new alarm** button based on **SuccessPercentage** and **Duration**. You can also select a **Simple Notification Services (SNS)** service to notify you of any additional actions you would like to take based on the alarms. It can create an incident in the ITSM system or action of remediation if required:

▼ **CloudWatch alarms - *optional*** Info
You can let Synthetics create alarms for your canary automatically, and customize these later.

Metric name	Alarm condition	Threshold		Period	
SuccessPer... ▼	Lower ▼	90	%	5 minutes ▼	Remove
Duration ▼	Greater ▼	30000	ms	5 minutes ▼	Remove

Add new alarm

▼ **Set notifications for this canary**
Select where to receive notifications when this canary reaches the level you define

Send a notification to the following SNS topic. Info
Define the SNS (Simple Notification Service) topic that will receive the notification

◉ Select an existing SNS topic
○ Create new topic

Select topic...

Q *Select an email list*

Only SNS topics for this account are available

Figure 8.16 – CloudWatch alarms

8. If your website is hosted within AWS VPC, you have the option to select the VPC as the location for executing the canary. However, in our current example, the website URL is hosted outside of our AWS account, hence we have selected **No VPC** in **VPC settings**:

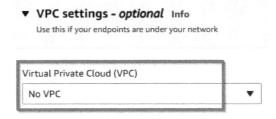

▼ **VPC settings - *optional*** Info
Use this if your endpoints are under your network

Virtual Private Cloud (VPC)	
No VPC	▼

Figure 8.17 – VPC settings

9. You can enable **Active tracing** (as we have selected Node.js script) to track how the Synthetics canary is executing, and click on the **Create canary** button:

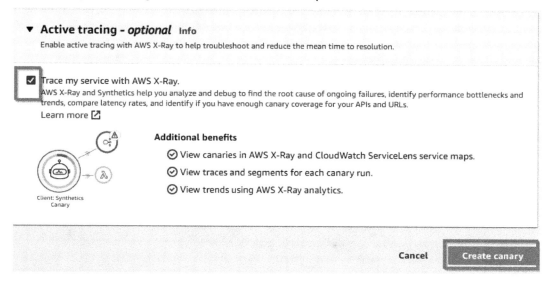

Figure 8.18 – Active X-Ray tracing

It will take a minute or two to create the Synthetics canary.

Now, let's explore the results of the Synthetics canaries created.

10. You can verify the availability of the website, the last state, and any issues as a **Summary** view:

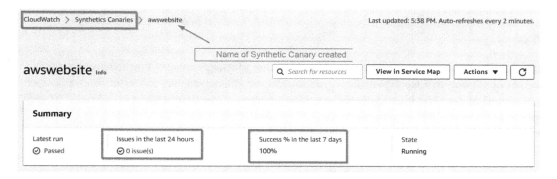

Figure 8.19 – Availability metrics for the Synthetics canary

11. When you navigate to the **Availability** tab and verify tabs at the bottom-right corner of the Synthetic canaries created, you can verify **Screenshots, Logs, HAR File**, and **Traces**. Additionally, HAR tells you where the greatest amount of time is spent in loading the website, which will be further useful to understand where the issues are and fine-tuning the page load times:

Figure 8.20 – Navigating details of Synthetics results

You can also navigate for metrics and alarms, as described in the *How CloudWatch Synthetics works* section in this chapter.

API canary

As a second example for using the Synthetics canary blueprints, let's set up a synthetic canary for the REST API endpoint that we deployed as a part of *Chapter 7, Observability for Serverless Application on AWS*, and understand how to use a canary to understand issues with the API endpoint. If you have not deployed CloudFormation as a part of *Chapter 7, Observability for Serverless Application on AWS*, you can deploy the same from the following Quickstart CloudFormation template and refer to *Chapter 7* for the application overview:

```
https://console.aws.amazon.com/cloudformation/home#/stacks/
new?stackName=serverless-app2&templateURL=https://insiders-guide-
observability-on-aws-book.s3.amazonaws.com/chapter-07/final/template.
yaml
```

Now, let's understand how to create a Synthetics canary for an API endpoint and understand the issues regarding APIs using the canary metrics:

1. Navigate to **CloudWatch | Synthetics Canaries | Use a blueprint | API canary** and enter the name apicanary_restapi:

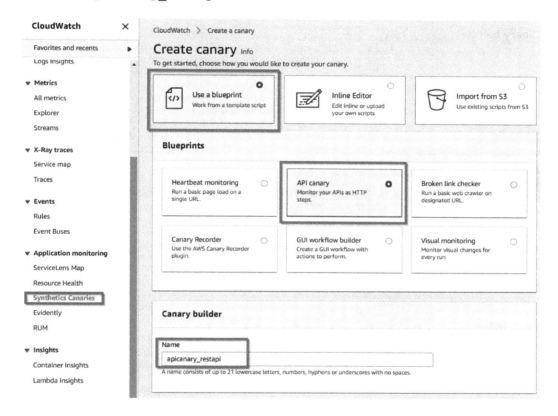

Figure 8.21 – API canary

2. If the API is a third-party service such as an external API (for example, Boomi or Apigee), you could directly select **Add HTTP request** and provide the URL along with the methods to invoke. As we are using an Amazon API Gateway API, select the **I'm using an Amazon API Gateway API** checkbox, then select **Choose API** | **Choose API and stage from API Gateway** | **serverless-app2**, and set the **Stage** option to **Prod**:

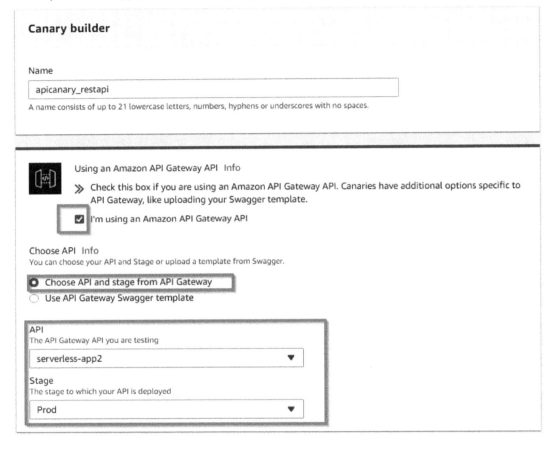

Figure 8.22 – Selecting Amazon API Gateway

3. Application or endpoint URLs are available to select automatically based on the Amazon API Gateway URL. Click on **Add HTTP request**:

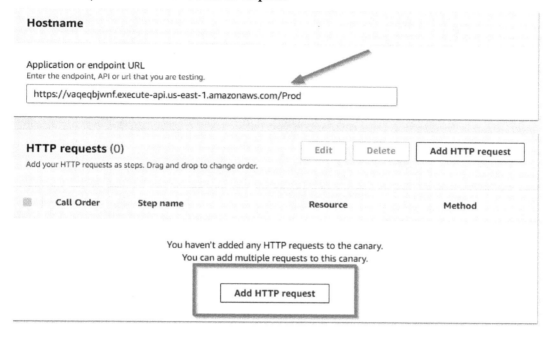

Figure 8.23 – Add HTTP request

4.   Select **/items** for **Resource**, set **Method** to **GET**, and click on **Save**:

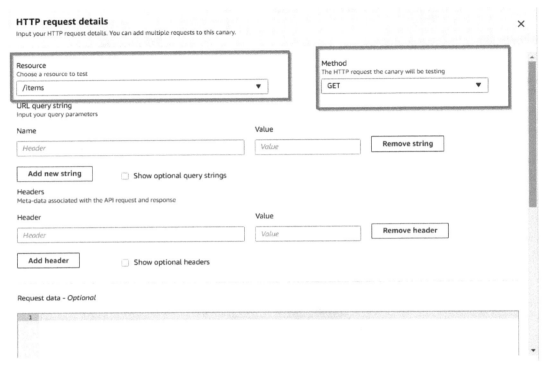

Figure 8.24 – GET method response verification

5.   Optionally, if you are looking to capture the headers and response body, you can select the same in **Reporting configuration**. This will be especially useful when you would like to analyze the response details of the API; click **Save**:

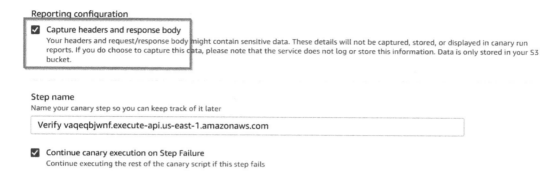

Figure 8.25 – Optional response capture

Now when you navigate to **Script editor**, the Synthetics canaries script has been auto-generated based on the selected options:

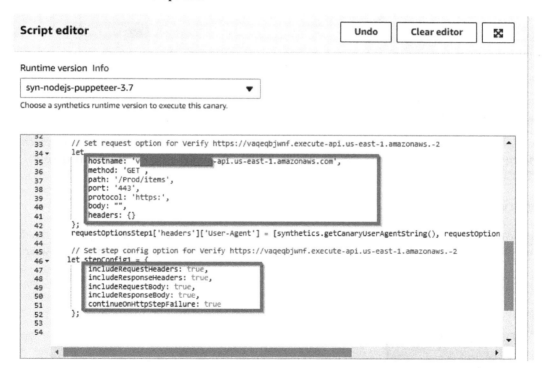

Figure 8.26 – API canary script

6.   Leave the remaining options as default and click **Create canary**:

▼ **Tags - *optional*** Info
    Add tags to canaries to help set permissions, organize, and search for them later

Key

blueprint

Value - *optional*

apicanary

Remove

Add new tag

You can add 49 more tag(s).

▶ **Active tracing - *optional*** Info
    Enable active tracing with AWS X-Ray to help troubleshoot and reduce the mean time to resolution.

Cancel   **Create canary**

Figure 8.27 – Create canary

If you are looking for an explanation of the remaining default ones and to understand the results, please refer to the *Heartbeat monitoring* subsection in this chapter.

The implementation of broken link monitoring and visual monitoring could follow a similar approach to heartbeat monitoring. If the default blueprints are inadequate for the task, we can explore the usage of the Canary Recorder.

Canary Recorder

CloudWatch Synthetics Recorder is a useful tool that can assist you in recording the steps performed on a website. This tool is available as a Chrome extension and can be easily installed on your browser:

Figure 8.28 – CloudWatch Synthetics Recorder in the Chrome Web Store

Once installed, you can simply click on the **Start recording** button and the recorder will begin capturing all the actions you perform on the website:

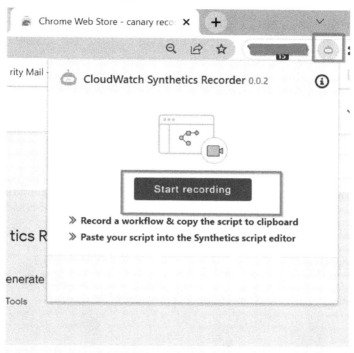

Figure 8.29 – Recording using CloudWatch Synthetics Recorder

The **Synthetics Recorder** is helpful when you need to generate Synthetics canary scripts quickly. By recording the steps you take on the website; the recorder auto-generates the Synthetics canary script for you. This feature saves time and effort, especially if you are not familiar with scripting.

Using the **Synthetics Recorder** can be an effective solution if the available default blueprints do not meet your requirements. With the recorder, you have the flexibility to customize the steps and actions you want to capture, enabling you to create more tailored and effective monitoring scripts for your website.

So far, we have gained a solid understanding of CloudWatch Synthetics, including how it functions and the various options for recording synthetic canaries. We have also explored two examples of using Synthetics for monitoring a website's heartbeat and API Gateway.

Now, let's dive into the topic of RUM: how it can be set up on AWS and how to analyze the RUM results.

CloudWatch RUM

CloudWatch RUM is a monitoring technology that captures and records all the user interactions with your website or application. With RUM, you can gain valuable insights into your application's frontend performance from the perspective of real users. It helps you correlate the client side and server side and provides an end-to-end view of the application performance.

CloudWatch RUM helps you gain a comprehensive understanding of your website's performance, identify areas for improvement, and continually optimize your web application for the best possible user experience. It will help you identify performance issues of the website, the **Application Performance Index** (**Apdex**) by country, and errors on the website. It also helps you understand the browsers/devices accessing your website and analyze the sessions and traffic on the website.

Besides these benefits, CloudWatch RUM offers all the benefits of Amazon CloudWatch, such as metrics, alarms, anomaly detection, and so on. It provides flexibility in configuration, allows you to customize session samples and capture response times of specific pages, and also add X-Ray tracing.

Now, let's understand how CloudWatch RUM works.

How CloudWatch RUM works

Here is the high-level workflow overview of how CloudWatch RUM works:

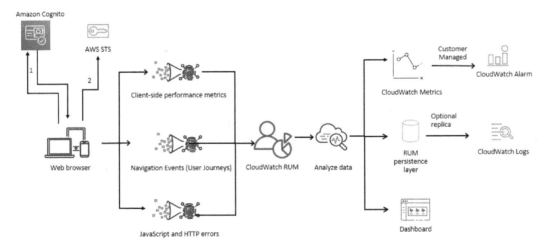

Figure 8.30 – How RUM works

If you look at the left side of *Figure 8.30*, to enable CloudWatch RUM monitoring for your web application, the first step is to include a small script in the application's frontend code. Once the web application loads in a user's browser, the script retrieves the credentials from the Cognito identity pool, which is configured as part of CloudWatch RUM.

With the credentials to post the data to CloudWatch RUM, the CloudWatch RUM script collects client-side performance telemetry, navigation events, and any JavaScript/HTTP errors that occur during user journeys on your website. This data provides valuable insights into how users interact with your application and can identify performance issues and other areas for improvement.

If you look at the right side of *Figure 8.30*, after receiving data from the CloudWatch RUM script, the service processes and publishes the collected metrics. You can choose to enable alarms as needed to notify you of any performance issues. Additionally, CloudWatch RUM provides the option to store logs in CloudWatch log groups from the RUM persistent data layer. This allows you to maintain a comprehensive record of user interactions and performance metrics for future analysis.

Finally, CloudWatch RUM automatically creates default dashboards that provide a clear and concise overview of performance issues based on end user experience. We can customize these dashboards to meet your specific monitoring and analysis needs, providing a valuable tool for identifying and addressing performance issues and optimizing your application for the best possible user experience.

The following are the default metrics provided by CloudWatch RUM:

- **Users**: Active user count, device, screen resolution, browser, and location
- **User interaction**: Sessions, entry point, exit point, user journey, and top viewed pages
- **Site behavior**: Page load times, latencies, and web vitals

The logs stored can generate insights by leveraging the CloudWatch Logs Insights service.

In the next section, we will look into how to set up CloudWatch RUM for an S3 static website and understand the real user performance metrics and issues.

Setting up CloudWatch RUM for an S3 static website

In this practical exercise, we will deploy a static website on Amazon S3 and configure CloudWatch RUM to monitor user interactions and provide insights into the performance of the **Single Page Application (SPA)** running on the S3. By setting up CloudWatch RUM, we can gain a deeper understanding of how users interact with our SPA, identify potential performance issues, and take steps to optimize the user experience.

As a part of the first step, let's deploy the static website and upload a simple HTML page, index.html:

1. Let's deploy the following CloudFormation template as a quick start, which will deploy an S3 bucket along with making it public using the ACLs. I do not recommend this configuration for production deployment, as the S3 bucket is made public:

```
https://us-east-1.console.aws.amazon.com/cloudformation/
home?region=us-east-1#/stacks/quickcreate?templateUrl=h
ttps%3A%2F%2Finsiders-guide-observability-on-aws-book.
s3.amazonaws.com%2Fchapter-08%2Fcreates3staticwebsite.
yaml&stackName=S3WebappRUM
```

2. Navigate to the output of the CloudFormation template deployment to confirm the creation of the S3 website URL. This URL will serve as our endpoint for the website and will be used to verify and observe the website's performance using CloudWatch RUM:

Figure 8.31 – S3 SPA

3. Download the index.html file from this URL: https://insiders-guide-observability-on-aws-book.s3.amazonaws.com/chapter-08/index.html.

    This is a simple plain HTML file with a test message. You can right-click on the file and save it as index.html.

    This has been purposefully kept for download and upload to the S3 bucket as RUM requires the addition of a script to capture the real user data.

4. Browse to the S3 website URL that was noted down in the **Outputs** tab during *Step 2*. By visiting this WebSite URL, we can verify that the S3 static web page is available to users. This step is essential to ensure that the website is properly deployed and functioning as intended before proceeding with the CloudWatch RUM setup.

5. The CloudWatch RUM setup is divided into three steps:

    - Add the app you want to monitor

    - Copy and paste the JavaScript code into the header of the application

    - Monitor and troubleshoot the issues:

Figure 8.32 – RUM setup

6. To set up CloudWatch RUM, there are two options available: configuring it through the AWS Console or using CloudFormation. For this exercise, we will utilize the AWS Console to set up CloudWatch RUM.

    Navigate to **AWS Console | CloudWatch | RUM | Add app monitor**:

Figure 8.33 – Add app monitor

7.   Provide S3RUM-App as the app monitor name and the website URL as noted in the *Step 2* output S3BucketSecureURL without http(s). We are selecting to capture performance telemetry, JavaScript errors, and HTTP errors, though we do not receive any JavaScript errors in this exercise.

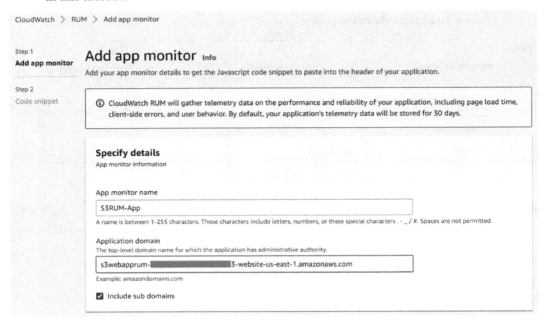

Figure 8.34 – Add app monitor

8. Select to allow cookies by selecting **Check this option to allow the CloudWatch RUM Web Client to set cookies** and type 100% in **Session samples,** which will allow you to sample 100% of the sessions. Finally, select **Check this option to store your application telemetry data in your CloudWatch Logs account** to log the request to CloudWatch Logs, as shown here:

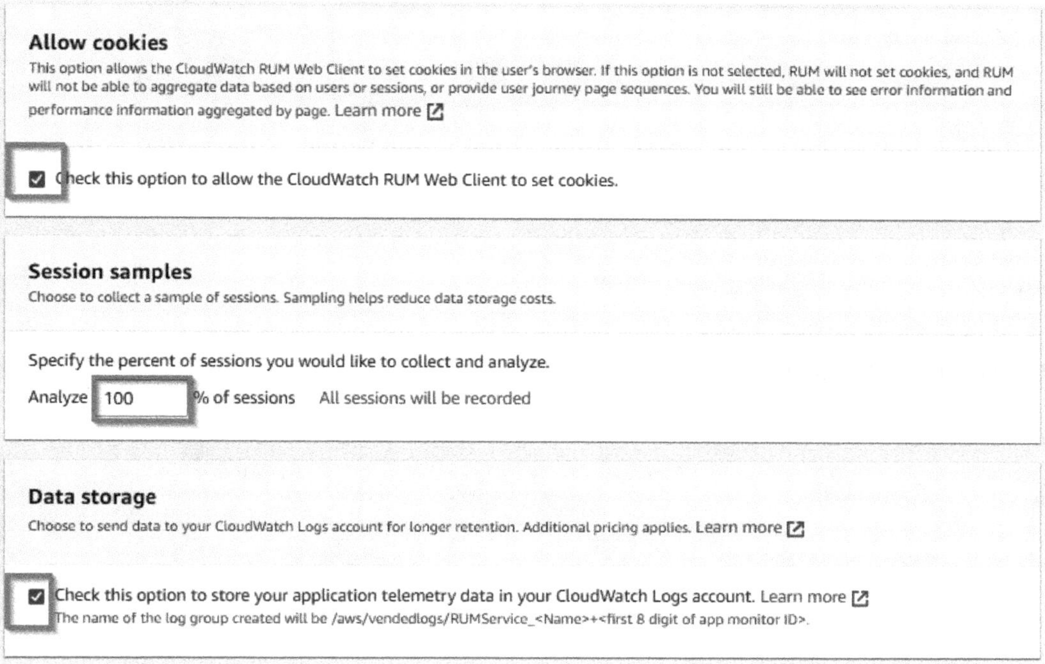

Allow cookies

This option allows the CloudWatch RUM Web Client to set cookies in the user's browser. If this option is not selected, RUM will not set cookies, and RUM will not be able to aggregate data based on users or sessions, or provide user journey page sequences. You will still be able to see error information and performance information aggregated by page. Learn more [↗]

☑ Check this option to allow the CloudWatch RUM Web Client to set cookies.

Session samples

Choose to collect a sample of sessions. Sampling helps reduce data storage costs.

Specify the percent of sessions you would like to collect and analyze.

Analyze 100 % of sessions   All sessions will be recorded

Data storage

Choose to send data to your CloudWatch Logs account for longer retention. Additional pricing applies. Learn more [↗]

☑ Check this option to store your application telemetry data in your CloudWatch Logs account. Learn more [↗]
The name of the log group created will be /aws/vendedlogs/RUMService_<Name>+<first 8 digit of app monitor ID>.

Figure 8.35 – CloudWatch logging

9.   Select **Create new identity pool** and configure it to capture information from **All pages**:

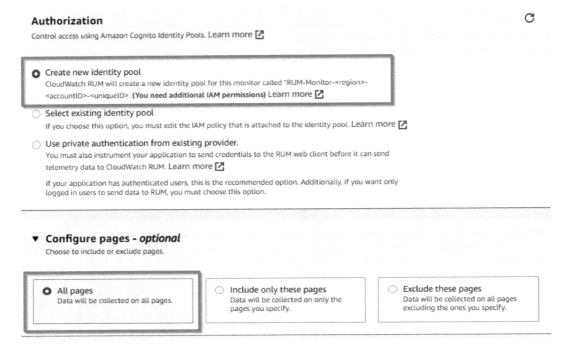

Figure 8.36 – Cognito pool creation and configuring pages

10. Click on **Add app monitor**:

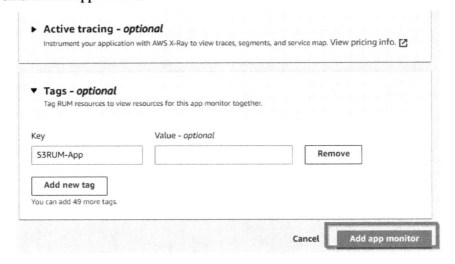

Figure 8.37 – Clicking on Add app monitor

11. The app monitor will generate a script that needs to be added to the SPA. This process may take a few minutes. It's worth noting that if the web application has multiple pages with disjoint domain names/URLs, the RUM monitor needs to be created separately for each page. This will ensure that the performance data for each page is captured accurately.

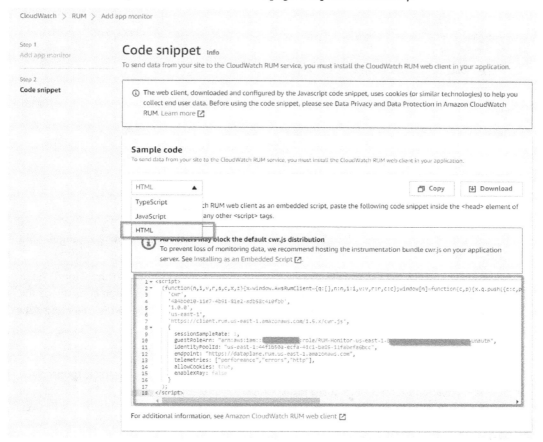

Figure 8.38 – Selecting the HTML script

12. Copy the script generated and edit `index.html` using any editor, such as Notepad on Windows or TextEdit on macOS, and add the script to the header of the HTML page after the HTML tag `<head>`:

```
<!-- saved from url=(0087)https://insiders-guide-observability-on-aws-book.s3.amazonaws.com/chapter-08/index.html -->
<html>
<head>
<script>
  (function(n,i,v,r,s,c,x,z){x=window.AwsRumClient={q:[],n:n,i:i,v:v,r:r,c:c};window[n]=function(c,p){x.q.push({c:c,p:p});};z=document.creat
   'cwr',
   '19b5dd5e-d8b3-4048-a91a-090f29d681df',
   '1.0.0',
   'us-east-1',
   'https://client.rum.us-east-1.amazonaws.com/1.5.x/cwr.js',
   {
     sessionSampleRate: 1,
     guestRoleArn: "arn:aws:iam::░░░░░░░░░░:role/RUM-Monitor-us-east-1-░░░░░░░░░░░░░░░░░░░░░░░-Unauth",
     identityPoolId: "us-east-1:ad8f7f25-c3ba-43cc-a5a4-ac6198281c6b",
     endpoint: "https://dataplane.rum.us-east-1.amazonaws.com",
     telemetries: ["performance","errors","http"],
     allowCookies: true,
     enableXRay: false
   }
 );
</script>
<meta http-equiv="Content-Type" content="text/html; charset=windows-1252"></head>
<body>
<h1>
```

Figure 8.39 – Changed index file

13. Upload the changed `index.html` file to the S3 bucket noted in *Step 2* by replacing the old `index.html` file.

14. To verify that the S3 static website is posting the user experience data to the CloudWatch RUM, you can browse the S3 static website URL and check the CloudWatch console for the RUM metrics. This will confirm that the RUM monitor works properly and is collecting data from the website.

15. Navigate to **CloudWatch | RUM** and verify the summary of the RUM results:

Apdex is an open standard solution used to measure user satisfaction with the response time of web applications and services. It is a ratio of total requests made over a period to the value of the number of satisfied and tolerable requests. The Apdex score measures customer satisfaction. It will be in the range of 0 to 1, with 0 being the worst and 1 being the best. If you look at the RUM output, in our case it is **1** as shown in *Figure 8.40*.

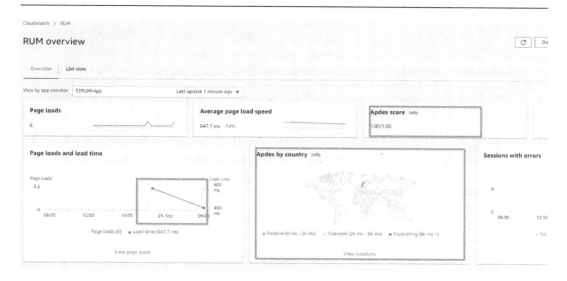

Figure 8.40 – RUM overview

16. Navigate to **CloudWatch | RUM | S3RUM-App** and verify the statistics provided by CloudWatch RUM:

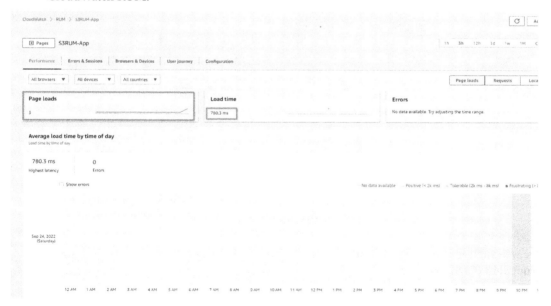

Figure 8.41 – CloudWatch RUM performance output

In *Figure 8.41*, you can see the overall number of page loads, load time, and any errors caused while browsing the application:

- **Page loads**: Page loads provides the number of page loads over the period of time selected.

- **Errors**: Here, you can find JavaScript and HTTP errors. You can also find the number of sessions and the number of sessions with errors.

To comprehend the results produced by CloudWatch RUM, it's important to note that the tool's performance output includes the term *web vitals*, which is a concept introduced by Google (https://support.google.com/webmasters/answer/9205520) to offer crucial data points for evaluating the user experience.

Let's refer to *Figure 8.42* to understand the user's experience details.

Web vitals is an initiative by Google that provides unified guidance for quality signals that are essential to delivering a great user experience on the web. It is a way to simplify the understanding of the performance landscape to focus on the metrics that matter the most.

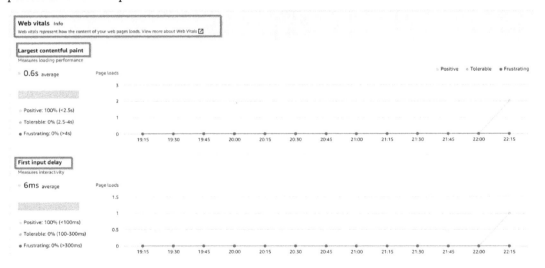

Figure 8.42 – Web vitals in CloudWatch RUM

Core web vitals include **Largest Contentful Paint (LCP)**, **First Input Delay (FID)**, and **Cumulative Layout Shift (CLS)**, and we could interpret the results as follows along with some possible remediations:

- The LCP metric reports the render time of the largest image or text block visible within the viewport, relative to when the page first started loading. LCP considers the largest image, video, and background images of the website, and block-level elements containing text. LCP issues and solutions include improving server performance, optimizing JS/CSS bundles, compressing/caching images, pre-fetch data, using **server-side rendering** (**SSR**) if possible, and using edge solutions for your data.

- The FID metric measures the time from when a user first interacts with a page (i.e., when they click a link, tap on a button, or use a custom JavaScript-powered control) to the time when the browser can begin processing event handlers in response to that interaction. The ideal score for this web vital is 100 ms or less. FID solutions include avoiding large blocking times (long tasks), reviewing your dependencies, removing unused polyfills, caching your assets, and lazy loading of third-party assets.

- CLS measures visual stability – the percentage of content shifting around your website, caused by images loading slowly, and new pieces of content pushing items. CLS solutions include adding size attributes to images/videos, reserving space for slow-loading content (skeleton), allowing your text to be visible while your fonts are loading, and preloading your fonts.

When you look at **Step and duration** in *Figure 8.43*, which is part of the CloudWatch RUM **Summary** tab, you can find the duration of the steps. One important metric to look into is **Time to first byte** (**TTFB**):

Figure 8.43 – Steps and duration of user experience

TTFB is a metric for determining the responsiveness of a web server. It measures the amount of time between creating a connection to the server and downloading the contents of a web page

If you are looking to understand the user performance by browser, navigate to the **Browsers & Devices** tab and verify the detail metrics by browser and by device:

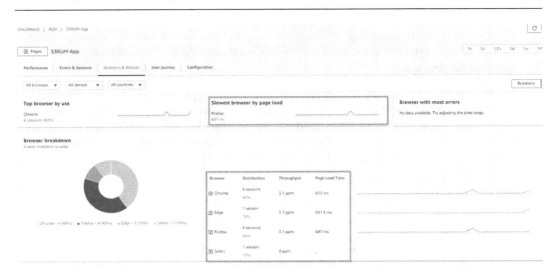

Figure 8.44 – Experience of the user by browser or devic

You can also analyze the user journey by navigating the **User Journey** tab.

In this section, we have understood what CloudWatch RUM is and how it works from the conceptual point of view. We have further gone ahead with the setup of CloudWatch RUM for an S3 static website and now have a grasp of the important performance statistics to focus on for analyzing the user experience on a website. We could adopt the same for any website where we would like to measure the performance experienced by the end users.

Summary

In this chapter, we have explored the significance of DEM in achieving observability and how it can help in identifying user issues. We have delved into the AWS services that can be used to measure DEM. Specifically, we have learned how CloudWatch Synthetics operates, its various use cases, and how to set it up using a blueprint and canary recorder. Additionally, we have examined the need for RUM and how it operates, and have gone through the process of setting up an end-to-end RUM for an S3 static website, including how to analyze the output.

In the next chapter, we will navigate into the open source observability landscape offered by AWS. We will explore collecting metrics and traces using the **OpenTelemetry (OTel)** open standard.

Questions

1. What are the services available from AWS for measuring the digital experience of the users?

2. What is the use of synthetic monitoring?

3. What are the different models available to set up CloudWatch Synthetics?

4. What is the use of CloudWatch RUM?

5. What is the importance of web vitals and how couild they be useful to understand the output from CloudWatch RUM?

Part 3:
Open Source Managed
Services on AWS

In this part, we will discuss the existing **software development kits (SDKs)**, APIs, and AWS services that support organizations looking for ways to implement observability but using the open source ecosystem. It shows how AWS services can easily integrate with existing practices, helping to reduce much of the heavy lift of deploying and managing those open source tools done by the infrastructure team.

This section has the following chapters:

- *Chapter 9, Collecting Metrics and Traces Using OpenTelemetry*
- *Chapter 10, Deploying and Configuring an Amazon Managed Service for Prometheus*
- *Chapter 11, Deploying the Elasticsearch, Logstash, and Kibana Stack Using Amazon OpenSearch Service*

Collecting Metrics and Traces Using OpenTelemetry

Regardless of whether you are a seasoned veteran or a beginner on your observability journey, since you are following this book and doing your homework, you already have all the pieces in place now: you can now collect metrics, traces, and logs from your application on AWS, whether using **Elastic Compute Cloud (EC2)** instances, containers, or Lambda functions.

But let's take a step back to see the big picture. To collect metrics for an EC2 environment, you need to install an agent on your virtual machine, a sidecar on your containerized application, or a Lambda layer on your serverless application. To collect application-specific metrics, you need to use a library as a dependency and write code to collect it and all the essential context around it. The same can be said about traces. You need to retrieve the trace ID in your code entry point, send it in every cascaded call, again using a library or component, and make changes to the code where you want to capture more granular information.

All of this work is realized using agents, sidecars, instrumentation, and SDKs specific to one observability provider. This work is required; it doesn't matter whether your application is in your data center or in the cloud.

Talking about vendors, now you may realize that on your quest to gather as much information as possible from your application, you have become highly coupled to a single provider, whether AWS or not. This does have some advantages, as your provider may evolve new features faster, so your coupling becomes not an investment but rather a liability. AWS is customer-obsessed, with a history of a fast development pace, so for an application born in the AWS cloud, this may be less of an issue, but if you are planning to migrate from/to different observability providers, this becomes a challenge.

In this chapter, we will talk about the OpenTelemetry project, a set of tools, APIs, and SDKs for collecting telemetry data from your application. OpenTelemetry is a Cloud Native Computing Foundation incubating project. As such, it is an open source, vendor-neutral project supported by the community and 23 vendors at the time of writing. OpenTelemetry provides SDKs to many languages and supports integration with a broad number of backends. OpenTelemetry is on a mission to standardize how we collect and export metrics. In this chapter, we will cover the following main topics:

- An open standard to collect metrics and traces using AWS Distro for OpenTelemetry
- How to instrument once for multiple monitoring destinations
- Instrumenting a container application running on AWS **Elastic Container Service** (**ECS**) using OpenTelemetry

Technical requirements

In this chapter, you will deploy an Amazon ECS workload using a CloudFormation template. So, knowledge of containers and how to write CloudFormation templates is needed.

We will see sample code in Python. Python syntax is easy to understand, so if you do not know Python but know C, C++, Java, or C#, you are well equipped to understand the source code.

All the source code for this chapter can be found here: `https://github.com/PacktPublishing/An-Insider-s-Guide-to-Observability-on-AWS/tree/main/chapter-09`.

An open standard to collect metrics and traces using AWS Distro for OpenTelemetry

OpenTelemetry works on different levels to provide a vendor-agnostic experience. It offers open-standard semantics so you can write vendor-agnostic code. We can deploy the vendor-neutral collector binary in various ways. It supports multiple open source and commercial protocols (see `https://github.com/open-telemetry/opentelemetry-collector-contrib/tree/main/receiver`) to export to a vendor-specific backend. It provides a vendor-agnostic library for these languages:

- .NET
- C++
- Erlang/Elixir
- Go
- Java
- Javascript
- PHP
- Python
- Ruby
- Rust
- Swift

So, in a typical OpenTelemetry-based observability deployment, you have an application using the OpenTelemetry SDK to manually or automatically instrument the application code. The collected metric is sent using an agnostic protocol to the OpenTelemetry Collector. The OpenTelemetry Collector receives, processes, and exports the telemetry data to the backend. The OpenTelemetry Collector implements modular architectures, where receivers, processors, and exporters can be organized into pipelines to export the collected telemetry to one or more backends. See the following figure:

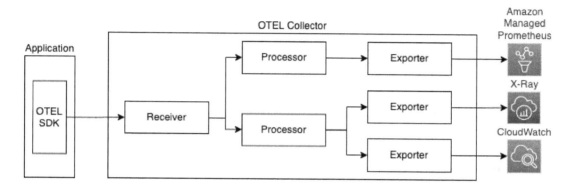

Figure 9.1 – A typical OpenTelemetry deployment

As you can see, you need to put together many pieces to make an OpenTelemetry deployment a success. You need to deploy the OpenTelemetry Collector in your infrastructure. You need to add configurations to connect the application to the OpenTelemetry Connector and from it to your backend of choice. And to do so, you need to authenticate to the backend, and mechanisms to authenticate to the backend are vendor-specific. That is where the OpenTelemetry distributions play a crucial role.

If you use any Linux operating system, you are familiar with the concept of distribution. In the Linux world, the distribution is a collection of open source software that makes up a cohesive operational system. Any end user could build their distribution and enjoy the freedom of doing so, but good luck with that. It is arduous to put together many pieces of software and all the configuration necessary. In the same way, OpenTelemetry provides many software and modules, and users can combine them in any way they desire, or trust the work already done by many different vendors. Vendors invest time to ensure end users can use OpenTelemetry effortlessly on their platforms. AWS provides the AWS Distro for OpenTelemetry (see https://aws.amazon.com/otel/), a secure, tested, production-ready OpenTelemetry distribution. AWS Distro for OpenTelemetry will automatically collect metadata from the AWS environment where your code is running. It supports AWS Lambda, Amazon EC2, Amazon ECS, and Amazon **Elastic Kubernetes Services** (**EKS**), the last two regardless of whether you use EC2 or AWS Fargate as the compute runtime for your worker nodes. AWS Distro for OpenTelemetry can export telemetry for all the usual suspects, such as AWS X-Ray, Amazon CloudWatch, Amazon Managed Service for Prometheus, and a list of third-party monitoring solutions as follows:

- AppDynamics

- Datadog

- Dynatrace

- Grafana

- Honeycomb

- Lightstep

- Logz.io

- New Relic

- Splunk

- Sumo Logic

In this section, you learned that to use OpenTelemetry, you need to instrument your code and deploy the OpenTelemetry Collector to collect the telemetry data and send it to one of the many backend services. AWS Distro for OpenTelemetry is the distribution created by AWS to help users to integrate effortlessly into AWS services and backends. In the next section, let us examine how to instrument your application in a vendor-agnostic way so you can later export the collected data to any vendor, including AWS.

How to instrument once for multiple monitoring destinations

As discussed in the previous section, OpenTelemetry exposes vendor-agnostic semantics and APIs to your application. All you need to do is to implement your code with a dependency on OpenTelemetry and configure where to send the telemetry data. With that, you abstract away any vendor-specific details. To achieve vendor-agnostic semantics, OpenTelemetry classifies the different telemetry data, or signals, into three distinct categories:

- **Traces**

- **Metrics**

- **Logs**

Let us discuss what they are and how to implement them one by one.

Traces

A trace, also known as **distributed trace**, contains information about all the steps to fulfill a single user request inside and among different services. Multiple spans make up a single trace.

A span is a single work unit. Spans can be nested, so a span can have a parent span, while the first span without a parent is known as the root span. A span also has a span context, which contains the trace ID, an essential piece of information carried around by the different services to aggregate all the different spans of a single request into a single trace. Many vendors and backends represent traces and spans as waterfall diagrams, as seen in the following figure:

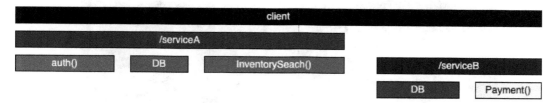

Figure 9.2 – Example of a trace representation using a waterfall diagram

See an example here of how to add spans to your code (you can see the full code example at `https://github.com/PacktPublishing/An-Insider-s-Guide-to-Observability-on-AWS/blob/main/chapter-09/basic_tracer/basic_tracer.py`):

```
...
// Removed code for clarity
11: tracer = trace.get_tracer(__name__)
12: with tracer.start_as_current_span("foo"):
13:     print("Hello world!")
```

In this example, we start a new trace, and inside this trace, we initialize the root span.

Let's execute this sample code. Go to the root folder of the companion repository you cloned before and execute the commands:

```
cd chapter-09/basic_tracer/
python3 -m venv .venv
source .venv/bin/activate
pip install -r requirements.txt
python3 basic_tracer.py | nl -w2 -s': '
```

We can see the structure of the trace as a JSON object on the console in the following figure:

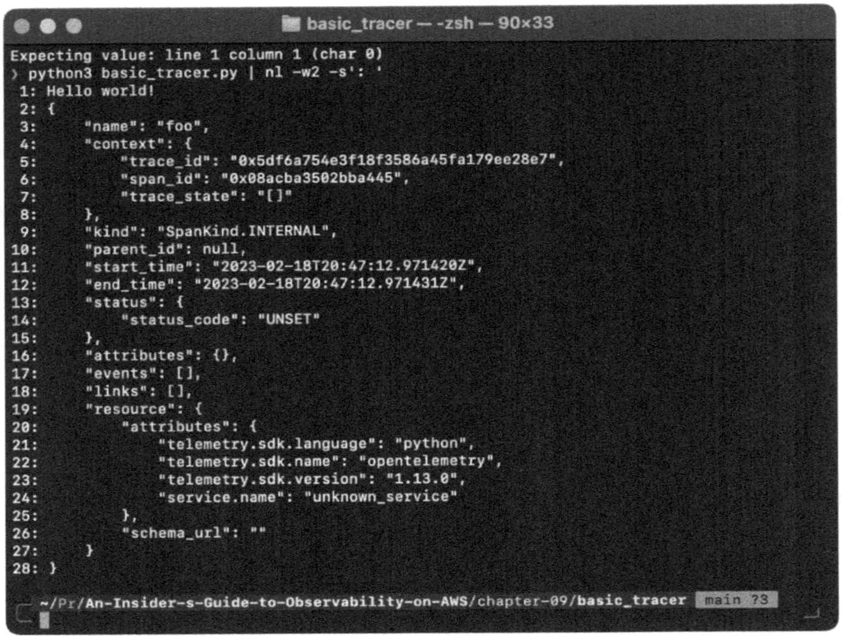

```
● ● ●                    📁 basic_tracer — -zsh — 90×33
Expecting value: line 1 column 1 (char 0)
) python3 basic_tracer.py | nl -w2 -s': '
 1: Hello world!
 2: {
 3:     "name": "foo",
 4:     "context": {
 5:         "trace_id": "0x5df6a754e3f18f3586a45fa179ee28e7",
 6:         "span_id": "0x08acba3502bba445",
 7:         "trace_state": "[]"
 8:     },
 9:     "kind": "SpanKind.INTERNAL",
10:     "parent_id": null,
11:     "start_time": "2023-02-18T20:47:12.971420Z",
12:     "end_time": "2023-02-18T20:47:12.971431Z",
13:     "status": {
14:         "status_code": "UNSET"
15:     },
16:     "attributes": {},
17:     "events": [],
18:     "links": [],
19:     "resource": {
20:         "attributes": {
21:             "telemetry.sdk.language": "python",
22:             "telemetry.sdk.name": "opentelemetry",
23:             "telemetry.sdk.version": "1.13.0",
24:             "service.name": "unknown_service"
25:         },
26:         "schema_url": ""
27:     }
28: }
  ~/Pr/An-Insider-s-Guide-to-Observability-on-AWS/chapter-09/basic_tracer  main ?3
  ▊
```

Figure 9.3 – JSON output representing a trace

We can see the following data in the preceding JSON:

- **Lines 4-8**: Here, we see the span context, with the trace ID, which is transferred between services to glue them together in the same chain of calls

- **Line 10**: The parent ID but in this case, as this is a root span, it is null

- **Lines 20-25**: Span attributes, a set of key-value pairs with context information at the time of the trace execution

Now that we have seen what a trace signal means in OpenTelemetry and how to collect and what information it contains, let's take a look at the next signal: metrics.

Metrics

The OpenTelemetry-agnostic definition of a metric is as follows:

"*A metric is a measurement about a service, captured at runtime.*"

The moment at which you capture a metric is called the *metric event*, which contains the metric value, the moment in time, and its metadata.

See examples of metric collections in the following source code (you can find the full source code at `https://github.com/PacktPublishing/An-Insider-s-Guide-to-Observability-on-AWS/blob/main/chapter-09/metrics/example.py`):

```
// Removed code for clarity
19: # Counter
20: counter = meter.create_counter("counter")
21: counter.add(1)

22: # UpDownCounter
23: updown_counter = meter.create_up_down_counter("updown_counter")
24: updown_counter.add(1)
25: updown_counter.add(-5)

26: # Histogram
27: histogram = meter.create_histogram("histogram")
28: histogram.record(99.9)
```

The important lines in this piece of code are as follows:

- **Lines 19-21**: We start a simple, monotonic counter, or in other words, it can only go up

- **Lines 22-25**: We have a non-monotonic counter, so this time, we can increment as well decrement it

- **Lines 26-28**: We create and add a single data point to a histogram, a collection of recorded metrics in a compressed format, divided into buckets

Let's see the code in action. In one terminal window, let's start an OpenTelemetry collector. Execute the following command:

```
docker run \
    -p 4317:4317 \
    -v $(pwd)/otel-collector-config.yaml:/etc/otel/config.yaml \
    otel/opentelemetry-collector-contrib:latest
```

In a second terminal window, run the following commands:

```
cd chapter-09/metrics
python3 -m venv .venv
source .venv/bin/activate
pip install -r requirements.txt
```

In the OpenTelemetry Collector standard output, you will see an output similar to the following figure:

```
                          metrics — -zsh — 80×50
> cat output.txt | nl -w2 -s': '
 1: Resource SchemaURL:
 2: Resource labels:
 3:      -> telemetry.sdk.language: STRING(python)
 4:      -> telemetry.sdk.name: STRING(opentelemetry)
 5:      -> telemetry.sdk.version: STRING(1.13.0)
 6:      -> service.name: STRING(unknown_service)
 7: ScopeMetrics #0
 8: ScopeMetrics SchemaURL:
 9: InstrumentationScope getting-started 0.1.2
10: Metric #0
11: Descriptor:
12:      -> Name: counter
13:      -> Description:
14:      -> Unit:
15:      -> DataType: Sum
16:      -> IsMonotonic: true
17:      -> AggregationTemporality: AGGREGATION_TEMPORALITY_CUMULATIVE
18: NumberDataPoints #0
19: StartTimestamp: 2023-02-18 21:12:28.156817 +0000 UTC
20: Timestamp: 2023-02-18 21:12:28.156972 +0000 UTC
21: Value: 1
22: Metric #1
23: Descriptor:
24:      -> Name: updown_counter
25:      -> Description:
26:      -> Unit:
27:      -> DataType: Sum
28:      -> IsMonotonic: false
29:      -> AggregationTemporality: AGGREGATION_TEMPORALITY_CUMULATIVE
30: NumberDataPoints #0
31: StartTimestamp: 2023-02-18 21:12:28.156836 +0000 UTC
32: Timestamp: 2023-02-18 21:12:28.156972 +0000 UTC
33: Value: -4
34: Metric #2
35: Descriptor:
36:      -> Name: histogram
37:      -> Description:
38:      -> Unit:
39:      -> DataType: Histogram
40:      -> AggregationTemporality: AGGREGATION_TEMPORALITY_CUMULATIVE
41: HistogramDataPoints #0
42: StartTimestamp: 2023-02-18 21:12:28.156854 +0000 UTC
43: Timestamp: 2023-02-18 21:12:28.156972 +0000 UTC
44: Count: 1
45: Sum: 99.900000
46: Min: 99.900000
~/Pr/An-Insider-s-Guide-to-Observability-on-AWS/chapter-09/metrics   main ?4
```

Figure 9.4 – OpenTelemetry Collector output after metric sample code execution

The preceding export shows the following details:

- **Lines 10-21**: Here, we see all the information captured by the collection about the first counter we saw in our code. You can see we don't have just the value, but also the timestamp and metadata for it.

- **Lines 22-33**: Here, we have the data about the second counter. As a non-monotonic counter, `IsMonotonic` is set to `false`.

- **Lines 34-51**: This shows the data from our histogram. As you can see, it keeps the information about `sum`, `minimum`, `maximum`, and the number of data points.

With that, we conclude the OpenTelemetry take on metrics and how the agnostic API is implemented and represented. Next stop: logs.

Logs

In OpenTelemetry terms, a log is "*a timestamped text record, either structured (recommended) or unstructured, with metadata*".

Let us see an example of how to log data using OpenTelemetry (you can find the full source code at https://github.com/PacktPublishing/An-Insider-s-Guide-to-Observability-on-AWS/blob/main/chapter-09/logs/example.py):

```
// Removed code for clarity

22:    log_emitter_provider = LogEmitterProvider(
23:      resource=Resource.create(
24:        {
25:          "service.name": "shoppingcart",
26:          "service.instance.id": "instance-12",
27:        }
28:      ),
29:    )
30:    set_log_emitter_provider(log_emitter_provider)

31:    exporter = OTLPLogExporter(insecure=True)
32:    log_emitter_provider.add_log_
processor(BatchLogProcessor(exporter))
33:    handler = LoggingHandler(
34:      level=logging.NOTSET, log_emitter_provider=log_emitter_provider
35:    )

36:    # Attach OTLP handler to root logger
37:    logging.getLogger().addHandler(handler)

38:    # Log directly
39:    logging.info("Jackdaws love my big sphinx of quartz.")

40:    # Create different namespaced loggers
41:    logger1 = logging.getLogger("myapp.area1")
42:    logger2 = logging.getLogger("myapp.area2")

43:    logger1.debug("Quick zephyrs blow, vexing daft Jim.")
44:    logger1.info("How quickly daft jumping zebras vex.")
45:    logger2.warning("Jail zesty vixen who grabbed pay from quack.")
46:    logger2.error("The five boxing wizards jump quickly.")
```

```
47:   # Trace context correlation
48:   tracer = trace.get_tracer(__name__)
49:   with tracer.start_as_current_span("foo"):
50:   # Do something
51:   logger2.error("Hyderabad, we have a major problem.")

52:   log_emitter_provider.shutdown()
```

The important sections of this code snippet are as follows:

- **Lines 22-37**: Configuration of the log handler and exporter

- **Lines 38-46**: Normal logging

- **Lines 47-51**: Attaching a log entry to a trace span

Let's execute the sample code and see how OpenTelemetry interprets the log data. Let's start an OpenTelemetry collector with the following command:

```
docker run\
    -p 4317:4317 \
    -v $(pwd)/otel-collector-config.yaml:/etc/otel/config.yaml \
    otel/opentelemetry-collector-contrib:latest
```

In another terminal, type the following commands:

```
cd chapter-09/logs
python3 -m venv .venv
source .venv/bin/activate
pip install -r requirements.txt
```

In the OpenTelemetry Collector standard output, you will see an output like that in the following figure:

```
) cat output.txt
01: Resource SchemaURL:
02: Resource labels:
03:      -> telemetry.sdk.language: STRING(python)
04:      -> telemetry.sdk.name: STRING(opentelemetry)
05:      -> telemetry.sdk.version: STRING(1.8.0)
06:      -> service.name: STRING(shoppingcart)
07:      -> service.instance.id: STRING(instance-12)
08: InstrumentationLibraryLogs #0
09: InstrumentationLibraryMetrics SchemaURL:
10: InstrumentationLibrary __main__ 0.1
11: LogRecord #0
12: Timestamp: 2022-01-13 20:37:03.998733056 +0000 UTC
13: Severity: WARNING
14: ShortName:
15: Body: Jail zesty vixen who grabbed pay from quack.
16: Trace ID:
17: Span ID:
18: Flags: 0
19: LogRecord #1
20: Timestamp: 2022-01-13 20:37:04.082757888 +0000 UTC
21: Severity: ERROR
22: ShortName:
23: Body: The five boxing wizards jump quickly.
24: Trace ID:
25: Span ID:
26: Flags: 0
27: LogRecord #2
28: Timestamp: 2022-01-13 20:37:04.082979072 +0000 UTC
29: Severity: ERROR
30: ShortName:
31: Body: Hyderabad, we have a major problem.
32: Trace ID: 63491217958f126f727622e41d4460f3
33: Span ID: d90c57d6e1ca4f6c
34: Flags: 1
```

Figure 9.5 – OpenTelemetry Collector output after sample code execution of logs

The previous output shows the following details:

- **Lines 11-26**: Log records without a span ID
- **Lines 27-34**: Log records with a span ID

With that, we have finished examining the vendor-agnostic vocabulary implemented by OpenTelemetry. With these building blocks, we can implement our application observability independent of the vendor, and OpenTelemetry will do the translation on a case-by-case basis.

In the next section, we will see how OpenTelemetry can be deployed to ingest data from your target application.

OpenTelemetry Collector deployment

The OpenTelemetry Collector makes it easy to decouple the application code and binary from the vendor's backend and reduces the potential overhead of processing, aggregating, connection handling, and retrying requests. There are two ways to deploy the OpenTelemetry Collector:

- **An agent**: The collector runs side by side with the application code in the same host or as a sidecar in the case of containers

- **A gateway**: An independent process running to collect telemetry from many applications, one per workload, cluster, or Availability Zone.

The two deployments have pros and cons. Deploying as an agent has the advantage of simplicity and resiliency, but it is more challenging to manage and scale the collector according to its resource requirements, and it also incurs higher overhead. A gateway deployment may be more complex, but it gives us the freedom to scale the collector independently and fine-tune the provided resources.

In this section, we saw how OpenTelemetry implements a vendor-agnostic API and semantics, taking care of the details of how to translate what the project calls signals and its vendor-specific implementation of them. This vendor-agnostic approach allows you to implement observability code once in your project and, with changes only in configuration, makes it possible to export the observability data to different vendor-specific backends.

Let's take a look at one example of how to implement observability using OpenTelemetry using the AWS-native container orchestrator: Amazon ECS.

Instrumenting a container application running on ECS using OpenTelemetry

In this section, we will go through all steps to instrument and deploy an application using AWS Distro for OpenTelemetry.

First, let us deploy the application:

1. Browse this link: `https://eu-central-1.console.aws.amazon.com/cloudformation/home?region=eu-central-1#/stacks/quickcreate?templateURL=https%3A%2F%2Finsiders-guide-observability-on-aws-book.s3.amazonaws.com%2Fchapter-09%2Faws-ecs-otel-main.yaml&stackName=OTELFlaskApp`. When you click on this link, you will be redirected to the CloudFormation stack creation page, with all the details pre-populated. You can leave all the default values. Just do not forget to check the two checkboxes at the end of the page before clicking on the **Create stack** button. Check the following figure:

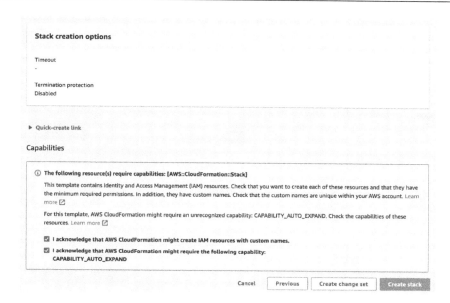

Stack creation options

Timeout
-

Termination protection
Disabled

▶ Quick-create link

Capabilities

ⓘ **The following resource(s) require capabilities: [AWS::CloudFormation::Stack]**

This template contains Identity and Access Management (IAM) resources. Check that you want to create each of these resources and that they have the minimum required permissions. In addition, they have custom names. Check that the custom names are unique within your AWS account. Learn more ☑

For this template, AWS CloudFormation might require an unrecognized capability: CAPABILITY_AUTO_EXPAND. Check the capabilities of these resources. Learn more ☑

☑ I acknowledge that AWS CloudFormation might create IAM resources with custom names.

☑ I acknowledge that AWS CloudFormation might require the following capability:
    CAPABILITY_AUTO_EXPAND

Cancel   Previous   Create change set   **Create stack**

Figure 9.6 – CloudFormation step 4, acknowledging the creation of IAM resources

2.   You will need to wait around 5-10 minutes for the infrastructure to be deployed. This CloudFormation template is split into one for the infrastructure, one for the application deployment, and the main template that orchestrates both. That is why, after deployment, you will see three Stacks: one being the main template, in which the deployment was initiated by you, and two more that are initiated by the main template. This technique is useful for creating modular, reusable templates and to help you handle the complexity of your infrastructure. You can create nested templates such as the ones you see here by following the instructions on the AWS documentation page (see https://docs.aws.amazon.com/AWSCloudFormation/latest/UserGuide/using-cfn-nested-stacks.html). You can easily identify the ones created by the main template as they are marked as **NESTED**. See the following figure:

Stack name	Status	Created time	Description
OTELFlaskApp-OTELFlaskAppStack-R3J41QJE1DC6 **NESTED**	⊘ CREATE_COMPLETE	2022-10-09 13:22:45 UTC+0200	CloudFormation template that represents a load balanced web service on Amazon ECS.
OTELFlaskApp-ECSInfraStack-1BAV7QCDIXG7K **NESTED**	⊘ CREATE_COMPLETE	2022-10-09 13:19:11 UTC+0200	CloudFormation environment template for infrastructure shared among ECS workloads.
OTELFlaskApp	⊘ CREATE_COMPLETE	2022-10-09 13:19:06 UTC+0200	-

Figure 9.7 – Main template (named OTELFlaskApp) and two nested templates

After deployment, you will have a load-balanced ECS-based web service deployed in your account. You can see the application architecture in the following figure:

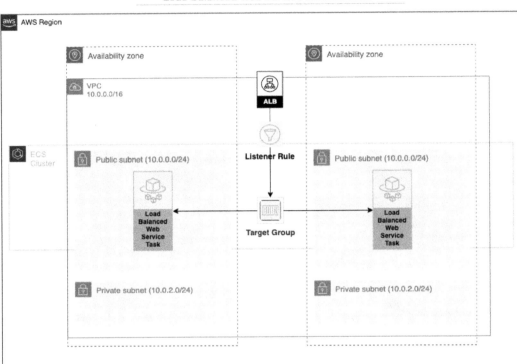

Figure 9.8 – Load-balanced web service deployed using Amazon ECS

3. Access the **Outputs** tab on the main template to find the public URL of the load balancer, as in the following figure:

Figure 9.9 – Main template | Outputs tab | load balancer URL

4. If you click on the public URL of the load balancer under the **PublicLoadBalancerDNSName** key name (in my case, `http://otelf-publi-1mn1ch1sjupfw-500899047.eu-central-1.elb.amazonaws.com/`, as you can see in the screenshot, but it can differ for every deployment), or right-click and open it in a new tab (better!), your browser will be redirected to the main application page, and you will receive a friendly **App running!** message, as in the following figure:

Figure 9.10 – Sample application main page

5. Our sample application is implemented in Python, and exposes four endpoints:

    *   `http://[load-balancer-URL]/`: The application main page you see when you click on the load balancer URL; it just returns a friendly **App running!** message.

    *   `http://[load-balancer-URL]/outgoing-http-call`: This endpoint makes a call to an external URL, mimicking a call to an external service, and returns the trace ID of this call chain.

    *   `http://[load-balancer-URL]/aws-sdk-call`: This endpoint makes a call to list all S3 buckets of your account, emulating a call to one AWS service, and returns the trace ID of this call chain.

    *   `http://[load-balancer-URL]/health`: This endpoint is used to check the application's health, and it is used by the load balancer to decide whether user traffic can be redirected to this node.

You can find the application source code and CloudFormation template here: `https://github.com/PacktPublishing/An-Insider-s-Guide-to-Observability-on-AWS/tree/main/chapter-09/manual_instrumentation`.

Let's navigate through the necessary components and configuration that allow this application to publish metrics using OpenTelemetry, or more specifically, AWS Distro for OpenTelemetry.

The OpenTelemetry Python SDK for traces

To configure exporting application traces, we need to configure the protocol we want to use to communicate and how often we want to do it. With the Python SDK, you have three main ways to export the application traces:

- **The console exporter**: This is easy to set up and is useful for development and debugging

- **An OTLP/gRPC exporter**: In other words, you use the agnostic protocol developed by the OpenTelemetry community to communicate to an OTLP-compatible endpoint (for example, the OpenTelemetry Collector), and use gRPC as the transport medium

- **An OTLP/HTTP exporter**: In other words, you use the same agnostic protocol developed by the OpenTelemetry community, but now use HTTP as the transport medium

In our Python application, we use OTLP/gRPC, as it is faster and recommended for production environments. We also accumulate and send trace information in batches by using the `BatchSpanProcessor` class to increase the export throughput. You can see how we configured it by checking the file at `https://github.com/PacktPublishing/An-Insider-s-Guide-to-Observability-on-AWS/blob/main/chapter-09/manual_instrumentation/docker/application.py`. The following is the most relevant code snippet:

```
// Removed code for clarityotlp_exporter = OTLPSpanExporter()
span_processor = BatchSpanProcessor(otlp_exporter)
trace.set_tracer_provider(
  TracerProvider(
    active_span_processor=span_processor,
    id_generator=AwsXRayIdGenerator(),
  )
)
```

With this configuration in place, and as the application load balancer adds the `X-Amzn-Trace-Id` trace ID header to all requests, we can access the trace information programmatically, as in the following code snippet:

```
# Test HTTP instrumentation
@app.route("/outgoing-http-call")
def call_http():
```

```
  requests.get("https://aws.amazon.com/")

  return app.make_response(
    convert_otel_trace_id_to_xray(
      trace.get_current_span().get_span_context().trace_id
    )
  )
```

In this section, we saw how we can use the OpenTelemetry API in our small sample application to collect traces. Next, let's check how we can do the same, but now for metrics.

The OpenTelemetry Python SDK for metrics

Similar to what we did in the Python application using the SDK for traces, to configure exporting application metrics, we need to configure how and how often we want to do so. In our sample application, we use the OTLP/gRPC exporter, and we configure a periodic batch processor, which exports the metrics at a regular cadence. See the following code snippet:

```
from opentelemetry import metrics
from opentelemetry.exporter.otlp.proto.grpc.metric_exporter import
OTLPMetricExporter
from opentelemetry.sdk.metrics import MeterProvider
from opentelemetry.sdk.metrics.export import
PeriodicExportingMetricReader

reader = PeriodicExportingMetricReader(
exporter=OTLPMetricExporter(),
export_interval_millis=5000
)

metrics.set_meter_provider(MeterProvider(metric_readers=[reader]))
```

With that configuration in place, we can create application metrics as follows:

```
apiBytesSentMetricName = "apiBytesSent"
latencyMetricName = "latency"

apiBytesSentCounter = meter.create_counter(
apiBytesSentMetricName, unit="1", description="API request load sent
in bytes"
)

apiLatencyRecorder = meter.create_histogram(
latencyMetricName, unit="ms", description="API latency time"
)
```

And then we can use the application metrics as shown in the following code:

```
@app.after_request
def after_request_func(response):
  if request.path == "/outgoing-http-call":
    apiBytesSentCounter.add(
      response.calculate_content_length() + mimicPayloadSize(),
      {
        DIMENSION_API_NAME: request.path,
        DIMENSION_STATUS_CODE: response.status_code,
      },
    )

  apiLatencyRecorder.record(
    int(time.time() * 1_000) - session[REQUEST_START_TIME],
    {
      DIMENSION_API_NAME: request.path,
      DIMENSION_STATUS_CODE: response.status_code,
    },
  )
  return response
```

In this section, we saw how to use the OpenTelemetry APIs to collect metrics from our sample application. Now that we have both traces and metrics, we need to preprocess them and send them to our backend of choice. We will see how to do so in the next section.

Deploying the OpenTelemetry Collector

We covered the two OpenTelemetry Collector deployment modes earlier, deploying as an agent and as a gateway. In our sample application, we decided to use the first one, to deploy it as an agent, which was the one that we implemented as a sidecar in Kubernetes. This is the simpler way to deploy for smaller cases – but remember that you can decide to deploy it just once in a group of microservices controlled by the same team, once per Availability Zone, or once per cluster, whatever makes the most sense for your architecture, reducing the total overhead.

To see how to set up a sidecar using Amazon ECS, you can check the CloudFormation template: `https://insiders-guide-observability-on-aws-book.s3.amazonaws.com/chapter-09/aws-ecs-otel-flask-application.yaml`. We have highlighted the relevant template snippet here.

In the container definition, we first define the application container:

```
ContainerDefinitions:
  - Name: !Ref WorkloadName
    Image: !Ref ContainerImage
```

```
Cpu: 256
Memory: 512
LogConfiguration:
  LogDriver: awslogs
Options:
  awslogs-region: !Ref AWS::Region
  awslogs-group: !Ref LogGroup
  awslogs-stream-prefix: ecs
PortMappings:
  - ContainerPort: !Ref ContainerPort
```

And then we define a second container as a sidecar using the OpenTelemetry Collector image:

```
- Name: aws-collector
  Image: 'amazon/aws-otel-collector:latest'
  Command:
    - '--config=/etc/ecs/container-insights/otel-task-metrics-
config.yaml'
  LogConfiguration:
  LogDriver: awslogs
  Options:
    awslogs-create-group: 'True'
    awslogs-group: /ecs/ecs-aws-otel-sidecar-collector
    awslogs-region: !Ref 'AWS::Region'
    awslogs-stream-prefix: ecs
```

You can see the collector being pulled from the public Amazon ECR repository, `amazon/aws-otel-collector`. Setting the right configuration is as important as downloading the right collector binary. AWS Distro for OpenTelemetry already comes with some default configurations backed in the Docker image, and you can see we select one to use in the `Command` directive, `/etc/ecs/container-insights/otel-task-metrics-config.yaml`. In the Docker image repository (`https://github.com/aws-observability/aws-otel-collector/tree/main/config`), you can see some of the configurations available. You usually select one based on the features you want to activate.

Checking the resulting application telemetry

After instrumenting our sample application to collect both traces and metrics using the OpenTelemetry API, and configuring the OpenTelemetry Collector to ingest the data and translate it into our AWS backend's format, our work is done. Time to check the results in Amazon CloudWatch Logs, Amazon CloudWatch metrics, and AWS X-Ray. Let's harvest the results of our hard work in the next sections.

Seeing container metrics in CloudWatch

Let's go to the CloudWatch console and select **Log groups** from the left-hand navigation. Search for our log group name – `/aws/ecs/containerinsights/{ClusterName}/performance` – and then click the log stream name, which is `task_id`. These logs use the CloudWatch **Embedded Metric Format** (**EMF**) (visit this link for more details: `https://docs.aws.amazon.com/AmazonCloudWatch/latest/monitoring/CloudWatch_Embedded_Metric_Format.html`) to generate CloudWatch metrics. The following is an example of received logs on the CloudWatch console.

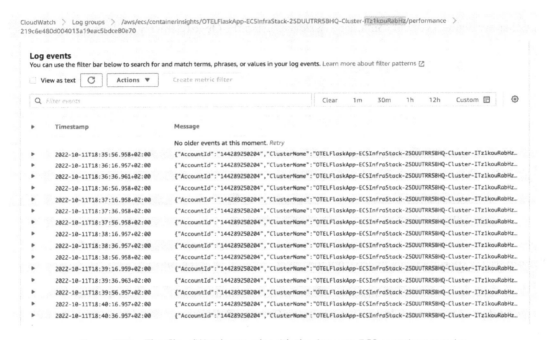

Figure 9.11 – The CloudWatch console with the Amazon ECS container metrics

Thanks to OpenTelemetry, we have metric data points converted into the Amazon CloudWatch EMF format, providing container metrics. Next, let's check application-specific metrics.

Seeing CloudWatch metrics

In the CloudWatch console, from the left-hand panel, select **Metrics**. You should be able to see the **ECS/ContainerInsights** namespace. These are metrics collected from the cluster and container automatically, ready to be used on your own dashboard. Click on it and you will see the expected metrics.

Figure 9.12 – CloudWatch Container Insights metrics

If you go to the CloudWatch console and select **Log groups** from the left-hand navigation and search for a log group named `/aws/ecs/application/metrics`, you will find the application-specific metrics we programmatically added before, such as the number of bytes sent when requesting the `/outgoing-http-call` path. See the following figure:

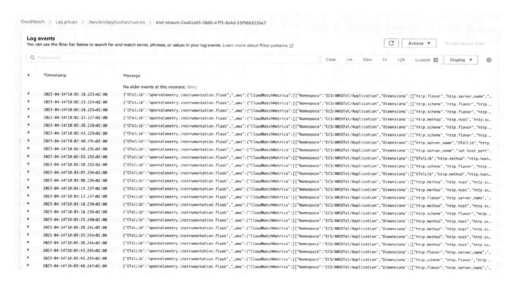

Figure 9.13 – CloudWatch application metrics

As we can see, we have the application-specific metrics published to Amazon CloudWatch using the Amazon CloudWatch EMF format. The last piece comes next: the application traces.

Seeing traces in AWS X-Ray

You can go to the **AWS X-Ray** console (`https://console.aws.amazon.com/xray/home`) and click **Traces** in the left-hand navigation. You can see the traces that were collected from the application and sent to AWS X-Ray in the following figure:

Figure 9.14 – List of application trace IDs

You can click on a trace ID to see the trace map as you can see in the following figure:

Figure 9.15 – Trace map from a single trace ID

You can click on the **Raw data** tab to see the same information in JSON format, as you can see in the following figure:

Figure 9.16 – Trace raw data

In this subsection, we saw the resulting metrics and traces of our sample application on Amazon CloudWatch after instrumenting it using OpenTelemetry. The last missing piece is logs, which we will see in the next subsection.

Seeing application logs in CloudWatch

You can go to the AWS CloudWatch console, and in the left-hand navigation, select the **/ecs/aws-otel-flask-app-test-aws-otel-flask-app** log group and then the latest log stream. You will see application logs collected from the application container. See the following figure:

Figure 9.17 – Application log data

OpenTelemetry captures the application logs and sends them to Amazon CloudWatch Logs for storage. You can see the log history of your infrastructure and your application in a single pane of glass.

In this section, we saw a sample application instrumented using OpenTelemetry, a sample deployment of this application, and the OpenTelemetry Collector as a sidecar. After generating a load, we could see application metrics, traces, and logs on the AWS backend services, demonstrating how we can use the standard vendor-agnostic OpenTelemetry libraries and tools to export observability signals to any backend, with just a change to the OpenTelemetry configuration.

Summary

In this chapter, you saw how OpenTelemetry approaches the problem of the standardization of SDKs, protocols, and tools to collect and export application telemetry, and how its implementation can make your application vendor-agnostic. We also saw how to implement it in a sample application, from the application SDK to the OpenTelemetry Collector deployment and backend.

With this information, you can now make a judgment call on when to apply OpenTelemetry, its benefits and drawbacks, and the value it brings to your architecture and observability objectives.

In the next chapter, we will continue our tour of the open source observability tools, talking about two important projects: Prometheus and Grafana.

10

Deploying and Configuring an Amazon Managed Service for Prometheus

Prometheus is a metrics-based monitoring and alerting system initially built at SoundCloud. It is now an open source project, and has been part of the Cloud Native Compute Foundation since 2016. It focuses on doing a few things well: it collects and stores metrics as time-series data, and it has a powerful, highly dimensional data model and query language.

In this chapter, we will discuss **Amazon Managed Service for Prometheus (AMP)** and **Amazon Managed Grafana (AMG)**. AMP is a Prometheus-compatible monitoring service, and you can use the same APIs and query language you are used to for monitoring your workloads. It is highly available, using multiple availability zones for deployment. And it automatically scales the ingestion, storage, and querying of operational metrics as your requirements go up and down. AMG is an open source visualization and analytics tool, often paired with Prometheus, to generate outstanding charts and dashboards.

In this chapter, we will see the following:

- Prometheus and Grafana overview
- Setting up Amazon Managed Service for Prometheus and Grafana
- Ingesting telemetry data
- Querying Prometheus metrics via APIs and Grafana
- Implementing container monitoring

By the end of this chapter, you will better understand what Prometheus and Grafana are, the value added by Amazon Managed Service for Prometheus and Grafana, and how to ingest, process, and query telemetry data using them in a real application. Let's get started!

Technical requirements

This chapter does not require programming skills, but a knowledge of using the terminal and understanding AWS CloudFormation templates will help you understand the code we use to deploy the infrastructure and the sample application.

Prometheus and Grafana overview

As I said in the introduction, Prometheus focuses on doing a few things well, and it doesn't provide rich charts or dashboards. Instead, it concentrates on collecting and storing metrics as time-series data and has a powerful, highly dimensional data model and query language.

Prometheus defines a simple text format that allows any application to expose metrics. Prometheus collects metrics from targets by scraping metrics from HTTP endpoints following the Prometheus-specified format. Since Prometheus exposes data in the same manner about itself and about the endpoints it consumes, it can also scrape and monitor its health or create a hierarchy of Prometheus systems to reduce the load on a single node and provide aggregated data. You can see an example of a Prometheus-compatible HTTP endpoint here:

```
# HELP http_requests A counter for the number of HTTP requests.
# TYPE http_requests counter
http_requests{method="put",code="200"} 329
http_requests_total{method="delete",code="400"}    5
```

In the Prometheus data model, time series are identified using a combination of metric names and key-value pairs known as labels. A metric name is a string that identifies the general type of data being collected, such as `http_requests` or `cpu_usage`. Labels are key-value pairs that provide more specific information about the data being collected, such as the endpoint that generated the request or the name of the process using the CPU.

Prometheus uses a combination of metric names and labels to create unique identifiers for time series. For example, the time series for HTTP requests to a particular endpoint might be identified by the metric name (`http_requests`) and the `method` (`GET`, `POST`, `PUT`, or `DELETE`) and `endpoint` (e.g., `/api/users`) labels.

When Prometheus scrapes metrics from a target, it organizes the data into time series based on their identifiers. Each time series represents a stream of data points, each with a timestamp and a numeric value. These data points represent the value of the metric at a particular point in time, and they can be used to visualize the performance of a system or to alert on anomalies.

There are Prometheus client libraries in all popular languages and runtimes to instrument code, including the following:

- C#/.Net
- Erlang
- Golang
- Haskell
- Java/JVM
- Node.js
- Python
- Ruby
- Rust

If third-party software exposes metrics in a non-Prometheus format, you can use exporters. Exporters act as bridges between these systems and Prometheus, translating their metrics into a format that Prometheus can understand and scrape.

Prometheus exporters work by exposing an HTTP endpoint that serves metrics in a format that Prometheus can understand. They collect data from a variety of sources, such as logs, system statistics, and application performance data, and they convert that data into a format that can be scraped by Prometheus.

There are many different types of Prometheus exporters available, each designed to collect metrics from specific systems or applications. Some popular examples of exporters include the following:

- **Node Exporter**: Collects metrics on system statistics such as CPU usage, memory usage, and disk I/O
- **Blackbox Exporter**: Collects metrics on network performance by running tests such as HTTP requests and DNS lookups
- **Redis Exporter**: Collects metrics on Redis, an open source, in-memory data structure store, including the number of connections, memory usage, and cache hits
- **Apache Exporter**: Collects metrics on the performance of the Apache web server, including request latency and status codes

Prometheus exporters can be run as standalone processes, or they can be run as sidecar containers alongside the systems they're monitoring. They provide a powerful way to collect metrics from a wide variety of systems and services, enabling users to gain deep insight into the performance and health of their entire infrastructure.

Prometheus is designed to work seamlessly with dynamic sources, such as those provided by container orchestration systems such as Kubernetes. To automatically discover and scrape new targets in a dynamic environment, Prometheus relies on a feature called service discovery.

Service discovery in Prometheus involves periodically querying a data source, such as a Kubernetes API server, to discover new targets that match a set of predefined criteria. These criteria can include things such as labels, annotations, and port numbers. Once new targets are discovered, Prometheus adds them to its list of targets and begins scraping them for metrics.

Prometheus supports a variety of dynamic service discovery mechanisms, including the following:

- **Kubernetes service discovery**: Prometheus can discover targets in a Kubernetes cluster using the Kubernetes API server. This can be done using Kubernetes' native service discovery mechanism, or by using the Prometheus-specific Kubernetes service discovery mechanism, which allows for more fine-grained control over target selection.

- **Consul service discovery**: Prometheus can discover targets in a Consul service mesh using the Consul API.

- **DNS service discovery**: Prometheus can discover targets by performing DNS lookups for targets that match a specified naming convention.

In addition to these built-in mechanisms, Prometheus also supports custom service discovery mechanisms through its pluggable service discovery API. This allows users to write their own service discovery mechanisms to integrate with any data source that can be queried via HTTP or DNS.

Overall, Prometheus provides a powerful set of features for dynamically discovering and scraping new targets in dynamic environments, making it an ideal monitoring solution for modern cloud-native infrastructure.

The following diagram is a typical Prometheus deployment architecture with some of the optional services:

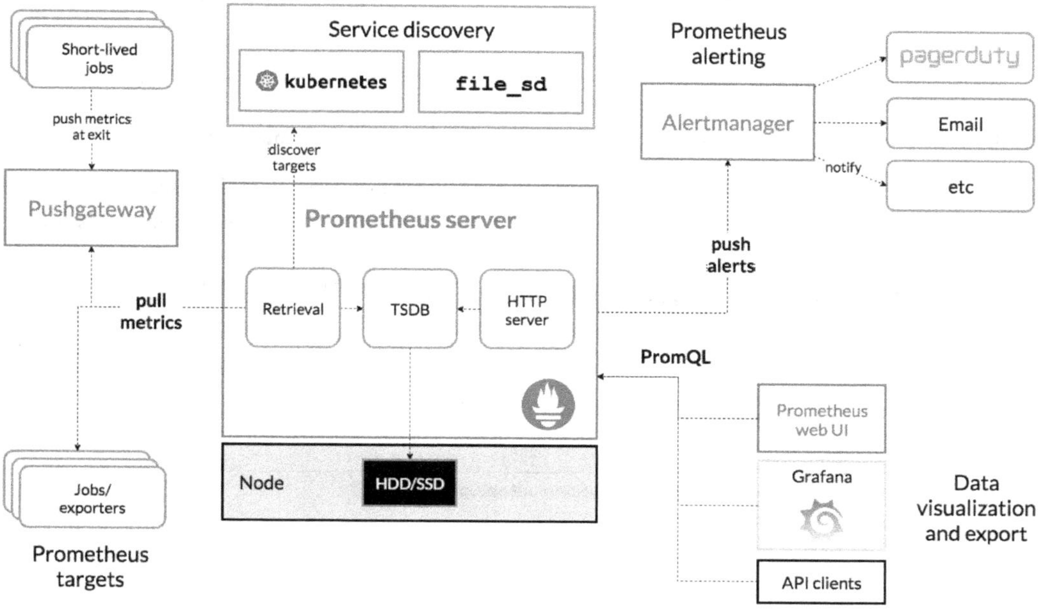

Figure 10.1 – Prometheus typical deployment

In the architectural diagram, we can see the following components:

- **Pushgateway**: This allows jobs and short-lived processes to send telemetry data to Prometheus. As the Prometheus server scrapes the HTTP endpoint at regular intervals, a job or a short-lived process is over before Prometheus can collect any information. With Pushgateway, jobs and short-lived processes push this information to it before the process is over, and the Pushgateway stores it until the next Prometheus scraping cycle.

- **Service discovery**: As mentioned, this is responsible for collecting information from dynamic environments such as the Kubernetes cluster, informing the Prometheus server of any application node that came online or was shut down to add/remove the new node from the scraping process.

- **Prometheus server**: This is the core Prometheus module, and is responsible for retrieving metrics using a pull process, storing them in an internal time-series database, and serving any query received by the Prometheus API.

- **Alertmanager**: This is responsible for handling the alerts sent by the Prometheus server, routing, deduplicating, and grouping the alerts, and sending them to one of the provided targets.

- **Prometheus web UI**: This provides a user interface that allows manual configuration of new endpoints to scrape, and lets you create simple dashboards and alerts using the PromQL.

Prometheus provides some basic dashboard features, but far from what Grafana provides.

In this section, we saw a brief overview of both Prometheus and Grafana. Next, let's see how to use Amazon Managed Service for Prometheus and Grafana to provision both with minimal overhead.

Setting up Amazon Managed Service for Prometheus and Grafana

The following figure illustrates the overall architecture of AMP and AMG and their interaction with other components:

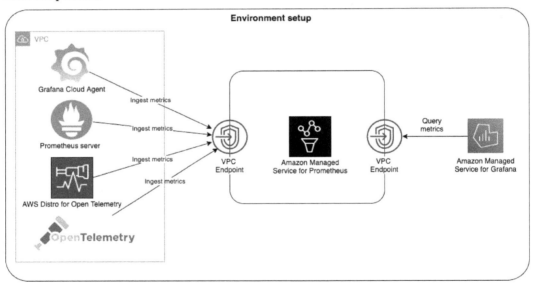

Figure 10.2 – Overall architecture of AMP and AMG solutions and components

Let's look at how to set up some of those components on AWS.

Setting up a Cloud9 development workspace

To have a standard **integrated development environment (IDE)** with the required set of tools, let's create an AWS Cloud9 environment:

1. Please click on the following link to start the deployment process:

   ```
   https://console.aws.amazon.com/cloudformation/home#/
   stacks/quickcreate?templateURL=https://insiders-guide-
   observability-on-aws-book.s3.amazonaws.com/common/cloud9.
   yaml&stackName=InsidersGuideCloud9Chapter10
   ```

2. After clicking on the link, you will be redirected to the AWS CloudFormation stack creation form. You can keep the default values and click on the checkbox asking for extra capabilities. Click on **Create Stack**, and in a few minutes, you will find the environment URL in the CloudFormation **Outputs** tab, as in the following figure:

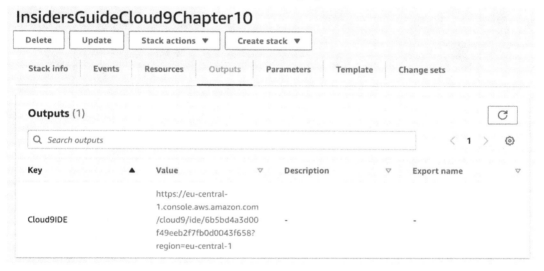

Figure 10.3 – CloudFormation Outputs tab, showing the AWS Cloud9 URL

3.   Click on the URL; this will take you to a newly configured environment, as in the following figure:

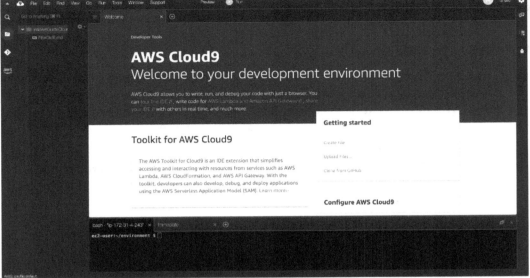

Figure 10.4 – AWS Cloud9 welcome page

With our web IDE environment in place, let's configure Amazon Managed Service for Prometheus.

Setting up an AMP workspace

Let's get started with setting up an AMP workspace:

1.   First, type `Prometheus` into the **Services** search box to create a new AMP workspace. The **Amazon Prometheus** service is the first one on the list. See the following figure:

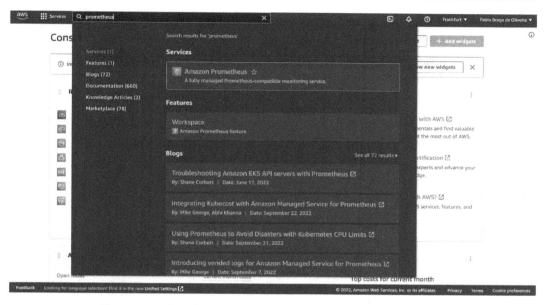

Figure 10.5 – Searching for Amazon Managed Service for Prometheus

2. After clicking on the Amazon Prometheus service in the search results box, you will be redirected to the Amazon Prometheus main page. Once on the main page, click on the **Create** button, as in the following figure:

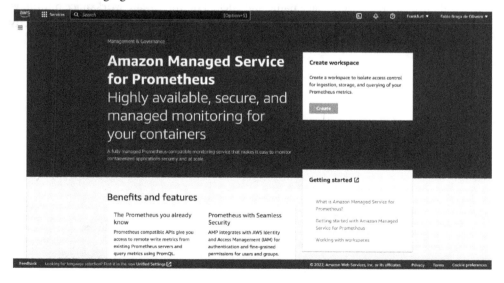

Figure 10.6 – Main Amazon Managed Service for Prometheus page

3. The page will redirect you to the creation form. The only information mandatory is the workspace name. See the following figure:

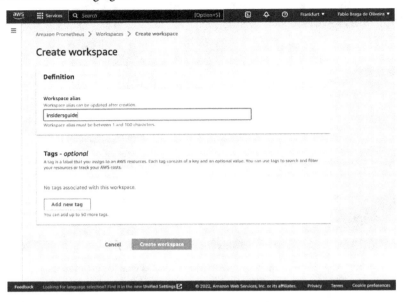

Figure 10.7 – AMP creation form

4. Once you have entered a workspace name and clicked on the **Create workspace** button, you will see the workspace details page. Wait a few minutes until the workspace status is **Active**. You will see a screen like the following:

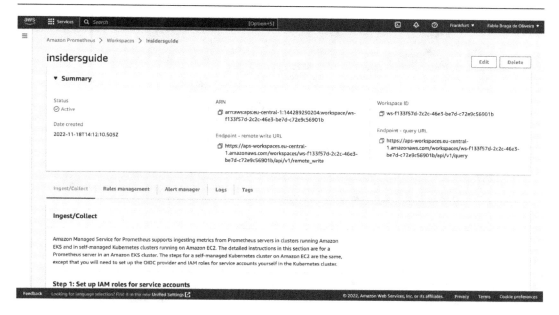

Figure 10.8 – AMP workspace details page

5.   Alternatively, if you prefer to run commands on the command line, you can use the following command to create a new Amazon Prometheus workspace:

```
aws amp create-workspace --alias insidersguide --region
$AWS_REGION
```

Setting up an AMG dashboard

Now let's configure a new Grafana workspace:

1.   First, let's search for the service. In the AWS **Console** search box, type Grafana, and the Amazon Managed Grafana (**Amazon Grafana**) service will appear as the first result. Check the following figure:

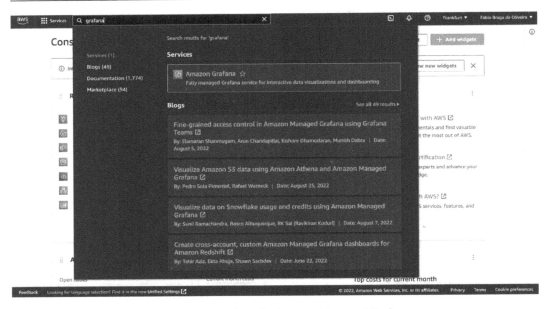

Figure 10.9 – Searching for Amazon Managed Grafana

2. After clicking on the service link on the search results page, you will see the main service page, as in the following figure. Click on the **Create workspace** button.

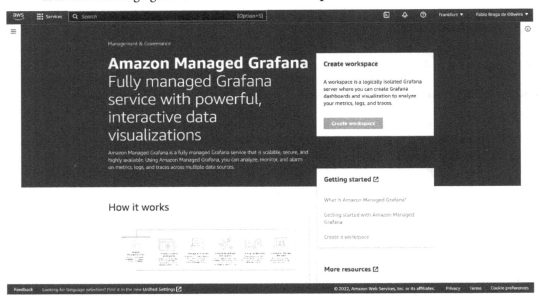

Figure 10.10 – Amazon Managed Grafana main page

This will redirect you to the AMG creation wizard.

3.   In **Step 1** (**Specify workspace details**), type in a workspace name, as in the following figure, and click on **Next**:

Figure 10.11 – AMG wizard, Step 1 – Specify workspace details

4.   In **Step 2** (**Configure settings**), you must select the form of authentication. You have two options: **AWS IAM Identity Center** (previously AWS SSO) and **Security Assertion Markup Language** (**SAML**). We use SAML when federating authentication to a third-party identity provider, which is outside the scope of this book. Let's use AWS IAM Identity Center for now, as it is easier to manage. Also, set **Permission type** to **Service managed** and click on **Next**. See the following figure:

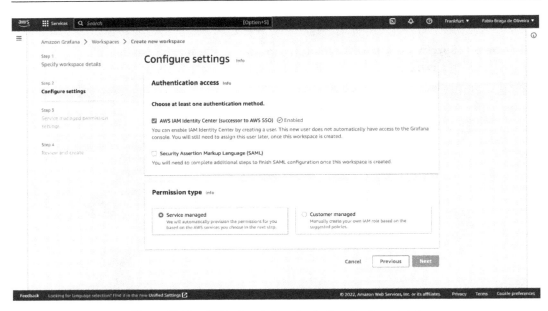

Figure 10.12 – AMG wizard, Step 2 – Authentication access

5. In **Step 3** (**Service managed permission settings**) of the AMG creation wizard, you select which accounts should access Grafana, and here we will choose the current account to keep it simple, and the data sources. Under **Data sources**, remember to select **Amazon Managed Service for Prometheus** to integrate the Prometheus workspace we configured in the previous section. Click on **Next**. Check the following figure:

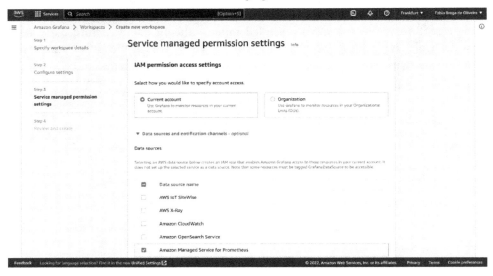

Figure 10.13 – AMG wizard, Step 3 – Service managed permissions settings

6. In **Step 4 (Review and create)**, you can review and confirm your selections. Review them and click on **Create workspace**, as in the following figure:

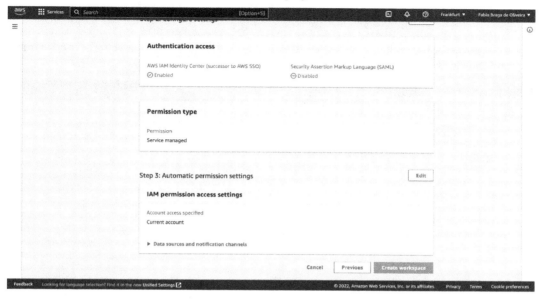

Figure 10.14 – AMG wizard, Step 4 – Review and create

The provisioning of a new workspace may take a few minutes. Wait until the status is **Active**. The service will warn you that you must assign users or groups before they can access the Grafana console, and that's what we will do next. Check the following figure:

Figure 10.15 – AMG workspace details; there's a warning in a blue box at the top

7.   The last step is to assign a user(s) or group(s) to access the Grafana graphical user interface. You can learn how to create new users or groups in the AWS documentation here: `https://docs.aws.amazon.com/singlesignon/latest/userguide/addusers.html`.

8.   If you, like me, already have one, you can click on the **Assign new user or group** button in the blue box at the top or, in the **Authentication** tab, click on the checkbox next to the user's name and then click on the **Assign users and groups** button. Check the following figure:

Figure 10.16 – AMG workspace, Assign user page

9.   After adding a new user or group, you can change the permissions by clicking on the user's checkbox and then on the **Action** button. Let's promote our newly assigned user to admin to give them enough powers to execute the actions described in the rest of this chapter. See the following figure:

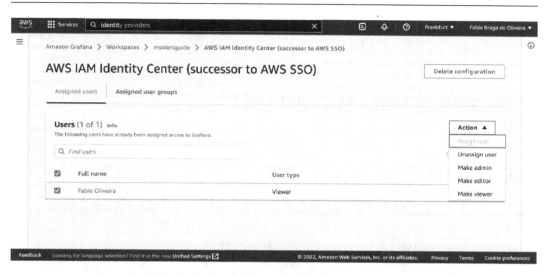

Figure 10.17 – AMG users and groups, changing the user permissions

After accessing your AWS SSO page and selecting the Grafana workspace, you should see the dashboard as in the following figure:

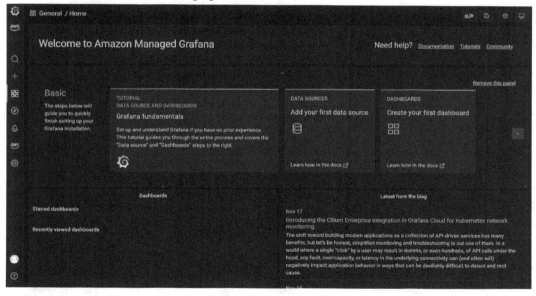

Figure 10.18 – Grafana dashboard

With the work done so far, we have both Amazon Managed Service for Prometheus and Grafana provisioned, and ready to be used. In the next subsection, we will create an Amazon EKS cluster as a sample workload to feed telemetry to both.

Setting up an Amazon EKS cluster and tools

We need a running environment to ingest data in our newly created AMP. We will configure and use a sandbox environment in the rest of this chapter. On the Cloud9 environment, run the following command:

```
curl -sSL https://insiders-guide-observability-on-aws-book.
s3.amazonaws.com/common/create-eks-ec2-eksctl.sh | bash
```

The preceding command will start the creation of a new EKS cluster. You can see the cluster status in the command output. The process of creating a new cluster may take a few minutes. Wait until the prompt comes back and you can type new commands. See the following figure:

Figure 10.19 – Amazon EKS cluster creation output

After creating the cluster, let's check the communication between the Cloud9 environment and the new cluster. Run the following command:

```
kubectl get svc
```

You should see an output such as this:

```
ec2-user:~/environment $ kubectl get svc
NAME         TYPE        CLUSTER-IP    EXTERNAL-IP    PORT(S)    AGE
kubernetes   ClusterIP   10.100.0.1    <none>         443/TCP    7m52s
```

Now, let's set up the service accounts in our Kubernetes cluster. Run the following command:

```
curl -sSL https://insiders-guide-observability-on-aws-book.
s3.amazonaws.com/chapter-10/setup-IAM-roles-service-accounts.sh | bash
```

The preceding command will execute the following actions:

- Generate an IAM role equipped with an IAM policy that grants authorization for remote writing to an AMP workspace
- Construct a Kubernetes service account and add annotations assigning it the IAM role
- Form a trust relationship linking the IAM role to the OIDC provider housed within your Amazon EKS cluster

In this section, we set up both Amazon Managed Service for Prometheus and Grafana, and an Amazon EKS cluster to be our source of telemetry data. In the next section, we will see the necessary configuration steps to send telemetry data to them.

Ingesting telemetry data

Now that we have the environment set up, it's time to set up the data ingestion software and permissions. We will see two ways to ingest data into AMP:

- Ingestion from a new Prometheus server
- Ingestion using AWS Distro for OpenTelemetry

Ingestion from a new Prometheus server

Let's configure the data ingestion using a Prometheus server deployed in our EKS cluster:

1. First, let's create a Prometheus namespace with the following command:

    ```
    kubectl create namespace prometheus
    ```

2. Now, let's execute the following command to install Prometheus:

    ```
    curl -sSL https://insiders-guide-observability-on-aws-book.
    s3.amazonaws.com/chapter-10/install-prometheus-eks.sh | bash
    ```

3. Now, let's go back and configure the AMP data source on Grafana. Access the Grafana dashboard and click on the AWS icon on the menu at the left. Select the **AWS services** option. Check the following figure:

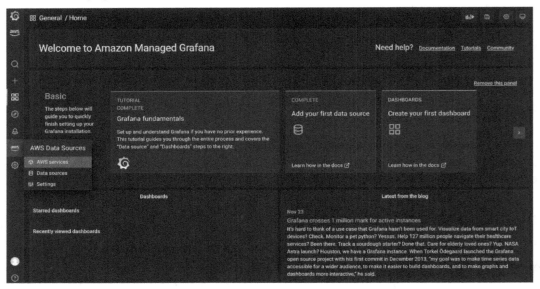

Figure 10.20 – Grafana dashboard, selecting the AWS services option on the left menu

4. On the next screen, select the **Amazon Managed Service for Prometheus** option. See the following figure:

Figure 10.21 – Grafana dashboard, selecting Amazon Managed Service for Prometheus

5. On the next screen, select the region where you have deployed AMP. The instance will appear on the list. Click on the checkbox and click on the **Add 1 data source** button to add the data source. See the following figure:

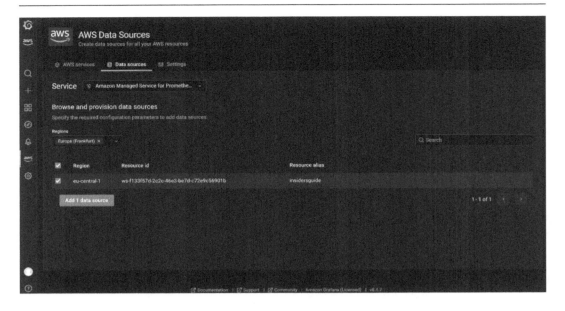

Figure 10.22 – Grafana dashboard, selecting the AMP instance as the data source

6.   Let's now import a public Grafana dashboard with metrics from Kubernetes environments. Hover your cursor over the plus (+) sign in the left navigation bar and select **Import**:

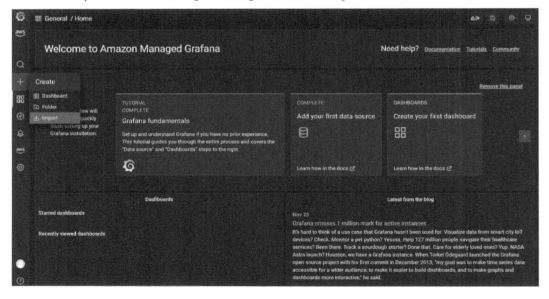

Figure 10.23 – Grafana dashboard, selecting the Import option on the left menu

7. On the **Import** screen, type 3119 into the **Import via grafana.com** textbox and click **Load**:

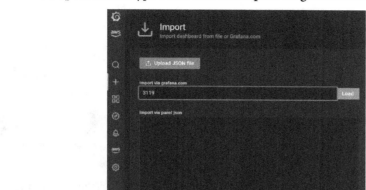

Figure 10.24 – Grafana dashboard, loading public dashboard ID 3119

8. Select the AMP data source in the drop-down box at the bottom and click on **Import**:

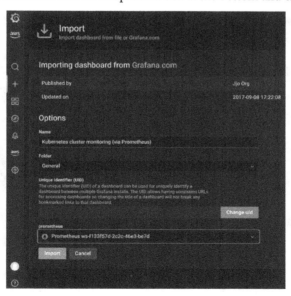

Figure 10.25 – Grafana dashboard, Import dashboard using data from AMP

9. After several minutes, you can find your dashboard using the **Dashboard** menu on the navigation bar. After selecting the correct data source at the top, you will see the cluster metrics in real time. See the following figures:

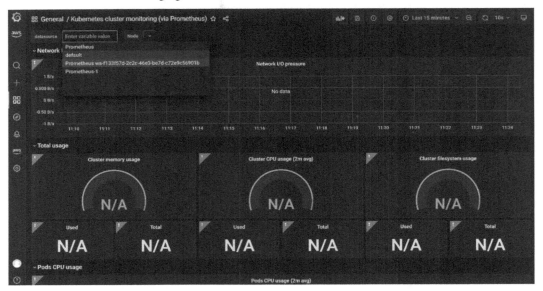

Figure 10.26 – Grafana dashboard, selecting the AMP data source

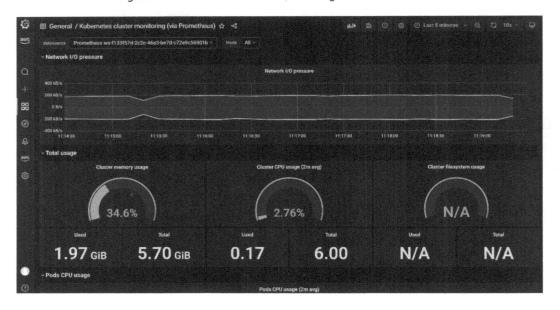

Figure 10.27 – Grafana dashboard, showing cluster metrics

Ingestion using AWS Distro for OpenTelemetry (ADOT)

Let's install a sample application to test the integration of ADOT with AMP:

1. To deploy the application pods in the existing Amazon EKS cluster, run the following command:

    ```
    curl -sSL https://insiders-guide-observability-on-aws-
    book.s3.amazonaws.com/chapter-10/install-sample-app.sh |
    bash
    ```

2. You can check whether the application installation went well by checking the status of pods in the application namespace:

    ```
    $ kubectl get pods -n aoc-prometheus-pipeline-demo
    NAME                                        READY
    STATUS      RESTARTS    AGE
    prometheus-sample-app-77b4c985db-fg6zv      1/1
    Running     0           11m
    ```

3. Now, let's install the AOT collector. The AOT collector will collect metrics for the application and forward them to AMP. Execute the following command:

    ```
    curl -sSL https://insiders-guide-observability-on-aws-
    book.s3.amazonaws.com/chapter-10/install-aot-collector.sh
    | bash
    ```

4. You can check the status of all pods to certify the collector installation using the following command:

    ```
    $ kubectl get pods -n adot-col
    NAME                    READY   STATUS    RESTARTS    AGE
    adot-collector-jx256    1/1     Running   0           26s
    adot-collector-qh9rw    1/1     Running   0           26s
    adot-collector-zc5mj    1/1     Running   0           26s
    ```

5. Let's check whether we are sending metrics to the AOT collector. Let's check the logs of one collector pod, as in the following example:

    ```
    $ kubectl get pods -A
    $ kubectl logs -n adot-col [name_of_your_adot_collector_
    pod]
    ```

You should see an output, as in the following figure:

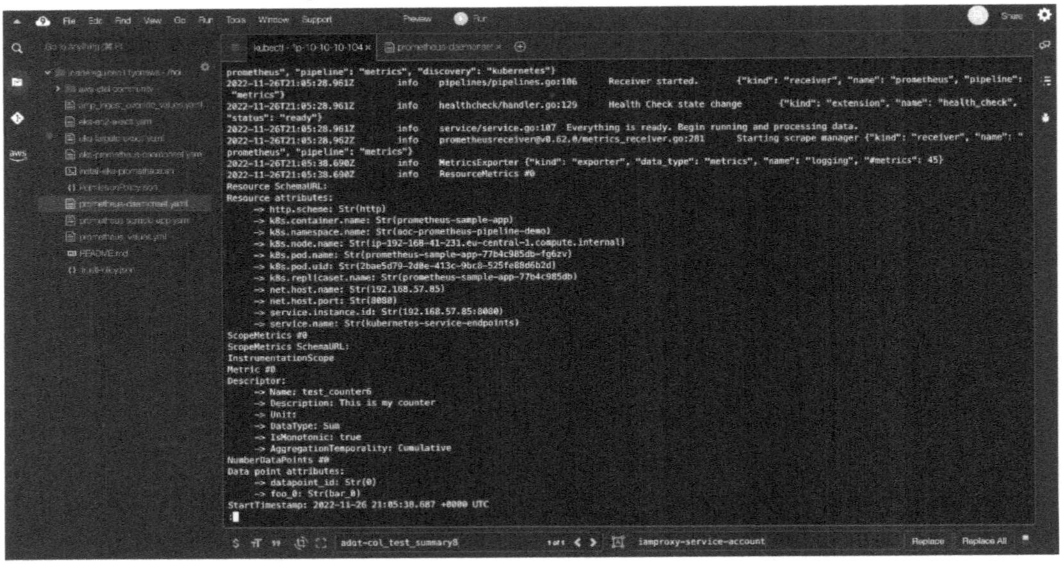

Figure 10.28 – ADOT collector output after receiving application metrics

In this section, we saw how to configure both a cluster and a sample application to send telemetry data to Amazon Managed Service for Prometheus and Grafana. In the next section, we will see different ways to query the stored information.

Querying Prometheus metrics via API and Grafana

You can use the Prometheus-compatible API and Grafana to query the metrics ingested to Prometheus. Let's look at how to do it in the following sections.

Querying Prometheus metrics using Prometheus APIs

To query AMP using the Prometheus-compatible APIs, you need to sign your requests using the AWS Signature Version 4 process (see https://docs.aws.amazon.com/general/latest/gr/signature-version-4.html). You can accomplish this requirement in many different ways, and between them, you can use a proxy such as the AWS Sig4 Proxy (see https://github.com/awslabs/aws-sigv4-proxy) or use the awscurl tool (see https://github.com/okigan/awscurl). In this section, we will use the awscurl tool.

So, let's get started with querying AMP using the Prometheus-compatible APIs:

1. On your Cloud9 shell, execute the following command to install the `awscurl` tool:

    ```
    $ pip3 install awscurl
    ```

2. To execute requests, you need the Prometheus query endpoint. You can find it on the Amazon Managed Service for Prometheus console page. Go to the console using `https://console.aws.amazon.com/prometheus/home`, click on the workspace we created in this chapter, and copy the information you find under the **Endpoint - query URL** section somewhere. Check the following figure:

Figure 10.29 – Prometheus workspace and the endpoint URLs

3. With this information, you can set an `AMP_ENDPOINT` environment variable, which you can use in subsequent commands, using the following command:

    ```
    export AMP_ENDPOINT=https://aps-workspaces.<Region>.
    amazonaws.com/workspaces/<Workspace-id>/api/v1/query
    ```

4. Remember to replace the `Region` and `Workspace-id` placeholders. You can see the command I used here:

    ```
    export AMP_ENDPOINT=https://aps-workspaces.eu-central-1.
    amazonaws.com/workspaces/ws-f133f57d-2c2c-46e3-be7d-
    c72e9c56901b/api/v1/query
    ```

5. With that variable in place, we can execute commands against the Prometheus API, as in the following example:

```
$ awscurl -X POST --region eu-central-1 --service aps
"$AMP_ENDPOINT?query=up"
{"status":"success","data":{"resultType":"vector",
"result":[]}}
```

We saw how to query the Amazon Managed Service for Prometheus using a command-line tool and the public API. Next, let's see how to use a graphical user interface, provided by the Amazon Managed Service for Grafana.

Querying Prometheus metrics using Amazon Managed Grafana

This section will continue from where we left off in the *Ingesting telemetry data* section. As soon as AMP receives telemetry data, we have everything necessary to explore metrics exposed by our systems and the powerful **Prometheus Query Language** (**PromQL**). Many users getting to know Prometheus need help to learn PromQL, and indeed, it can be a challenge to learn a new language to plot and aggregate your metrics as you see fit. But the Grafana team has implemented a handy query builder, which supports you in writing the correct query with a few clicks.

Let's look at how to do it:

1. First, open the AMG console we deployed a few sections ago (you can find your Grafana URL at https://console.aws.amazon.com/grafana/home). On the left navigation menu, select the **Explore** option. See the following figure:

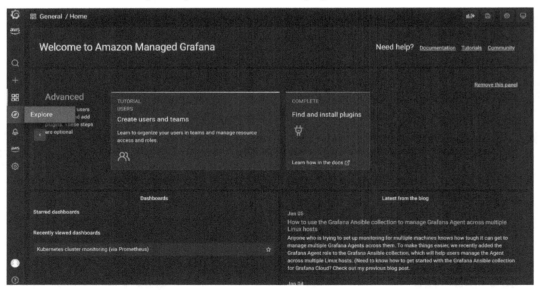

Figure 10.30 – AMG console, left menu, Explore option

2.   The browser will redirect you to the **Explorer** window, where you can type queries against your data sources. But first, you need to select the correct data source, which we created in the *Ingesting telemetry data* section. In the drop-down menu at the top, select the data source named **Prometheus ws-***, as shown in the following figure:

Figure 10.31 – AMG console, Explore window, selecting the Prometheus data source

3.   We are ready to build our first query. Click on the **Metrics browser** link, and you will see all the options to start building your query. See the following figure:

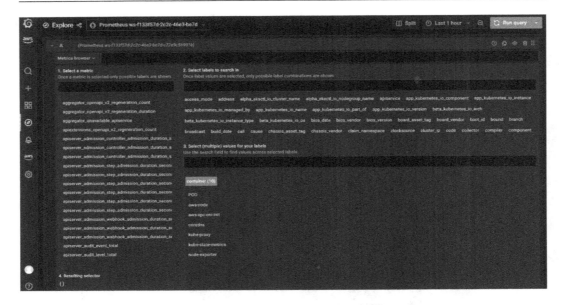

Figure 10.32 – AMG console, Explore window, metrics browser open

4. You can follow the steps suggested in this window, building your query in this order:

    I.   Select a metric.

    II.   Select a label to search in.

    III. Select (potentially) multiple values for your label.

5. Let's build our first query as an example. You can type or scroll to select the `container_memory_usage_bytes` metric, the `container` label, and the `prometheus-server` label value. Your screen should look like the following:

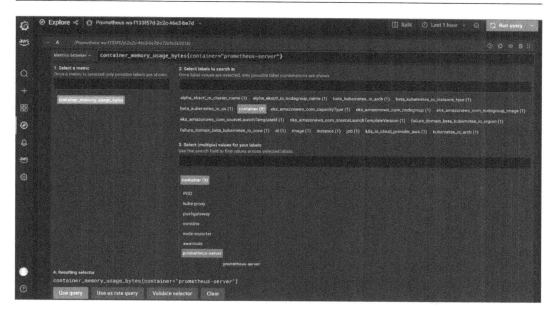

Figure 10.33 – AMG console, Explore window, building a query

With that, we are ready to plot the graph. Click on the **Use query** button at the bottom left to see a graph as in the following figure:

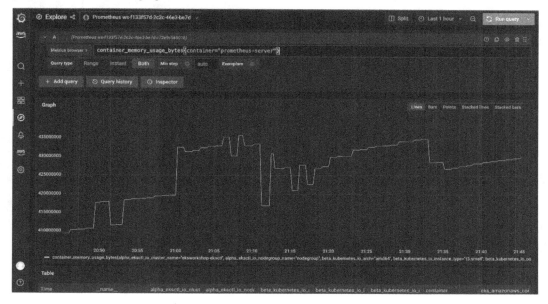

Figure 10.34 – AMG console, Explore window, PromQL built, and a graph plotted

Congratulations, you have queried the Prometheus API visually using Grafana.

In this section, we saw how to query the telemetry data stored in Amazon Managed Service for Prometheus using the API and the Amazon Managed Service for Grafana graphical user interface. In the next section, let's see how the community uses both to implement container monitoring.

Implementing container monitoring

Once your metrics are published to Prometheus, implementing your monitoring strategy involves combining the right charts into relevant dashboards. This task is very business- and application-specific and, as such, hard it is hard to give prescriptive guidance on how to do it. But one Grafana feature can help you to start from a good place and customize from there as you see fit: the capability to export and import dashboards.

You can easily export your dashboards and make them available to other teams in your organization or make them publicly available to the community. There are plenty of dashboards available for you to use at `https://grafana.com/grafana/dashboards/` URL. We will check some of the most useful if you plan to monitor your containerized application.

First, to import a new dashboard, on the AMG console, you need to navigate to the **Create** | **Import** menu on the left-hand side, as you can see in the following figure:

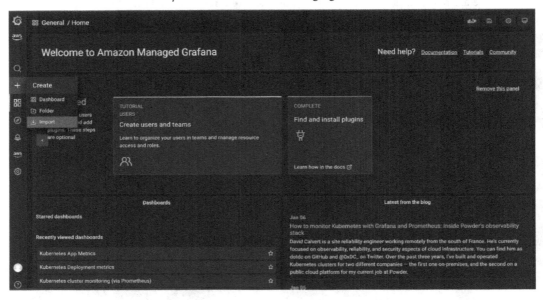

Figure 10.35 – AMG console, Create menu, Import

You can use the Grafana dashboard URL or ID to import a new dashboard. For our example, we will use only the ID, as we can see in the following figure:

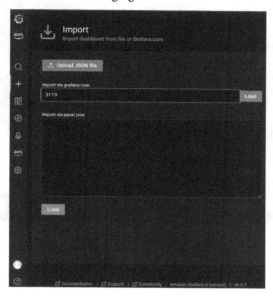

Figure 10.36 – AMG console, import dashboard ID 3119

Let's see some of the most useful (according to your author) community-provided dashboards available:

* **ID 3119, cluster monitoring** (see `https://grafana.com/grafana/dashboards/3119-kubernetes-cluster-monitoring-via-prometheus/`): With this dashboard, you can see some cluster-wide metrics such as CPU, memory, filesystem, and network usage, helping you identify more direct bottlenecks with which your application may suffer. See a sample dashboard here:

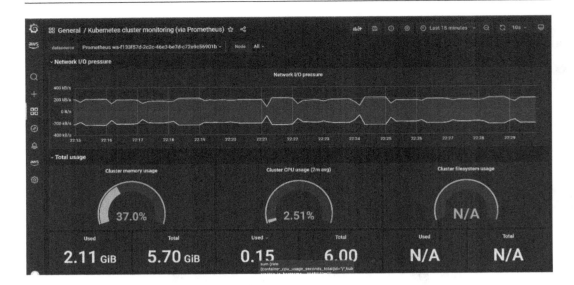

Figure 10.37 – Community dashboard 3119

- **ID 741, deployment metrics** (see `https://grafana.com/grafana/dashboards/741-deployment-metrics/`): This shows the usage of CPU/memory per deployment and the number of replicas running versus the maximum allowed. This level helps find issues within a single application. See the following figure:

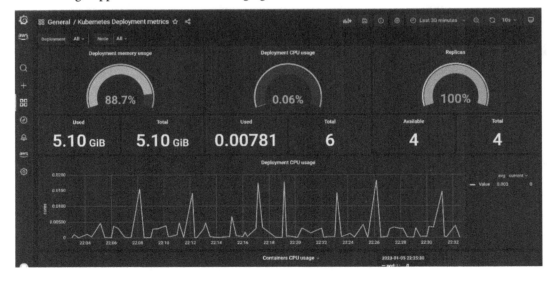

Figure 10.38 – Community dashboard 741

- **ID 747, pod metrics** (see `https://grafana.com/grafana/dashboards/747-pod-metrics/`): This shows the CPU, memory, filesystem usage, and other statistics, now per pod. Check the following figure:

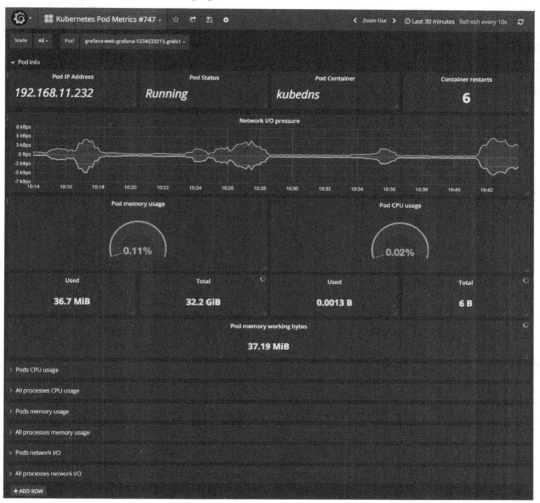

Figure 10.39 – Community dashboard 747

- **ID 1471, application metrics** (see `https://grafana.com/grafana/dashboards/1471-kubernetes-apps/`): This shows a rich collection of application-level metrics, such as the request rate, error rate, response times, pod count, CPU, and memory usage. See the following figure:

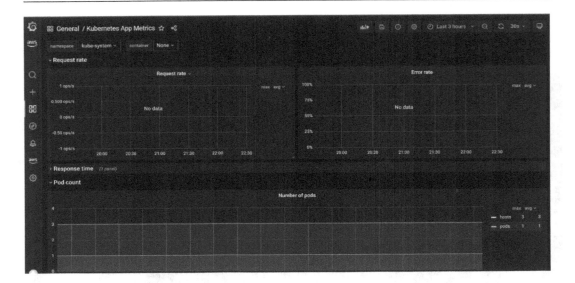

Figure 10.40 – Community dashboard 1471

As you can see, by using the dashboards described in this section, you have a wealth of data in an easy-to-use view at different levels to help you understand your infrastructure and application behavior and trends over time. And again, these are some generic dashboards. With time, you and your team will and should figure out more relevant metrics for your particular use case and add to/create new dashboards with the appropriate metrics.

Summary

In this chapter, you saw how to set up both Amazon Managed Services for Prometheus and Amazon Managed Grafana, how to set up the necessary components to monitor containerized workloads, how to reuse the community-provided dashboards, and how to create your own dashboards.

With the skills learned in this chapter, you can build your observability stack on AWS using open source tools and protocols, if your team already uses them on-premises and you want to leverage the team knowledge or if your organization has an open source first strategy toward aimed at telemetry.

In the next chapter, we will continue navigating the available AWS open source solutions to support your team with your observability needs.

11

Deploying the Elasticsearch, Logstash, and Kibana Stack Using Amazon OpenSearch Service

In the previous chapter, we understood how to gather metrics and visualize them using Prometheus and Grafana. Let's look at Amazon OpenSearch Service in this chapter. The ELK stack comprises **Elasticsearch**, **Logstash**, and **Kibana** (**ELK**). You might have used it on-premises for popular use cases such as log aggregation, observability, and SIEM and want to deploy a managed version on AWS. AWS used to offer Elasticsearch as a managed service on AWS until 2021.

OpenSearch is the successor of Elasticsearch and is a community-driven open source *search and analytics suite* derived from Apache 2.0-licensed Elasticsearch 7.10.2 and Kibana 7.10.2. It is a distributed search engine powered by Apache Lucene under the hood and provides data visualization and a user interface called **OpenSearch Dashboards**. OpenSearch includes a series of add-on tools and plugins. It includes all the advanced functionality ported over from Open Distro from Elasticsearch such as security capabilities, machine learning, alerting functionalities, and so on.

We will also dig into **Amazon OpenSearch Service** (**OSS**), which is a fully managed service that runs OpenSearch by making it easy to deploy, manage, and run cost-effectively by providing industry-leading reliability, scalability, and security functionalities. It provides all the functionalities securely with no overhead of infrastructure management and helps you focus on your use cases.

In this chapter, we will cover the following main topics:

- Amazon OpenSearch Service overview
- Setup and configuration of Amazon OpenSearch Service

- Observability of the application traces and logs using Amazon OpenSearch Service
- Security for Amazon OpenSearch Service
- Anomaly detection in Amazon OpenSearch Service

Technical requirements

To follow along with the chapter, you need the following:

- A working AWS account
- An understanding and fundamental knowledge of EC2 operations
- Fundamental knowledge of Docker and Docker Compose operations
- An understanding of how to deploy CloudFormation templates

Amazon OpenSearch Service overview

Amazon OpenSearch Service (**OSS**) provides dedicated clusters of a community-driven open source search and analytics service called **OpenSearch,** providing well-tuned and optimized deployments of search services with no management overhead of infrastructure provisioning, patching, installation, or ongoing maintenance. AWS monitors Amazon OSS 24/7 and takes care of ongoing maintenance. It also supports simple scaling and cluster topology changes with a single click. AWS handles cross-region replication to support high reliability with no downtime for updates or version upgrades.

Amazon OSS provides **Security Assertion Markup Language** (**SAML**) integration, encryption across use cases with AWS **Key Management Service** (**KMS**), fine-grained access control, detailed security auditing, backward-compatible security patches for all supported versions to minimize required upgrades, and compliance with the HIPAA, FEDRAMP, SOC, PCI, ISO, and CSA STAR.

Amazon OSS presents numerous opportunities to optimize your search and analytics workflows. Here are some examples of the diverse use cases where OpenSearch can be effectively leveraged:

- **Business Insights**: Provides data analytics to improve the user experience by providing personalized recommendations
- **Document Portal**: This helps in a fast and relevant document search experience
- **Observability**: Monitors and debugs applications and infrastructure, log analytics for infrastructure, application, IoT, and trace analytics for observability
- **Security Monitoring**: Detects potential threats to systems based on machine learning and alerting

The observability use case, which we will discuss in this chapter, provides a combination of log and trace analysis. It also provides integrated alerting across use cases with Amazon CloudWatch, scales up to 3 Pb in a single cluster, alerts in real time, and includes ML-powered anomaly detection for real-time adjustments. Additionally, it only charges for consumed infrastructure and comes bundled with a dashboard visualization tool called OpenSearch Dashboards.

One of the key benefits of using Amazon OSS is its cost-effectiveness. With Amazon OSS, you can easily stream data from various log sources using a range of services such as Amazon S3, Amazon Kinesis Data Streams, Amazon DynamoDB, Amazon Kinesis Data Firehose, Amazon CloudWatch, and AWS IoT. The best part is that there are no additional software licensing fees, making it a budget-friendly solution for businesses of all sizes. This allows you to leverage powerful data streaming capabilities without worrying about costly licensing fees, which can be a significant expense for many organizations.

The advantages of using Amazon OSS include the following:

- Support for multiple query languages (OpenSearch, SQL, and PPL)
- Fast access to virtually unlimited pre-indexed data via cold storage in S3
- Free automated backups for 14 days
- 90% cost reduction with UltraWarm compared to hot storage

Now, with this being the last chapter focused on open source observability, let's see how it will fit into all the open source tools using the following figure:

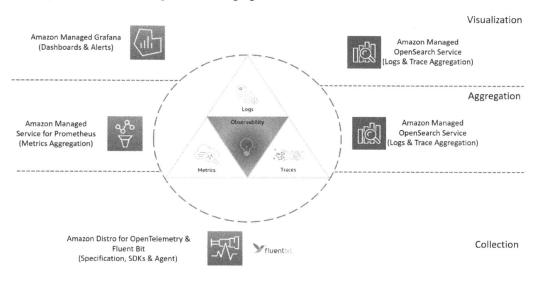

Figure 11.1 – OpenSearch

The left side of *Figure 11.1* shows that while you are collecting metrics, logs, and traces using **AWS Distro for OpenTelemetry** (**ADOT**) or using **OpenTelemetry Collector** (**OTEL**) and Fluent Bit, metrics are aggregated using Prometheus and can be visualized using Amazon Managed Grafana, which we covered in *Chapter 10, Deploying and Configuring an Amazon Managed Service for Prometheus*. The right side of *Figure 11.1* shows that we can use Amazon OpenSearch for the aggregation of logs and traces. Amazon OSS also provides a visualization layer called OpenSearch Dashboards, which could support building operational dashboards.

Metrics functionality is not yet available at the time of writing. Currently, metrics are not directly supported for ingestion/collection into OpenSearch. There is a separate project available on GitHub as a sample: `https://github.com/aws-samples/amazon-opensearch-service-monitor`.

In the next section, let's understand the fundamental components of Amazon OSS and learn how to set up an Amazon OSS standalone cluster practically, using the AWS Console.

Setup and configuration of Amazon OpenSearch Service

Before going into the setup and configuration of Amazon OSS, let's try to understand the fundamental concepts of the OpenSearch Service on AWS. They are as follows:

- **Domain or Cluster**: A domain or cluster is a collection of nodes that share the same `cluster. name` attribute. As nodes are added or removed from the cluster, the system reorganizes itself to distribute the data across the remaining nodes.

- **Nodes**: A node is a single server that forms part of a cluster, stores your data, and contributes to the cluster's indexing and search capabilities.

- **Leader/Master Nodes**: Amazon OpenSearch Service master nodes are the ones that will provide a control plane for managing the cluster. Master nodes perform routine management tasks such as monitoring the health of all the nodes, tracking the number of indexed documents in the cluster and shards in each index, and so on.

- **Data Nodes**: Data nodes hold the data responding to indexes and search queries, and can be a distributed data cluster. For development workloads, you can combine leader and data nodes in the same instance, where one of the data nodes will act as a leader node. But for production workloads, a dedicated master node helps in performing the cluster management tasks without holding any data.

- **UltraWarm Nodes**: UltraWarm nodes are data nodes that use S3 as a data store to reduce the cost of storing older data. You can store large amounts of read-only data on Amazon OSS using the UltraWarm nodes.

- **Indexes**: An index is a collection of documents that have somewhat similar characteristics. It can be thought of as similar to a database.

- **Cold Storage Indexes**: Cold storage also uses S3 as a data store to store any amount of infrequently accessed or historical data on Amazon OSS. This storage is appropriate for doing periodic research or forensic analysis. When there is a requirement to query the cold data, it can be selectively attached to existing UltraWarm nodes.

You can understand the relationship between the cluster/domain, leader nodes, data nodes, UltraWarm nodes, indexes, and cold storage indexes from the following diagram:

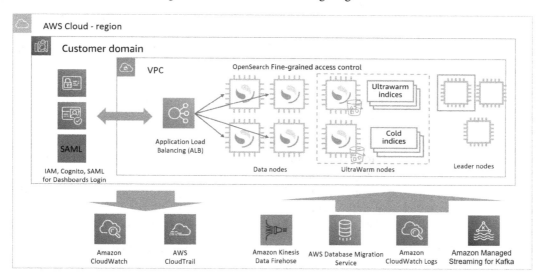

Figure 11.2 – Components of OpenSearch

You can securely install the Amazon OSS domain inside a VPC and use a VPC private endpoint to ingest traffic from various sources such as Amazon CloudWatch Logs, Amazon Kinesis Data Firehose, and so on. You can monitor the performance of the Amazon OSS cluster using CloudWatch and monitor any API activities using AWS CloudTrail. Amazon OpenSearch provides various integration mechanisms with identity providers such as IAM and Cognito, and SAML providers for providing the logging for OSS Dashboards.

As we understand the technical services that the Amazon OpenSearch domain consists of, let's look at the logical components of Amazon OpenSearch Service:

- **Shards**: OpenSearch provides the ability to subdivide your index into multiple pieces called shards.

- **Replica**: A replica is a copy of every single shard.

- **Document**: A document is a fundamental unit of information that can be indexed. It can be thought of as similar to a record in a relational database.

- **Segment**: Each shard comprises multiple segments, where a segment is an inverted index. A search within a shard will iteratively search each segment, then combine the results from each segment to produce the final results for that shard.

Let's understand the additional components that are required when you are looking to deploy Amazon OpenSearch Service in your AWS account:

- **Authentication**: To log in to OSS Dashboards and also access the data nodes/master nodes, you can integrate Amazon OSS with IAM, Cognito, or SAML-based external authentication services.

- **Amazon EC2**: Amazon **Elastic Compute Cloud (EC2)** instances will be used to set up master and data nodes and are managed automatically by AWS to provide high availability based on the domain configuration provisioned.

- **Amazon VPC**: Amazon **Virtual Private Cloud (VPC)** will be used to communicate and understand the heartbeat between the master and data nodes.

- **Application Load Balancer**: An **application load balancer (ALB)** is required to distribute incoming search requests to multiple search instances. The ALB also provides features such as SSL/TLS termination and request routing based on the path or host of the incoming request. Additionally, the ALB can also be used for health checking of the search instances, and automatically redirecting traffic to healthy instances in case of failures. It can also be made public/private based on the configuration for accessing OSS Dashboards.

- **AWS CloudFormation/AWS Console**: The OSS cluster/domain can be set up either using AWS CloudFormation or using the AWS Console. You can also leverage other forms of **infrastructure as code (IaC)** tools to build the OSS domain.

Overall, the AWS components that would be part of deployment when setting up an Amazon OSS cluster are depicted in the following figure:

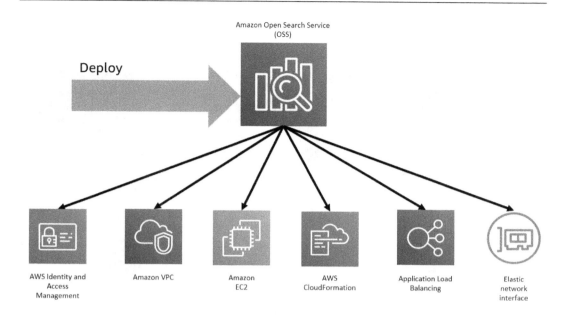

Figure 11.3 – AWS components to deploy Amazon OpenSearch Service

We have reviewed the logical and physical components involved in setting up Amazon OSS, and we have also reviewed the dependent services required for the setup. Next, we will examine the deployment of a standalone cluster of Amazon OSS through the AWS Console.

Installation of a standalone cluster of Amazon OpenSearch Service

Let's proceed with a standalone cluster to understand how to deploy Amazon OpenSearch Service using the AWS Console. Although this is not recommended for production installations, to keep the costs low, in this setup, we will not segregate data nodes and leader nodes. This section is hands-on, so log in to your AWS account and be prepared for the ride. AWS will charge you for the resources deployed in this section, but for short periods. When you finish, remember to clean up the used resources.

Let's get started with the installation:

1.   Navigate to the AWS Console and search for Amazon OpenSearch Service:

Figure 11.4 – Searching for Amazon OpenSearch Service

2.   Click on **Create domain**:

Figure 11.5 – Create domain

3.   Provide the domain name as observabilitydomain:

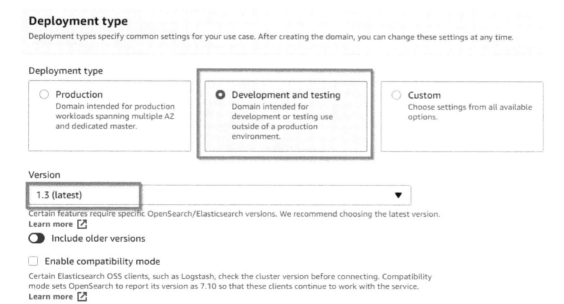

Amazon OpenSearch Service > Domains > Create domain

Create domain Info

Name

Domain name

observabilitydomain

The name must start with a lowercase letter and must be between 3 and 28 characters. Valid characters are a-z (lowercase only), 0-9, and - (hyphen).

Custom endpoint

Each Amazon OpenSearch Service domain has an auto-generated endpoint, but you can also add a custom endpoint using AWS Certificate Manager (ACM). **Learn more** 🗗

☐ Enable custom endpoint

Figure 11.6 – Providing the domain name

4.   Select **Development and testing**, and for **Version**, select **1.3**:

Deployment type

Deployment types specify common settings for your use case. After creating the domain, you can change these settings at any time.

Deployment type

○ Production
Domain intended for production workloads spanning multiple AZ and dedicated master.

● Development and testing
Domain intended for development or testing use outside of a production environment.

○ Custom
Choose settings from all available options.

Version

1.3 (latest)   ▼

Certain features require specific OpenSearch/Elasticsearch versions. We recommend choosing the latest version.
Learn more 🗗

🔘 Include older versions

☐ Enable compatibility mode

Certain Elasticsearch OSS clients, such as Logstash, check the cluster version before connecting. Compatibility mode sets OpenSearch to report its version as 7.10 so that these clients continue to work with the service.
Learn more 🗗

Figure 11.7 – Deployment type and Version

5. Select **1-AZ** for **Availability Zones** under **Data nodes** and **t3.small.search** for **Instance type**. Please note, I do not recommend this for production installation, and also, the t3 series does not support UltraWarm nodes and cold storage functionality.

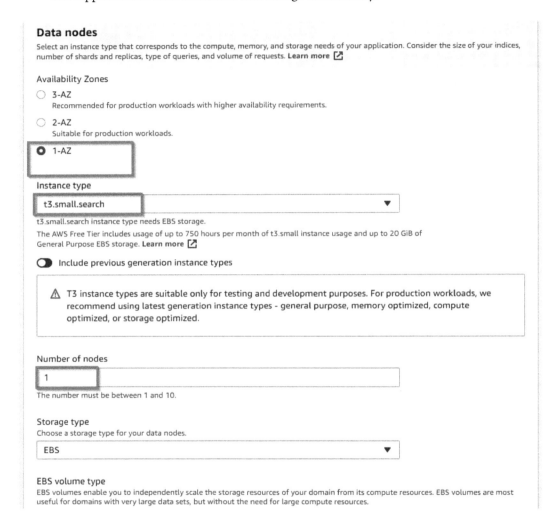

Figure 11.8 – Data nodes and Instance type

6. Set **EBS storage size per node** to **10** GB and uncheck **Enable dedicated master nodes**:

Storage type
Choose a storage type for your data nodes.

EBS ▼

EBS volume type
EBS volumes enable you to independently scale the storage resources of your domain from its compute resources. EBS volumes are most useful for domains with very large data sets, but without the need for large compute resources.

General Purpose (SSD) - gp3 ▼

🔘 Include previous generation EBS volume types

EBS storage size per node

10

EBS storage size per node in GiB. Minimum 10 GiB and maximum 100 GiB.

▶ **Advanced settings**

Warm and cold data storage

Enable UltraWarm to store even more data on Amazon OpenSearch Service. You can economically retain large amounts of data while keeping the same interactive analysis experience. **Learn more** 🔗

Enable cold storage to further reduce storage costs for data you rarely access. To view data in cold storage, you must first move it to warm storage. **Learn more** 🔗

ⓘ UltraWarm data nodes feature is not supported by the data instance type you selected.

Dedicated master nodes

Dedicated master nodes improve the stability of your domain. For production domains, three is recommended. **Learn more** 🔗

☐ Enable dedicated master nodes

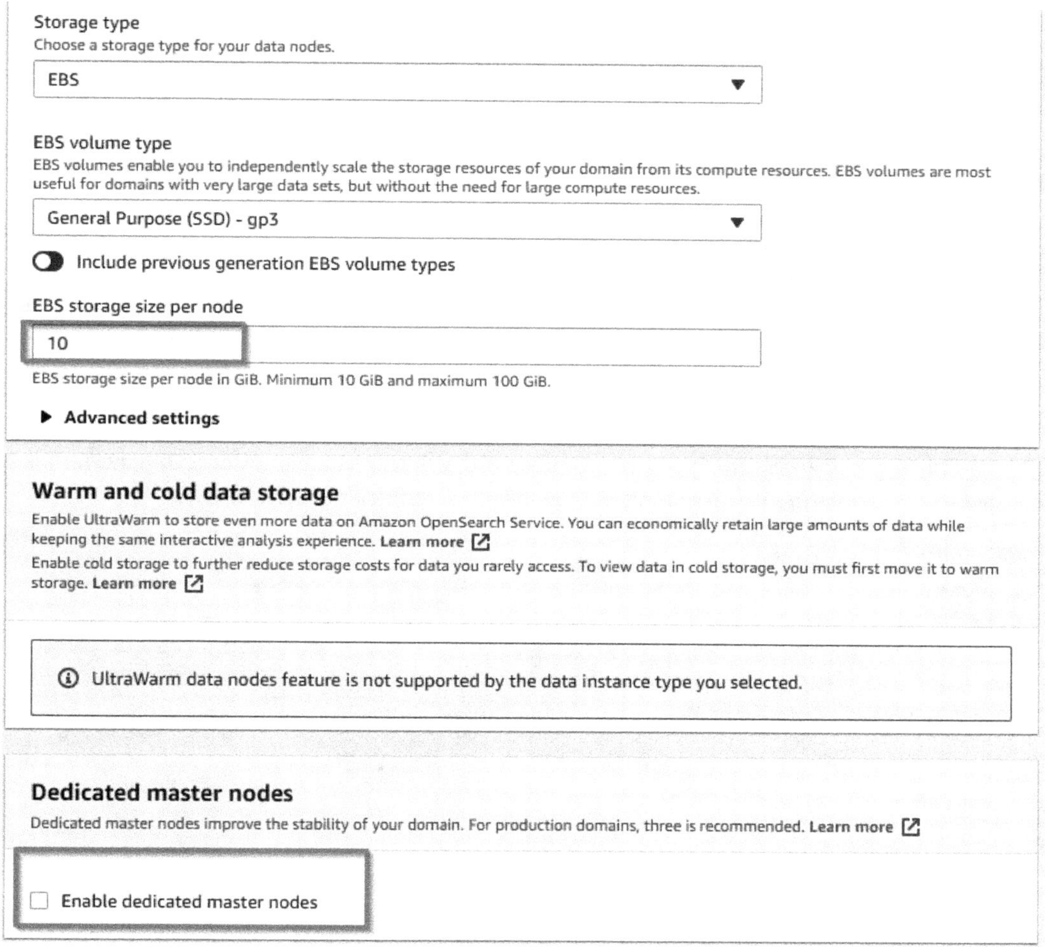

Figure 11.9 – EBS volume and master nodes

7.   For **Network**, select **Public access** (not recommended for production):

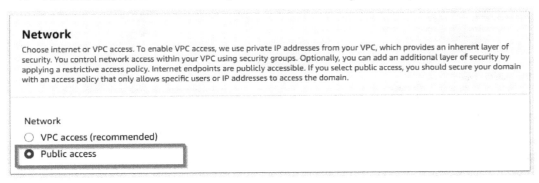

Figure 11.10 – VPC or Public access

8.   For **Fine-grained access control**, you can embed with either IAM, Cognito, or SAML integration or a local account in OSS. We are going ahead with creating a new master user:

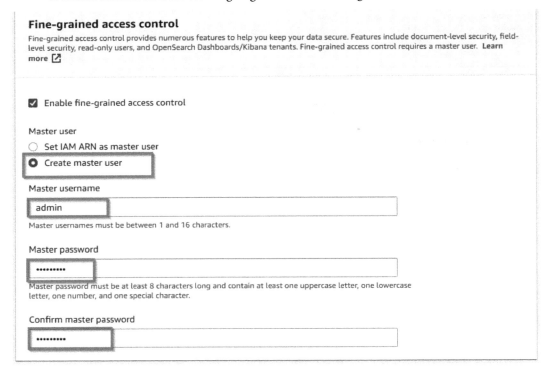

Figure 11.11 – Local master user

9. For **Domain access policy**, select **Only use fine-grained access control**, where we are allowing open access to the domain:

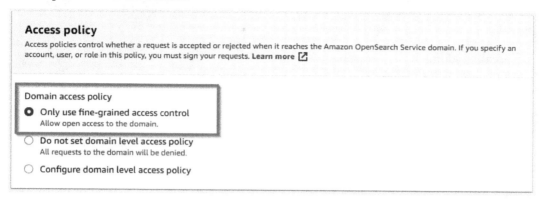

Figure 11.12 – Open access to domain

10. Leave the remaining settings as the defaults and click **Create**:

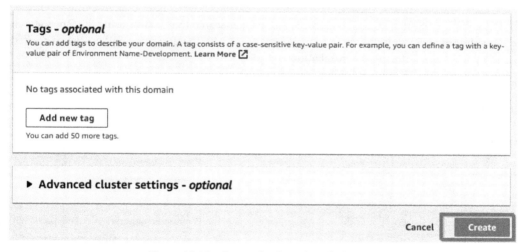

Figure 11.13 – Create the OpenSearch domain

The process will take around 10-15 minutes to complete.

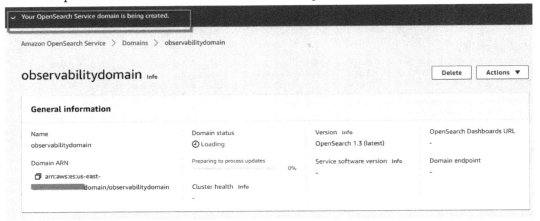

Figure 11.14 – Amazon OSS creation in progress

Metrics related to the OpenSearch performance and the API access will be logged to Amazon CloudWatch and CloudTrail, respectively.

11. Once the OpenSearch domain is active, log in to OpenSearch Dashboards using the **OpenSearch Dashboards URL** as highlighted in the *Figure 11.15* with the username and password created as a part of *step 7* during the installation:

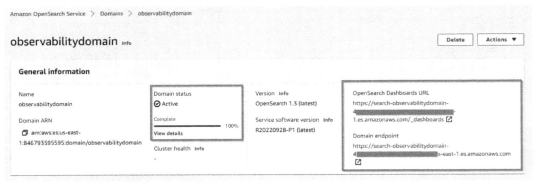

Figure 11.15 – Domain active and log in

Let's log in to OpenSearch Dashboards with our username and password:

Figure 11.16 – OpenSearch Dashboards login

12. Let's select the tenant as **Global**. This is ideal when you would like to share the data with all users:

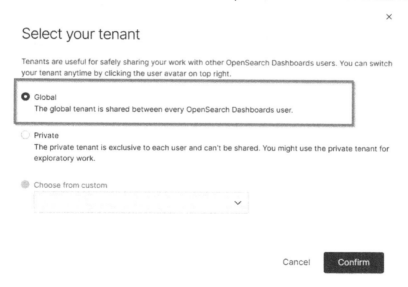

Figure 11.17 – Selecting the Global tenant

13. You can add the sample data by clicking **Add sample data** and selecting **Sample flight data**:

Add sample data

Figure 11.18 – Sample flight data

14. You can click on **View Data** and navigate through the sample dashboard provided by Amazon OpenSearch Service.

Let's navigate and verify the Amazon OSS service domain. Amazon OpenSearch Service has several pre-installed plugins, and you can see them when you click on the navigation bar.

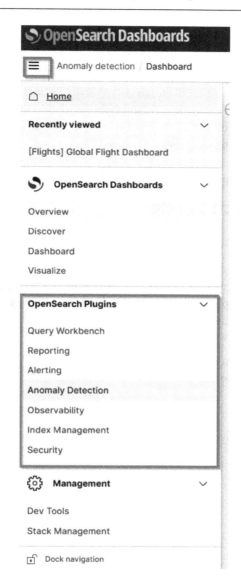

Figure 11.19 – OpenSearch Plugins

In this section, we have successfully deployed an Amazon OpenSearch Service domain and added sample data as a part of the exercise. If you are looking to deploy Amazon OSS for production, you can look into the best practices at `https://docs.aws.amazon.com/opensearch-service/latest/developerguide/bp.html`.

In the next section, we will look at deploying Amazon OpenSearch Service using CloudFormation and also setting up the same within the VPC. We will also set up an app to understand the observability of the application and visualize the same using Amazon OSS.

Observability of the application traces and logs using Amazon OpenSearch Service

Let's deploy the Quick Start CloudFormation template by clicking this URL: `https://insiders-guide-observability-on-aws-book.s3.amazonaws.com/chapter-11/ossdeploy.json`. Alternatively, you can download the CloudFormation template and deploy the template. You will need a *key pair* to be available in your AWS account to deploy the CloudFormation template. If you want to create a key pair, please look at `https://docs.aws.amazon.com/AWSEC2/latest/UserGuide/create-key-pairs.html`. It will take approximately 40 minutes to complete the full deployment of the test application.

Once you deploy the application and Amazon OSS using the CloudFormation template, you can navigate to the **Outputs** tab and should be able to see the OpenSearch Proxy URL and Sample Application URL, as shown here:

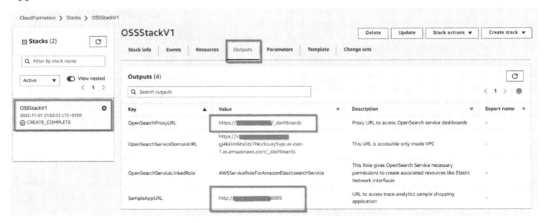

Figure 11.20 – CloudFormation output URLs

The application website will look as follows:

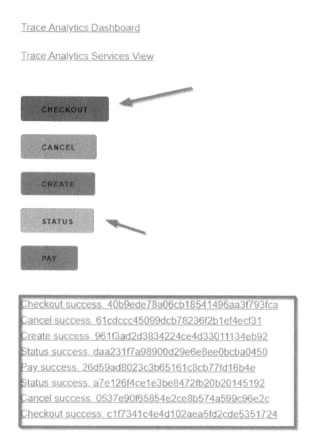

Figure 11.21 – Sample application website

Once you navigate and click on the listed items on the web app, you should be able to see success/failure messages at the bottom. You generate application traces by clicking on them.

Now, let's explore the deployment and high-level architecture of the application and the data flow of the application. The sample application has multiple containers running inside an EC2 using Docker. You can see the high-level application and the data flow to OpenSearch Service in *Figure 11.22* . The application is instrumented using OTEL. We covered OTEL in *Chapter 10, Deploying and Configuring an Amazon Managed Service for Prometheus.*

Now let's understand how to send the data gathered using OTEL into OpenSearch Service. Please refer to *Figure 11.22*. It consists of two paths, one for gathering the traces and the other for gathering the log files. In path 1 (**Traces**), we are collecting the application traces and transforming the data from OTEL using **Data Prepper version 2.0** and ingesting the data using the HTTP/HTTPS endpoint of OpenSearch Service. The second path (**Logs**) is for ingesting application logs into Amazon OSS. The application logs are captured using the **Fluent Bit** logs processor and are sent to the `sample_app_logs` index using the HTTP/HTTPs endpoint of OpenSearch Service.

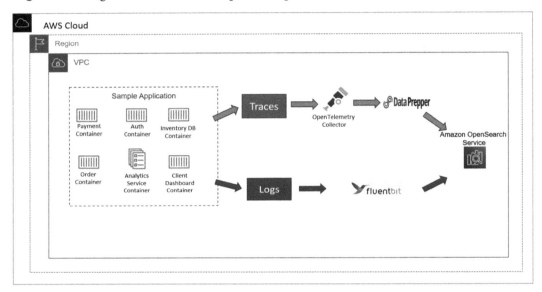

Figure 11.22 – Application architecture

Let's deep-dive into the configuration of Data Prepper and Fluent Bit. The source code for the application, along with the `dataprepper/fluentbit` configurations, is available at this GitHub URL: `https://github.com/PacktPublishing/An-Insider-s-Guide-to-Observability-on-AWS/tree/main/chapter-11/traceanalytics`.

Important files to pay attention to are `trace_analytics_no_ssl.yml`, `fluentbit.conf`, and `docker-compose.yml`. Let's understand the configuration and why they are relevant in the next two sections.

Application traces

Let's start with understanding the trace pipeline and how the data is transformed using Data Prepper and the relevant configuration in the application deployment. Here are the components of application traces:

- **Trace pipeline:** This refers to the traces generated from the application leveraging OTEL instrumentation and captured and segregated into two different formats – **raw traces** and **service map traces**. They are transformed by the Data Prepper pipeline, as shown in the following figure. The raw pipeline transforms the individual traces and their relevant metrics, such as delay. The service map pipeline provides you with the relationship between different containers in the sample application that could be visualized in the Amazon OSS console.

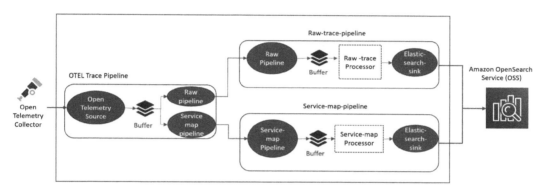

Figure 11.23 – Trace pipeline

- **Trace configuration:** The configuration for the traces in Data Prepper is shown in the following figure and is part of the trace_analytics_no_ssl.yml file, where you can see the entry from OpenTelemetry and segregate application traces into two different pipelines (raw-pipeline and service-map-pipeline), and both the pipelines send the data into Amazon OpenSearch Service into two different indices (trace-analytics-raw and trace-analytics-service-map) using the HTTPS Amazon OSS endpoint.

```
entry-pipeline:
  delay: "100"
  source:
    otel_trace_source:
      ssl: false
  processor:
  sink:
    - pipeline:
        name: "raw-pipeline"
    - pipeline:
        name: "service-map-pipeline"
raw-pipeline:
  source:
    pipeline:
      name: "entry-pipeline"
  processor:
    - otel_trace_raw:
  sink:
    - opensearch:
        hosts: [ "https://OSS_DOMAIN:443" ]
        username: "OSSDOMAIN_USERNAME"
        password: "OSSDOMAIN_PASSWORD"
        index_type: trace-analytics-raw
service-map-pipeline:
  delay: "100"
  source:
    pipeline:
      name: "entry-pipeline"
  processor:
    - service_map_stateful:
  sink:
    - opensearch:
        hosts: [ "https://OSS_DOMAIN:443" ]
        username: "OSSDOMAIN_USERNAME"
        password: "OSSDOMAIN_PASSWORD"
        index_type: trace-analytics-service-map
```

Figure 11.24 – Trace configuration

- **Data Prepper deployment**: To deploy Data Prepper along with the dependencies, we have used Docker Compose to build the containers. Let's look at the Docker configuration. You can see that we are using a Data Prepper 2.0 container image and have used the pipelines as shown in *Figure 11.23* and configuration as shown in *Figure 11.25*:

```
data-prepper:
  restart: unless-stopped
  container_name: data-prepper
  image: opensearchproject/data-prepper:2
  command: sh data-prepper-wait-for-odfe-and-start.sh
  volumes:
    - ./data-prepper-wait-for-odfe-and-start.sh:/usr/share/data-prepper/data-prepper-wait-for-odfe-and-start.sh
    - ./trace_analytics_no_ssl.yml:/usr/share/data-prepper/pipelines/pipelines.yaml
    - ./data-prepper-config.yaml:/usr/share/data-prepper/config/data-prepper-config.yaml
    - ./root-ca.pem:/usr/share/data-prepper/root-ca.pem
  ports:
    - "21890:21890"
  networks:
    - my_network
  logging:
    driver: fluentd
```

Figure 11.25 – Data Prepper deployment

Now let's log in to the **OpenSearch Dashboard Proxy** URL as shown in the CloudFormation **Outputs** tab (*Figure 11.20*) to visualize the traces and service map generated by the application flow. Once you log in, you can navigate to **Amazon OSS Console | Observability | Trace analytics**. This will provide a list of traces captured by OTEL and visible on the Amazon OSS Dashboard.

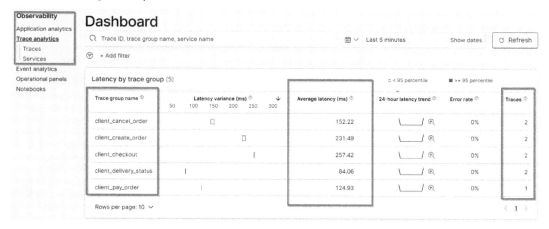

Figure 11.26 – Amazon OSS Trace analytics dashboard

You can click on **Amazon OSS Console | Observability | Trace analytics | Services** and view the overall traces by service, as shown in the following figure:

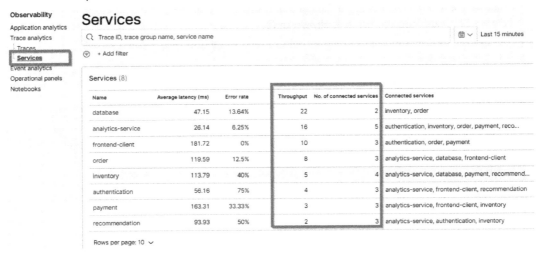

Figure 11.27 – Amazon OSS Services view

Additionally, you can also see the **Service map** view, as shown in the following figure:

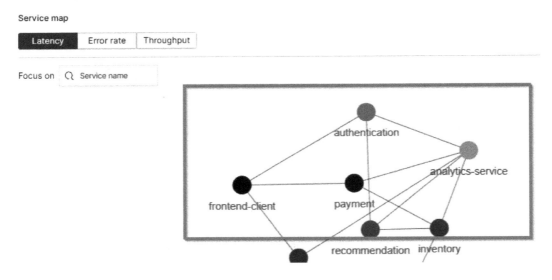

Figure 11.28 – Service map view of the traces

Application logs

We learned about the trace pipeline and configuration files in the last section. Let's understand the components of the application logs pipeline into Amazon OSS and the relevant configuration in this section. We are using a Fluent Bit log processor tool to send the log data from the application into Amazon OSS. Fluent Bit is an open source and multi-platform log processor tool. It can collect and aggregate different types of log data and send them to one or more log destinations. You can see the high-level Fluent Bit log flow in the following figure:

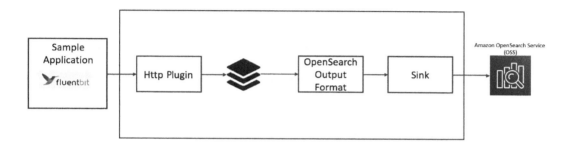

Figure 11.29 – Amazon OSS logs pipeline

As shown in *Figure 11.29*, Fluent Bit uses the HTTP output plugin to send data to Amazon OSS along with the required transformation. You can see the HTTP plugin details in the `fluentbit.conf` file in *Figure 11.30*. The `fluentbit.conf` file provides the configuration to forward the data to Amazon OpenSearch Service utilizing the HTTP endpoint. You can observe the configuration of the output to Amazon OSS in the configuration file and note the HTTP endpoint and the type of log file (`docker` in this case, as the application is installed as a Docker container):

```
[SERVICE]
    Flush       5
    Daemon      Off
    Log_Level   debug

[INPUT]
    Name    forward
    Listen  0.0.0.0
    Port    24224

[OUTPUT]
    Name    es
    Match   *
    Host    OSS_DOMAIN
    Port    443
    Index   sample_app_logs
    Type    docker
    HTTP_User OSSDOMAIN_USERNAME
    HTTP_Passwd OSSDOMAIN_PASSWORD
    tls On
    tls.verify Off
```

Figure 11.30 – Amazon OSS Fluent Bit configuration

Let's look into the Fluent bit deployment and configuration and verify the logs sent to Amazon OSS:

- **Fluent Bit Deployment**: Fluent Bit is deployed as a container and the relevant logs are forwarded to the Fluent Bit container and transformed into an Amazon OSS-understandable format. We can see relevant configuration details for Docker in *Figure 11.31*:

```
version: "3.8"
services:
  fluent-bit:
    image: fluent/fluent-bit:latest
    ports:
      - '24224:24224' # logging port
    volumes:
      - ./fluent-bit.conf:/fluent-bit/etc/fluent-bit.conf
```

Figure 11.31 – Fluent Bit deployment

- **Verification of Logs**: To match the log format of the sample logs and query them, let's navigate to the Amazon OSS URL from the CloudFormation **Outputs** tab and carry out the following steps:

    I.   Navigate to Amazon OSS | **Management** | **Stack Management** | **Create index pattern**:

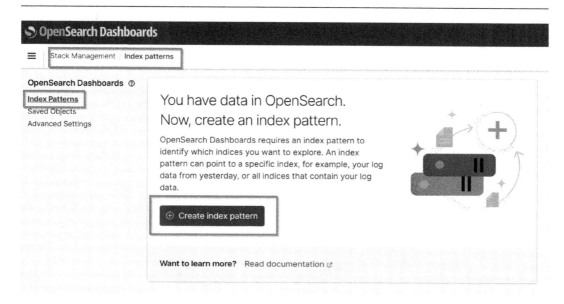

Figure 11.32 – Amazon OSS Create index pattern

II.   Input `sample_app_logs*` and you should see an index matching one source:

Figure 11.33 – Matching index

III. Select [@timestamp] as the primary field to filter and click on **Create index pattern**:

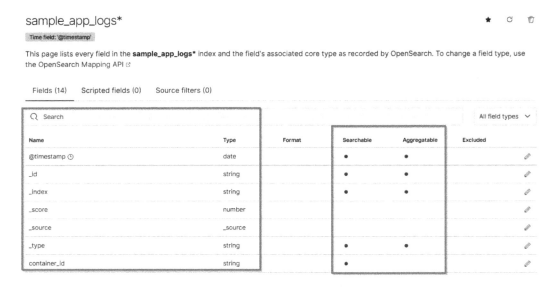

Create index pattern

An index pattern can match a single source, for example, `filebeat-4-3-22`, or **multiple** data sources, `filebeat-*`.
Read documentation ⤴

Step 2 of 2: Configure settings

Specify settings for your **sample_app_logs*** index pattern.

Select a primary time field for use with the global time filter.

Time field | Refresh

@timestamp

@timestamp

I don't want to use the time filter

‹ Back | Create index pattern

Figure 11.34 – Global filter

It will discover the log fields and the type automatically:

sample_app_logs*

Time field: '@timestamp'

This page lists every field in the **sample_app_logs*** index and the field's associated core type as recorded by OpenSearch. To change a field type, use the OpenSearch Mapping API ⤴

Fields (14)   Scripted fields (0)   Source filters (0)

Q Search   All field types ∨

Name	Type	Format	Searchable	Aggregatable	Excluded	
@timestamp ⏱	date		•	•		✎
_id	string		•	•		✎
_index	string		•	•		✎
_score	number					✎
_source	_source					✎
_type	string		•	•		✎
container_id	string		•			✎

Figure 11.35 – Discovered fields

IV.   Navigate to **OpenSearch Dashboards** and click on **Discover**:

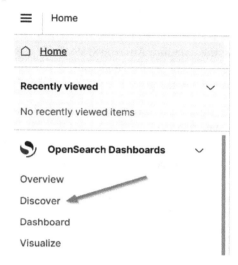

Figure 11.36 – Searching the data using the Discover option

V.   Once you click, you can see the time series-based logs generated from the containers in Amazon OSS. You can further use the OpenSearch Query language and **Piped Processing Language (PPL)** to query the log data ingested.

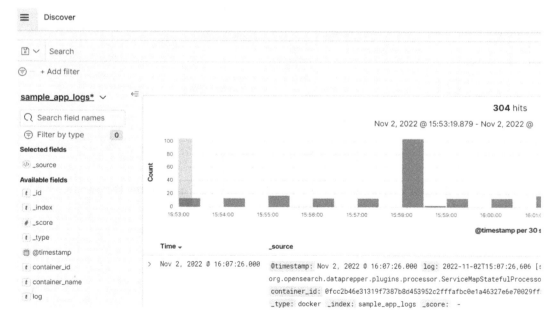

Figure 11.37 – Searchable logs

In this section, we have learned how to deploy Data Prepper and Fluent Bit and forward the logs and traces into OSS, and how to query them and visualize them in Dashboards.

In the example shown in this section, we have only discussed how to use two tools, namely Data Prepper and Fluent Bit, to ingest the data into Amazon OSS. There is a larger number of use cases such as log analysis, security management, and various sources of ingesting the data from various AWS-native services that are feasible, but those are beyond the scope of the book.

In the next section, we will understand how to leverage the anomaly detection feature in Amazon OSS to observe anomalies in the data ingested.

Anomaly detection in Amazon OpenSearch Service

Amazon OSS's **anomaly detection** feature uses the **Random Cut Forest** (**RCF**) algorithm to detect anomalies in OpenSearch log data in real time. RCF is an unsupervised machine learning algorithm that models a sketch of your incoming data stream. It calculates an anomaly score and level of confidence for each incoming data point. Anomaly detection then uses these scores to distinguish abnormal data from normal variations. You can also utilize the pre-created anomaly detection features such as **monitoring HTTP responses**, **monitoring e-commerce orders**, and **monitoring host health**.

Let's go ahead and configure anomaly detection for the container logs deployed from the sample application:

1.   Navigate to Amazon OSS | **OpenSearch Plugins** | **Anomaly detection** | **Create detector**:

Figure 11.38 – Create detector

2.   Let's set the detector's **Name** field to `ContainerAppLogs`, and for the **Data Source** field, select **sample_app_logs** from the dropdown:

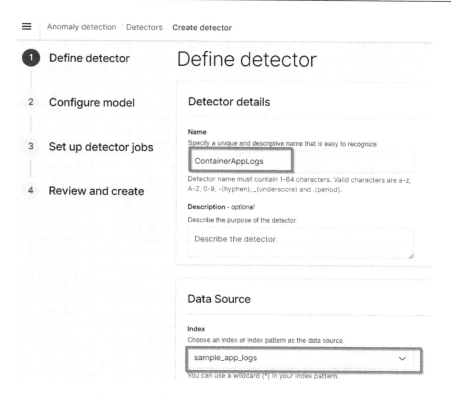

Figure 11.39 – Select Data Source to detect anomalies

3. Set **Timestamp field** to @timestamp:

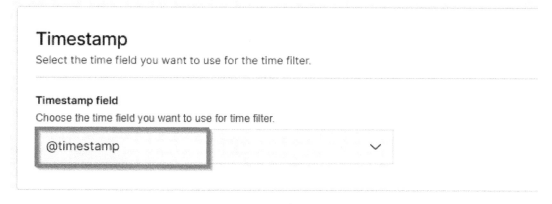

Figure 11.40 – Select the Timestamp field options to use

4. Leave **Operation settings** as the defaults (**Detector interval** as `10` minutes, and **Window delay** as `1`):

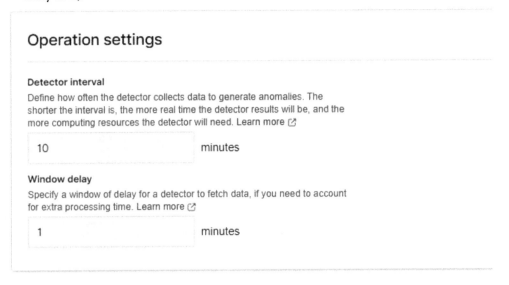

Figure 11.41 – Anomaly detection operation settings

5. Select **Enable custom result index** and add `sample_logs` as the suffix of the index, which will enable creating a custom index to store the output of anomaly detection, as shown here:

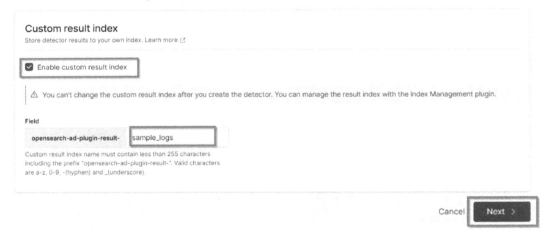

Figure 11.42 – Provide the name for the anomaly detection result index for sample logs

6. Configure the model to include the detection of the count of `log.keyword` received over some time. You need to set **Feature name** to `LogData`, **Aggregation method** to `count()`, and **Field** to `log.keyword`, and then click **Next**:

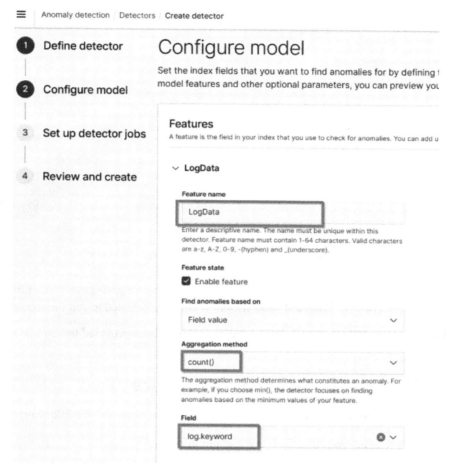

Figure 11.43 – Configure model

7. Enable both real-time detection and historical detection by selecting the **Start real-time detector automatically** and **Run historical analysis detection** checkboxes, then click **Next** to create the job:

Figure 11.44 – Detector job schedule

In the current implementation, we detect the anomalies for the data ingested every 10 minutes and also run anomaly detection for historical data. The limits could be adjusted to suit the requirements.

In the next section, we will look at high-level security functionalities provided by AWS in securing OpenSearch Service.

Security for Amazon OpenSearch Service

Securing Amazon OpenSearch Service at a high level could be classified into the following types:

- **Encryption**: Keeping your data secure at rest and in transit
- **Authentication**: Leveraging authentication infrastructure to authenticate to the OpenSearch domain
- **Authorization**: Granular authorization can be used to control user actions in your cluster
- **Auditing**: Auditing functionality allows you to track and record all user actions, helping you to meet compliance requirements such as the HIPAA and PCI

AWS offers various services to meet the objectives of security in Amazon OpenSearch Service:

- **Encryption**: For encryption of data during transit, you can enable node-to-node encryption and also enforce HTTPS for the web URL using certificates.

  For encryption of data at rest, you can use AWS Key Management Service to store and manage keys. You can create your own or use the one that is provided by AWS. You could protect the data at rest for indices, OpenSearch logs, swap files, automated snapshots, and all other data in the application directory.

- **Authentication**: Authentication to Amazon OSS could be done either using local authentication within the cluster or leveraging your existing authentication infrastructure:

  - **Basic authentication**: Amazon OSS provides a basic authentication service using an internal DB (which we used in the chapter) with a user and password configured in a local DB on the service.

  - **IAM authentication**: You can configure Amazon OSS to use IAM authentication, which uses STS and can integrate with Amazon Cognito.

  - **SAML authentication**: You can configure your third-party identity provider. Authentication will be done by your identity provider and Amazon OSS provides a secure way to integrate with your SAML provider.

- **Authorization**: You can get granular access control to control the user's actions on the OSS cluster:

  - **IAM policy**: IAM policies can be used to provide permissions on the Amazon OSS control plane and data plane. You could also use IAM to deploy as short-term credentials.

  - **Domain policy**: In Amazon OpenSearch Service, the domain policy enables you to define IP address-based permissions for both the control plane and data plane. Similar to the IAM policy, this policy provides a way to control access to the cluster and its resources. With the domain policy, you can specify which IP addresses are allowed or denied access, as well as what actions can be performed by users or applications associated with those addresses. This level of control is essential for ensuring the security and integrity of your OpenSearch cluster.

- **Auditing**: You can track, end to end, all the user actions using Amazon CloudTrail. Additionally, you can also meet the compliance requirements for the HIPAA and PCI using Amazon Managed OpenSearch Service.

Overall, the security features for Amazon OpenSearch Service can be summarized as follows:

Figure 11.45 – Security features for Amazon OSS

In this section, we have understood various AWS services and options available for securing your Amazon OpenSearch cluster.

Summary

In this chapter, we understood how to set up Amazon OpenSearch Service and the options for deploying OSS in our AWS account. Further, we have implemented instrumentation of end-to-end sample applications for ingesting traces and logs into Amazon OSS and enabled anomaly detection to understand anomalies in the log data ingested. In the final section, we understood the security features available in AWS to secure our Amazon OSS implementation.

We learned how Amazon OSS could be helpful in meeting our organization's observability goals when we are looking for open source solutions as a part of the observability strategy.

In the next chapter, we will look at the Cloud Adoption framework and how AWS observability services will support accelerating cloud adoption.

Questions

1. What are the storage options available in Amazon OSS for optimizing the storage costs in AWS?
2. What is Data Prepper and its use case in Amazon OSS?
3. What are the advantages of using Amazon OSS?
4. What is the algorithm used for anomaly detection in Amazon OSS?

Part 4: Scaled Observability and Beyond

In this final part, we will look at how to scale the observability techniques and practices we have seen so far. You will understand observability's role in an organization's journey to the cloud while applying the best practices described by the Cloud Adoption Framework. We will also learn about the relationship between the Well-Architected Framework and the Management and Governance Lens to reach operational excellence. And finally, we will discuss the limits and challenges users face in highly complex environments, and what the future may hold in terms of addressing those complexities.

This section includes the following chapters:

- *Chapter 12, Augmenting the Human Operator with Amazon DevOps Guru*
- *Chapter 13, Observability Best Practices at Scale*
- *Chapter 14, Be Well-Architected for Operational Excellence*
- *Chapter 15, The Role of Observability in the Cloud Adoption Framework*

12

Augmenting the Human Operator with Amazon DevOps Guru

Today's applications are becoming increasingly distributed and complex. We learned in the previous chapters that we need the three pillars of *Metrics*, *Logs*, and *Traces* to achieve good observability. To visualize the data that's been collected, we need dashboards that can correlate data and provide a drill-down view of the application, such as the CloudWatch service map. While this model is effective for less complex systems, as the volume and diversity of data increase, it becomes challenging to identify and troubleshoot issues manually. Developers or administrators may face difficulties in locating and resolving problems as they need to correlate information manually from multiple sources and tools. The constant alerts and notifications from different tools can also lead to alarm fatigue and difficulty in determining the most pressing issue. That's where DevOps Guru steps in and comes to the rescue.

DevOps Guru is a **machine learning (ML)**-powered service that learns from your operational data, such as metrics, application logs, and events, to offer **Artificial Intelligence for IT Operations (AIOps)**-based root-cause analysis. This streamlines the process for developers and operators to identify problems automatically, increasing application uptime and minimizing downtime by linking issues and offering suggestions for a fast resolution. We can achieve all of this without users having any prior ML experience. We discovered the observability landscape of AWS in *Chapter 2, Overview of the Observability Landscape on AWS*, and DevOps Guru is part of the AI and ML insights. In this chapter, we are going to do the following:

- Get an overview of Amazon DevOps Guru for enhanced application availability
- Understand RDS database performance issues using Amazon DevOps Guru
- Review Amazon DevOps Guru insights for resources in AWS
- Understand other AI and ML insight services in AWS

Technical requirements

To follow along with this chapter, you will need the following:

- A working AWS account

- A fundamental understanding of AWS Lambda and Amazon DynamoDB

- An understanding of Amazon RDS and basic performance issues

- A fundamental understanding of Python and running Python scripts

The source code for this chapter can be downloaded from `https://github.com/PacktPublishing/AWS-Observability-Handbook/tree/main/Chapter12`.

Overview of Amazon DevOps Guru

Amazon DevOps Guru is an easy-to-use service with no configuration or no prior ML experience requirements for delivering anomaly detection and observability insights. It helps with continuously analyzing the streams of metrics and logs from disparate data sources and understanding the application's behavior in an automated way by leveraging ML. Amazon DevOps Guru helps in accelerating the resolution of issues quickly by providing ML-powered insights and recommendations. It also helps reduce alarm fatigue by automatically correlating and grouping related anomalies. It is easy to scale and maintain with minimum or no intervention when new AWS resources are added.

DevOps Guru Insights, a component of the DevOps Guru service, automates the process of setting alarms and thresholds and provides clear, actionable guidance to aid developers and operations teams in quickly identifying and addressing the underlying cause of an issue.

The DevOps Guru console provides five different views:

- **Dashboard**: A simple interface that provides a summary of the overall system health, which consists of a summary of the number of resources analyzed, the impacted cloud formation stacks, and ongoing reactive and proactive insights.

- **Insights**: DevOps Guru Insights offers both *Reactive* and *Proactive* insights into issues affecting your analyzed resources. Reactive insights provide recommendations for improving the performance of your application and reducing the **mean time to recover** (**MTTR**), which is the average time it takes to fix a problem after it has occurred. Proactive insights provide you with potential issues that may affect your application in the future, allowing you to address them before they cause disruptions.

- **Settings**: Settings allows you to set up Amazon DevOps Guru either at the organization level for resources if you would like to do so in a multi-account environment or at the current account level. You can also integrate the DevOps Guru insights with AWS System Manager OpsCenter and enable log anomaly detection to provide insights about logs, not just metrics and traces.

- **Analyzed resources**: The **Analyzed resources** view in Amazon DevOps Guru provides an overview of the analyzed resources and shows their estimated cost.

- **Integrations**: The **Integrations** view will help you if you would like to understand the code-level performance issues using the Amazon CodeGuru profiler and integrate the findings into Amazon DevOps Guru:

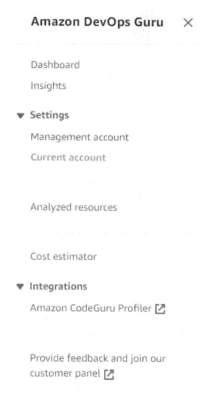

Figure 12.1 – View of Amazon DevOps Guru

The appearance of the service and the views available when accessing DevOps Guru in your account may differ from the views shown in *Figure 12.1*. These variations are determined by the organization-level configuration, which part of the multi-AWS account management settings.

Enabling Amazon DevOps Guru

Amazon DevOps Guru can be enabled by navigating to **Settings**, as shown in *Figure 12.1*, and enabling the service. You can add the resources later as a second step. This process will create the required **Identity and Access Management** (**IAM**) roles and permissions required by DevOps Guru to analyze the resources to be deployed to the current account.

The pricing for analysis in DevOps Guru can be found at `https://aws.amazon.com/devops-guru/pricing`.

Let's enable DevOps Guru and add the necessary resources:

1. Navigate to **Amazon DevOps Guru | Settings | Current account**:

Figure 12.2 – Getting started with DevOps Guru

2. Select **Monitor applications in the current AWS account**:

Figure 12.3 – Selecting the scope of resources to be monitored by DevOps Guru

3. DevOps Guru will automatically create and display **Amazon DevOps Guru_Role** with the required permissions to evaluate AWS resources:

Figure 12.4 – Automated IAM role creation

4. In the next section, *Reviewing Amazon DevOps Guru insights for serverless applications in AWS*, we will enable the analysis of resources. For now, let's skip the resources to analyze by selecting **Choose later**, as shown in the following figure, for Amazon DevOps Guru analysis coverage:

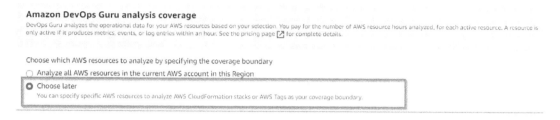

Figure 12.5 – Choose later

5. Click on **Enable** to enable the DevOps Guru service in your AWS account for usage:

Figure 12.6 – Enable

Once enabled, you will see a message stating **Amazon DevOps Guru has been successfully enabled for this account**, and you will also see integration with **Amazon CodeGuru Profiler** visible in the view:

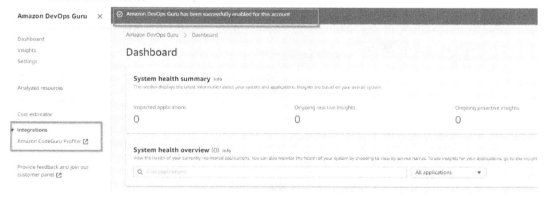

Figure 12.7 – A view of the DevOps Guru dashboard

Analyzing resources using Amazon DevOps Guru

Amazon DevOps Guru lets you discover and analyze resources across different boundaries based on the properties shown in the following diagram:

Figure 12.8 – DevOps Guru boundaries for resource discovery and analysis

Let's determine the optimal locations to enable each of these boundaries:

- **CloudFormation Stacks**: When you have adopted **Infrastructure as Code (IaC)** and you deploy your application as a CloudFormation stack, DevOps Guru can be utilized to analyze the resources generated by the stack.

- **Tags**: When you have a mix of IaC and console usage to deploy your applications and are also using third-party tools to deploy your infrastructure, you could use tags as a mechanism to discover the resources and onboard them to Amazon DevOps Guru.

- **Account level**: If you would like to analyze all your resources in a **specific region** and aren't worried about missing out on anomalies in any of your workloads, this would be an ideal way to onboard at the account level.

- **Organization level using a management account**: When you are working in a multi-account organizational structure and would like to enable DevOps Guru across the organization in multiple AWS accounts in a specific region, you can use this functionality. This is not visible by default and is available in an organizational view/management account view.

If you would like to enable resources to be analyzed by DevOps Guru, navigate to **Analyzed resources** and click on **Edit analyzed resources**:

Figure 12.9 – Edit analyzed resources

You could select one of the methods to discover based on the CloudFormation stack, tags, or the entire AWS account:

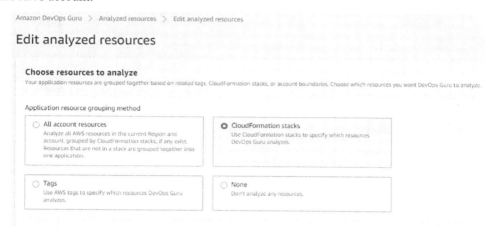

Figure 12.10 – Resource analysis

How DevOps Guru works

Let's understand how Amazon DevOps Guru works. As you deploy applications in your AWS account, you can use one of the options discussed in the *Analyzing resources using Amazon DevOps Guru* section to discover the resources and analyze them. Amazon DevOps Guru analyzes anomalies based on ML models by applying them to the data stores generated by the application (such as metrics, logs, and traces) to AWS-native services such as AWS CloudWatch and AWS X-Ray, as well as the events from AWS CloudTrail and RDS using RDS Performance Insights. Furthermore, Amazon DevOps Guru uses CloudTrail events to understand stack changes applied as part of CloudFormation and maps them to generate insights.

You can enable the generated insights to be created as an AWS Systems Manager OpsItem, which will help you visualize actions on all insights from a unified console. Alternatively, you can send notifications using SNS and integrate them with additional third-party collaboration tools such as Slack or IT service management systems such as ServiceNow for incident creation as required.

The overall behavior of Amazon DevOps Guru is summarized in the following figure:

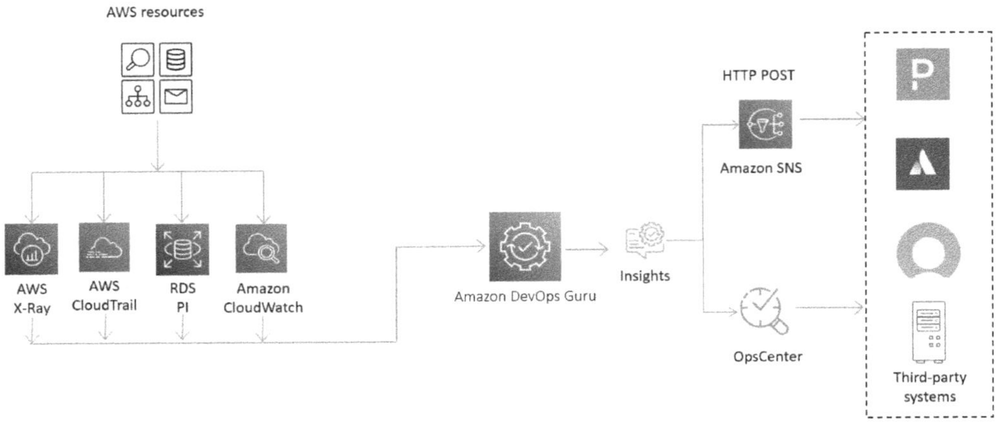

Figure 12.11 – DevOps Guru workflow

In this section, we covered the basics of Amazon DevOps Guru, including its uses and how to enable it in your AWS account. We also discussed the basic configuration for DevOps Guru in an AWS account and the overall workflow and life cycle of the service. Now, we will explore how to deploy a serverless application in an AWS account and see how DevOps Guru provides root cause analysis insights for any issues that may arise.

Reviewing Amazon DevOps Guru insights for serverless applications in AWS

In *Chapter 7, Observability for Serverless Application on AWS*, we deployed a serverless application using a CloudFormation template. We will use the same application to troubleshoot and provide root cause analysis using DevOps Guru. If you have come to this chapter directly, then execute the following Quick Start CloudFormation template to deploy the application and *insert a few records*, as explained in *Chapter 7*: https://console.aws.amazon.com/cloudformation/home#/stacks/new?stackName=serverless-app2&templateURL=https://insiders-guide-observability-on-aws-book.s3.amazonaws.com/chapter-07/final/template.yaml.

We will simulate the load on DynamoDB by continuously reading the records that have been inserted as a part of this example scenario to generate the load on the application and understand reactive insights.

Let's enable DevOps Guru for the CloudFormation's Serverless-app2 stack and try to understand the anomalies and insights generated by DevOps Guru here.

Discovering and analyzing resources

1. Navigate to **Analyzed resources | CloudFormation stacks | Serverless-app2** and click **Save**:

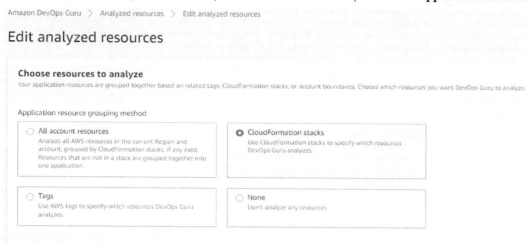

Figure 12.12 – Adding resources to DevOps Guru for analysis

2. It will prompt you for confirmation. Click on **Confirm** to save the changes:

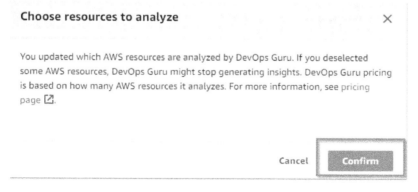

Figure 12.13 – Confirming the changes for analysis

3.   You will need to wait approximately 2-3 hours for DevOps Guru to show the analyzed services/ resource and provide insights:

Figure 12.14 – Resources analyzed

The list of resources in the **Analyzed resources** view, as shown in *Figure 12.13*, shows that DevOps Guru has discovered the resources that have been deployed by the CloudFormation stack.

Decreasing DynamoDB capacity

Next, let's go ahead and decrease the table capacity for DynamoDB for both **Read capacity** and **Write capacity** to 1:

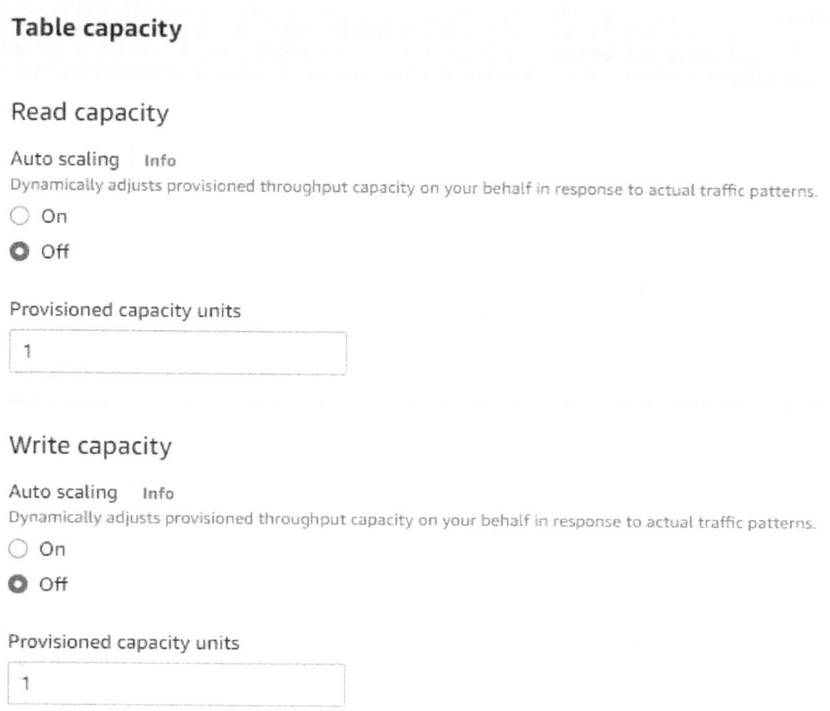

Table capacity

Read capacity

Auto scaling   Info
Dynamically adjusts provisioned throughput capacity on your behalf in response to actual traffic patterns.

○ On

● Off

Provisioned capacity units

1

Write capacity

Auto scaling   Info
Dynamically adjusts provisioned throughput capacity on your behalf in response to actual traffic patterns.

○ On

● Off

Provisioned capacity units

1

Figure 12.15 – DynamoDB capacity reduction

Generating traffic to create anomalies

Now, let's generate some traffic to create anomalies and understand the insights:

1. To generate load and simulate the traffic on Amazon DynamoDB, which is part of the application, you can download the Python script from the following URL or this book's GitHub Chapter12 folder: https://insiders-guide-observability-on-aws-book. s3.amazonaws.com/chapter-12/sendAPIRequest.py.

2. Replace the variable URL in the Python script with your API from the **CloudFormation outputs** section tab, as shown in the following screenshot:

```
#!/usr/local/python/3.3.2/bin/python3.3
#script-version: 1

import requests

url = 'https://replaceme/Prod/items/'

def main():

  #SEND API Requests
  while (True):
      print("\n\n  Iterating sending requests...+++++")
      response = requests.get(url)
      result = response.text
      code = response.status_code
      print (response, result, code)

if __name__ == "__main__":
    main()
```

Figure 12.16 – Replacing the variable URL in the Python script

3. You need to download Python and install it on your computer based on your OS from `https://www.python.org/downloads`. Ensure that you set up environment variables such as `path` on Windows and `env` on macOS/Linux.

4. Once Python has been installed, you can execute the following command to generate traffic:

    `python sendAPIRequest.py`

    Once traffic has been generated for some time (10 minutes), you can view the anomalies in the **Aggregated metrics** section and look at the DevOps Guru-provided recommendations for us to investigate and resolve issues. We have received both proactive and reactive insights for the application.

Reactive insights

To check the reactive insights, do the following:

1. Navigate to **Insights | Reactive** and check for any reactive insights that have been generated.

    You should have one reactive insight describing an issue with the Lambda duration. This is an anomaly that's been detected in the **Serverless-app2** application:

Figure 12.17 – The Reactive insights dashboard

2. When you click on the insights provided, you will see that DevOps Guru has analyzed the metrics and logs and provided an overview of the anomalies and recommendations for remediation, as shown in the following screenshot:

Figure 12.18 – Analysis carried out by DevOps Guru

3. DevOps Guru captured the timeline of events that occurred as a part of the analysis. You can see the CloudTrail events in the following screenshot:

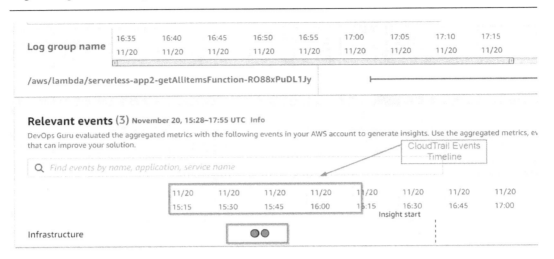

Figure 12.19 – CloudTrail events timeline view

Based on the analysis, DevOps Guru provided three recommendations for **Serverless-app2**:

- **Rollback the Amazon Dynamo DB table update** capacity changes based on the event changes reported as part of CloudTrail:

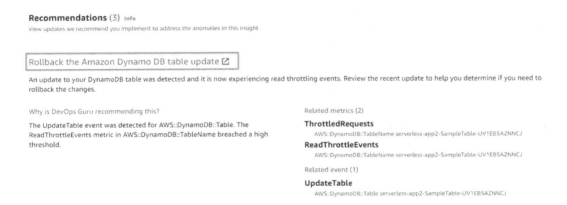

Figure 12.20 – Recommendation 1 – Rollback the Amazon Dynamo DB table update

As we have decreased the capacity of DynamoDB from 5 to 1, it is recommended that we revert the changes to the earlier state.

- **Troubleshoot throttling in Amazon DynamoDB**:

Troubleshoot throttling in Amazon DynamoDB ⬀

Read operations, write operations, or both on your DynamoDB table are being throttled. To learn how to fix throttle events, see Troubleshoot throttling in Amazon DynamoDB ⬀ .

Why is DevOps Guru recommending this?

The ReadThrottleEvents metric in DynamoDB breached a high threshold.

Related metrics (3)

ThrottledRequests

⊞ Show more resources

DynamoDB serverless-app2-SampleTable-UV1EB5AZNNCJ

Duration

AWS::Lambda::FunctionName serverless-app2-getAllItemsFunction-RO88xPuDL1Jy

ReadThrottleEvents

DynamoDB serverless-app2-SampleTable-UV1EB5AZNNCJ

Figure 12.21 – Recommendation 2 – Troubleshoot throttling in Amazon DynamoDB

DevOps Guru has recommended resizing the throttling capacity for DynamoDB and included a knowledge article on how to achieve that.

- **Configure provisioned concurrency for AWS Lambda**:

Configure provisioned concurrency for AWS Lambda ⬀

Your Lambda function is having trouble scaling. To learn how to enable provisioned concurrency, which allows your function to scale without fluctuations in latency, see Configure provisioned concurrency for AWS Lambda ⬀ .

Why is DevOps Guru recommending this?

The Duration metric in Lambda breached a high threshold.

Related metric (1)

Duration

Lambda serverless-app2-getAllItemsFunction-RO88xPuDL1Jy

Figure 12.22 – Recommendation 3 – Configure provisioned concurrency for AWS Lambda

DevOps Guru has recommended increasing the provisioned concurrency for the Lambda function since it has been left at its default concurrency.

When you navigate to **Graphed anomalies**, you will see the relevant metrics and anomalies:

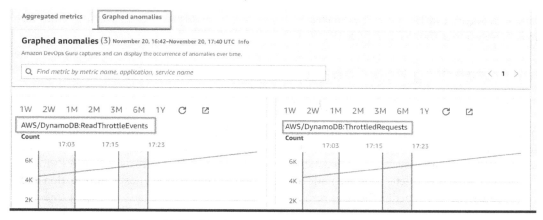

Figure 12.23 – Metric anomalies detected by DevOps Guru

Proactive insights

DevOps Guru also provides proactive insights. In this example, it has provided the recommendation to *enable DynamoDB point-in-time recovery for the DynamoDB*:

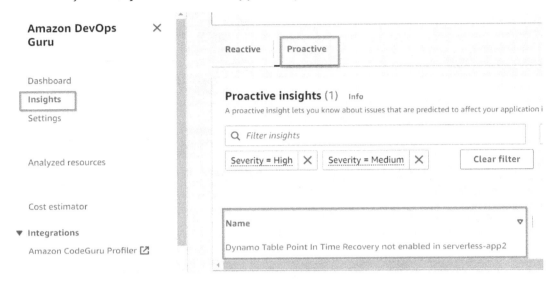

Figure 12.24 – Proactive insights

With no configured thresholds or correlation, it's apparent that Amazon DevOps Guru can identify the likely root cause of the problem and offer suggestions for resolving it.

In this section, we learned how DevOps Guru can help provide root cause analysis by examining the metrics and logs generated by the underlying services in your application stack and generating ML-based insights that enable you to take quick action.

In the next section, we'll explore the database Performance Insights provided by DevOps Guru for Amazon **Relational Database Service (RDS)**.

Understanding Relational Database Service (RDS) performance issues using DevOps Guru

Amazon DevOps Guru for RDS is an ML-powered capability that assists developers and DevOps engineers in detecting, troubleshooting, and resolving issues related to Amazon RDS. DevOps Guru for RDS identifies issues related to databases, such as excessive resource usage, suggests index creation for certain keys, detects problematic SQL queries, and delivers diagnostic information and recommendations that expedite the issue resolution process.

Enabling **Performance Insights** on RDS is a prerequisite for Amazon DevOps Guru to provide database performance analysis. **Performance Insights** is a feature in RDS that provides database performance tuning insights and a detailed view of which SQL statements are causing the load, along with key performance metrics such as the active transaction count, deadlocks, and more. The following screenshot shows an example **Performance Insights** dashboard for an AWS RDS instance:

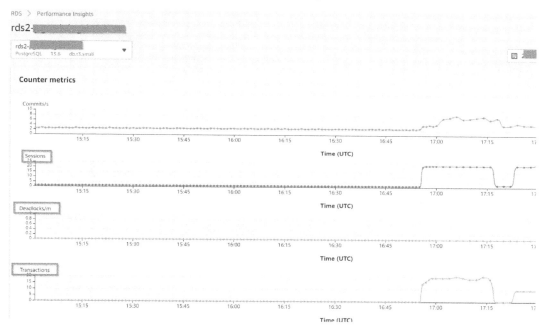

Figure 12.25 – Metric view in RDS Performance Insights

While we did not specifically explore an example of deploying an RDS database and analyzing it using DevOps Guru in this chapter, the process is identical once you have an RDS database in your AWS account. You can enable DevOps Guru for RDS databases using one of the methods described in the *Analyzing resources using Amazon DevOps Guru* section.

Once you have enabled it, DevOps Guru will collect, baseline, and understand the performance issues that are impacting your application and provide recommendations for resolving any issues. DevOps Guru reports two different types of anomalies based on the Performance Insights metrics:

- **Casual anomalies**: A casual anomaly is a top-level anomaly with insight into, for example, database load.

- **Contextual anomalies**: A contextual anomaly is a finding within the database load such as the instance size being small, the CPU capacity being exceeded, and so on, due to which the database is having performance issues.

The following screenshot shows a casual anomaly caused due to the database load:

Figure 12.26 – DevOps Guru DB load insights

This casual anomaly is high database load, while the contextual anomaly is the high CPU utilization metric of the database load. The analysis also suggests a solution – upgrading the RDS database instance to the next size:

Figure 12.27 – Database recommendation from DevOps Guru

In this section, we learned that Amazon DevOps Guru can be utilized to identify performance issues in Amazon RDS and offer suggestions to optimize your application's databases by utilizing insights from RDS Performance Insights.

AI and ML insights

In *Chapter 2*, *Overview of the Observability Landscape on AWS*, we discussed two additional services alongside Amazon DevOps Guru in terms of AI and ML insights: Amazon CodeGuru and Amazon Lookout for Metrics. Let's briefly look at these two services since we don't have an explicit chapter concerning them.

Amazon CodeGuru

Amazon CodeGuru is comprised of two different services: **Amazon CodeGuru Reviewer**, which is a static analysis tool that helps improve code quality by scanning for critical issues, identifying hard-to-find bugs, and recommending how to remediate them, and **Amazon CodeGuru Profiler**, which helps developers visualize their application to find the most expensive lines of code that impact application performance.

The relationship between these two services in terms of the software flow can be understood as follows:

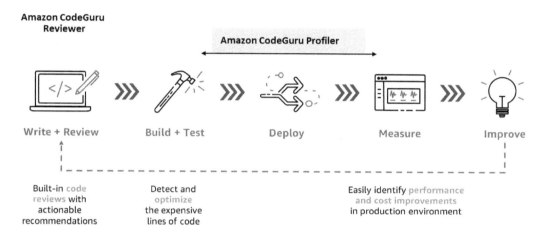

Figure 12.28 – Amazon CodeGuru – Software Lifecycle

While writing code, you can use CodeGuru Reviewer to provide recommendations about code quality, and you can use Amazon CodeGuru Profiler while building, deploying, and measuring the performance of your code to understand the expensive lines that will impact application performance.

You can integrate the recommendations from Amazon CodeGuru Profiler into DevOps Guru, as shown in the following figure:

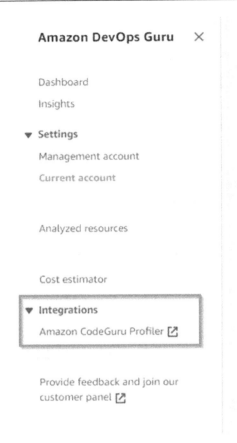

Figure 12.29 – CodeGuru Profiler integration with DevOps Guru

The integration between these two services allows DevOps teams to leverage CodeGuru's expertise in identifying performance issues in code and use DevOps Guru's ML capabilities to quickly detect anomalies and provide actionable insights to improve the application performance. This integration enables teams to quickly diagnose performance issues and apply best practices to optimize the performance of their applications. Overall, this integration helps teams save time, increase efficiency, and improve the reliability and availability of their applications.

Amazon Lookout for Metrics

Amazon Lookout for Metrics provides a similar service to DevOps Guru, but from the perspective of business metrics for different datasets. The life cycle of Amazon Lookout for Metrics can be seen in the following figure:

Create a detector

A detector monitors your dataset, finds anomalies and analyzes their impact.

Add a dataset

Select the data that you want to monitor.

Activate detector

Activate the detector to start monitoring the data for anomalies.

Add alerts - *optional*

Send automated anomaly alerts to Lambda functions, Webhooks, cloud applications like Slack, PagerDuty, and DataDog, or to SNS topics with subscribers that use SMS, email, or

Figure 12.30 – Stages in Amazon Lookout for Metrics

Let's look at these stages in more detail:

1. **Create a detector**: Detectors are ML models that find outliers in your data. First, you set up a detector and configure the anomaly detection interval so that it meets your use case.

2. **Add a dataset**: Lookout for Metrics supports analyzing different types of datasets. It supports over 19 data sources, including S3, CloudWatch, Salesforce, ServiceNow, Marketo, and others.

3. **Activate detector**: When you're ready, activate the detector to begin data analysis. You can view the detector's progress in real time.

4. **Set up alerts**: Setting up alerts is optional and can be done based on your requirements.

Configuring Amazon Lookout for Metrics

In Amazon Lookout for Metrics, creating a detector is a crucial step that enables you to monitor your metrics effectively. By creating a detector, you can specify the data source, time range, and frequency of data ingestion to generate anomaly detection results. This process involves specifying the data to analyze, such as metrics from Amazon CloudWatch, and setting up the data frequency to ensure that the detector is continually updated with new data:

1. Navigate to **Amazon Lookout for Metrics | Create detector**:

Figure 12.31 – Creating a detector in Amazon Lookout for Metrics

2. Set the name of the detector to my-detector1 and set **Interval** to **5 minute intervals**:

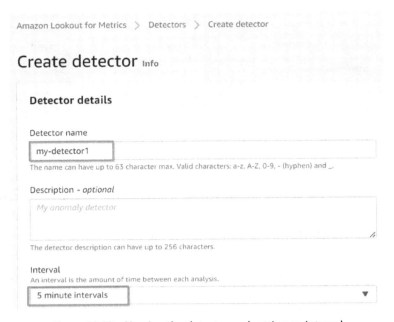

Figure 12.32 – Naming the detector and setting an interval

3. Then, click **Create**:

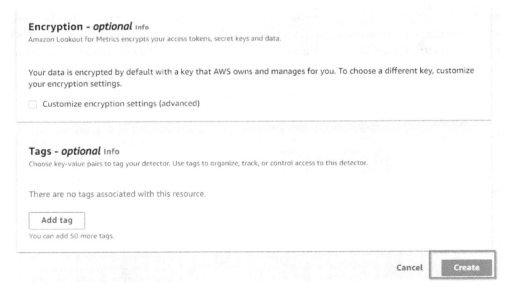

Figure 12.33 – Creating a detector

Adding a dataset

The next step is to add the dataset. Amazon Lookout for Metrics supports different datasets. For this example, we will use CloudWatch as the dataset:

1. Click **Add a dataset**:

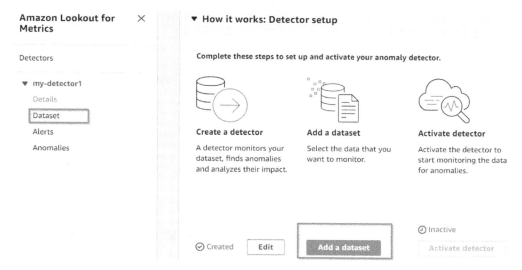

Figure 12.34 – Adding a dataset to the detector

2. Call the dataset `my-dateset1`:

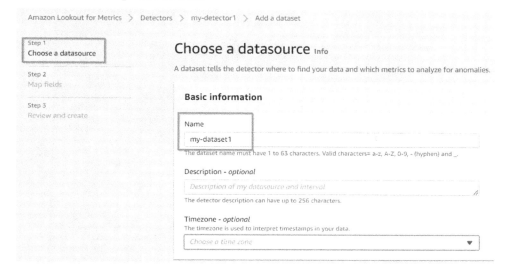

Figure 12.35 – Naming the dataset

3. Set **Datasource** to **Amazon CloudWatch** and set **Detector mode** to **Continuous** to monitor anomalies in real time:

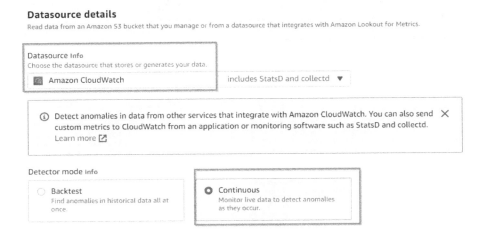

Figure 12.36 – Configuring Datasource and selecting a Detector mode option

4. Select **Create and use a new service role** so that you have access to CloudWatch data for Lookout for Metrics. Then, click **Next**:

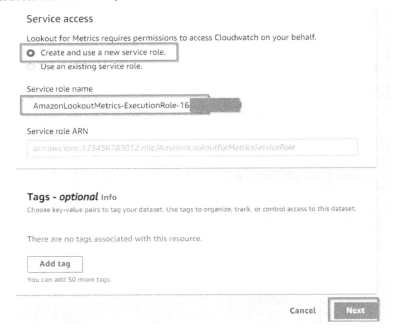

Figure 12.37 – Setting up an IAM role to gain access to CloudWatch data

It will take some time to create the role and load the data:

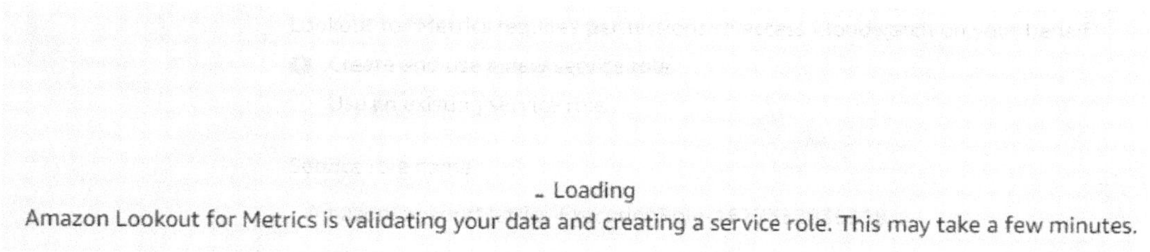

_ Loading

Amazon Lookout for Metrics is validating your data and creating a service role. This may take a few minutes.

Figure 12.38 – Validating the data and creating an IAM role

5. Select **AWS/EC2** as the metric's **Namespace** and select **InstanceId** under **Dimensions**. If you don't see the **AWS/EC2** namespace, then you must create an EC2 instance (https://docs. aws.amazon.com/AWSEC2/latest/UserGuide/EC2_GetStarted.html):

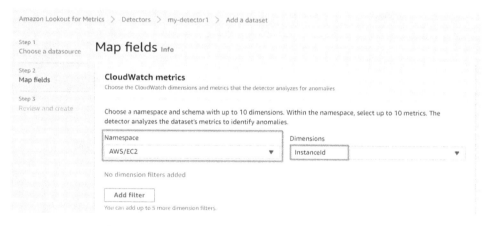

Figure 12.39 – Selecting a CloudWatch metrics Namespace

6. Select **CPUUtilization** under **Metric** and **AVG** under **Aggregation Function**. We are analyzing anomaly detection when the average CPU utilization is beyond the specified boundaries:

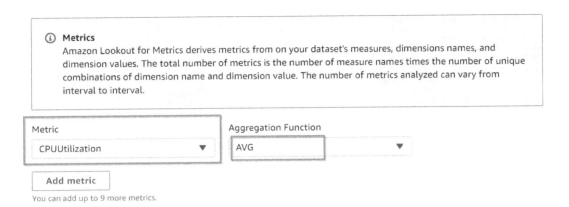

Figure 12.40 – Selecting Metric and Aggregation Function

7. Ensure you understand the cost of Amazon Lookout for Metrics for the selected metrics and click **Save dataset**:

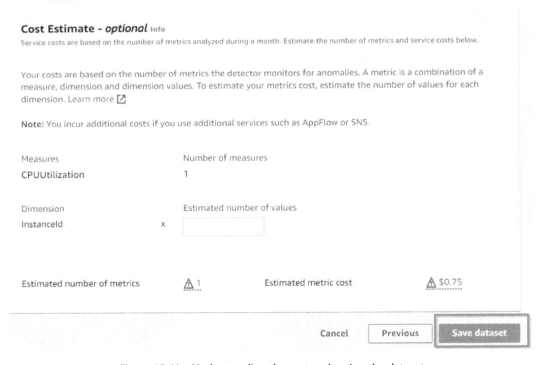

Figure 12.41 – Understanding the cost and saving the dataset

Next, you activate the detector to understand the anomalies in the data in real time:

1. Click on **Activate detector**:

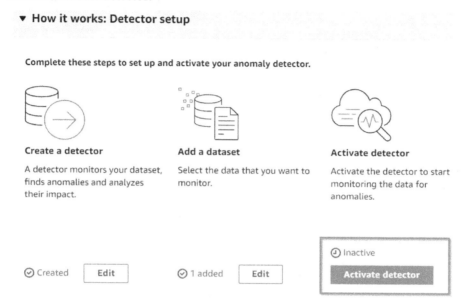

Figure 12.42 – Activate detector

2. Select **Activate**:

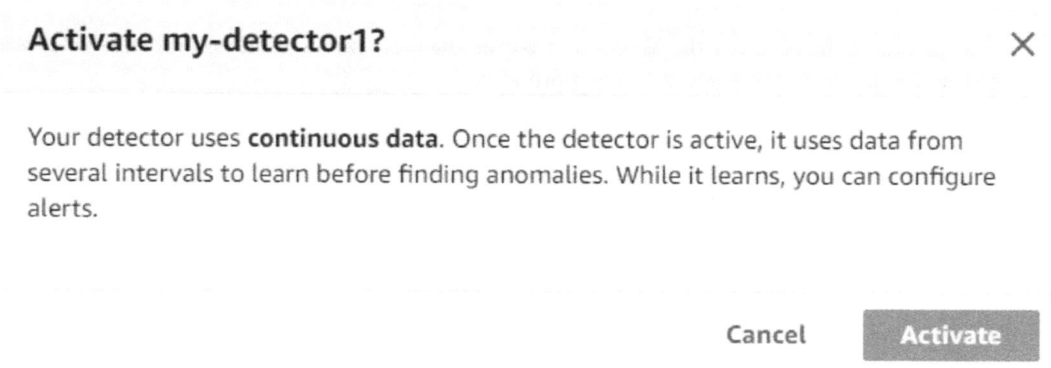

Activate my-detector1?   ✕

Your detector uses **continuous data**. Once the detector is active, it uses data from several intervals to learn before finding anomalies. While it learns, you can configure alerts.

Cancel   Activate

Figure 12.43 – Activating the detector

The detector will import the selected metric data and activate the detector so that it can understand anomalies:

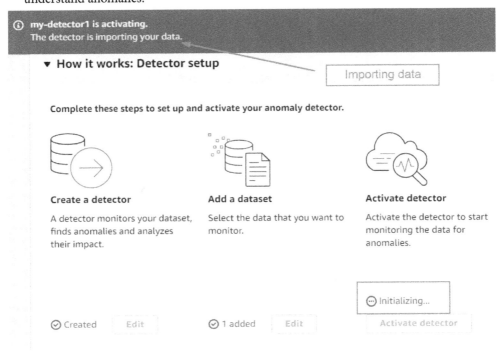

Figure 12.44 – Initializing and activating the detector

3. You can adjust the anomaly threshold to fine-tune the recommendations. Simply navigate to **Anomalies** and adjust the **Severity score** property:

Figure 12.45 – Fine-tuning anomaly threshold

Some of the use cases that you can leverage Amazon Lookout for Metrics for include web response times, a sudden increase/decrease in the number of orders, and more.

In this section, we learned about Amazon Lookout for Metrics and how to configure it and understood its use cases.

Summary

In this chapter, we examined the difficulties faced in modern application operations and explored how Amazon DevOps Guru, with its AIOps capabilities, can aid in addressing these challenges. We covered the process of activating Amazon DevOps Guru in your AWS account and the available options. We also delved into identifying performance issues in serverless applications and how DevOps Guru can assist in resolving these issues via AWS's RDS. We saw how the tool's recommendations can facilitate faster resolution of database problems.

We also delved into Amazon CodeGuru and its integration with DevOps Guru. In addition, we examined Amazon Lookout for Metrics and its various use cases.

In the next chapter, we will explore best practices for observability in a multi-account, multi-region environment at scale. Businesses operating across multiple geographic regions face challenges in managing a distributed system, making it crucial to have a comprehensive observability strategy in place to quickly detect and diagnose issues, regardless of the region or account in which they occur, by monitoring key metrics, logs, and events across all regions and accounts.

Questions

Answer the following questions to test your knowledge of this chapter:

1. What is AIOps?
2. How can Amazon DevOps Guru help in identifying issues in applications?
3. What are Performance Insights in RDS?
4. What options are available in AWS so that you can enable DevOps Guru for your workloads?

13

Observability Best Practices at Scale

As organizations adopt cloud technology, they must have a comprehensive observability strategy in place to ensure their applications are running smoothly and efficiently. In this chapter, we will look into observability best practices when applications are at scale and are spread across multiple **Amazon Web Services** (**AWS**) accounts and Regions. As organizations adopt multi-account and multi-Region strategies, this will create additional complexies in terms of observability and troubleshooting. In this chapter, we will take a deep dive into the various aspects of managing observability solutions in such a scenario.

In this chapter, we will look into these main topics:

- Observability best practices at scale
- Understanding cross-account cross-Region CloudWatch
- Beyond observability

Observability best practices at scale

An organization may adopt a multi-account and multi-Region strategy to improve security, manage costs, comply with regulations, prepare for disaster recovery, reduce latency, and improve performance. By creating separate accounts for different parts of the organization, sensitive data can be isolated, reducing the risk of a data breach and improving security. A multi-Region strategy helps organizations meet compliance requirements and ensure business continuity in the event of an outage or disaster. This strategy also helps to minimize the impact or **blast radius** of any potential security incidents or failures by limiting their scope to a specific account or Region. Let's look at the common deployment model of applications in multi-account, multi-Region topologies and how organizations manage them.

Understanding multi-account and multi-Region topologies

Let's have a look at how multi-account and multi-Region topologies are created in AWS. **AWS Organizations** is a powerful tool for managing multiple AWS accounts, and it can help customers simplify AWS management, improve their security and compliance posture, and better manage their costs. Whether you are a small business or a large enterprise, AWS Organizations is an essential tool for managing your AWS resources effectively and efficiently. AWS Organizations helps customers to manage multiple AWS accounts from a single parent account, called the root account, and then manage those accounts as an individual entity. This helps customers simplify multi-account management and ensures that all their AWS accounts are aligned with their overall IT governance policies.

The AWS Organizations service helps you manage multiple AWS accounts simultaneously. The organization is made up of two types of accounts: a management account that creates and has complete control over the organization and one or more member accounts that can either join the organization or be created within it. You can visualize the management and member accounts in the organization, as shown in *Figure 13.1*:

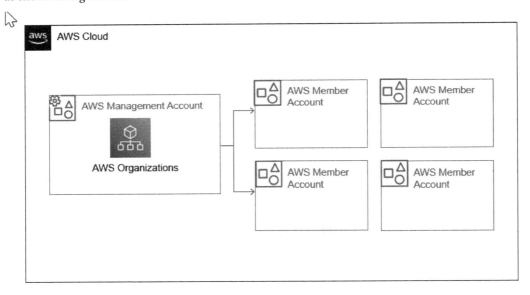

Figure 13.1 – Management account and member accounts in AWS Organizations

We can arrange the accounts in a tree-like structure with root and organizational units. The AWS Organizations console allows us to view the details and policies of the accounts in the organization. To access a member account, you can switch **identity access management** (**IAM**) roles in the AWS Management console using IAM user credentials. You can visualize the tree structure in AWS Organizations in *Figure 13.2*:

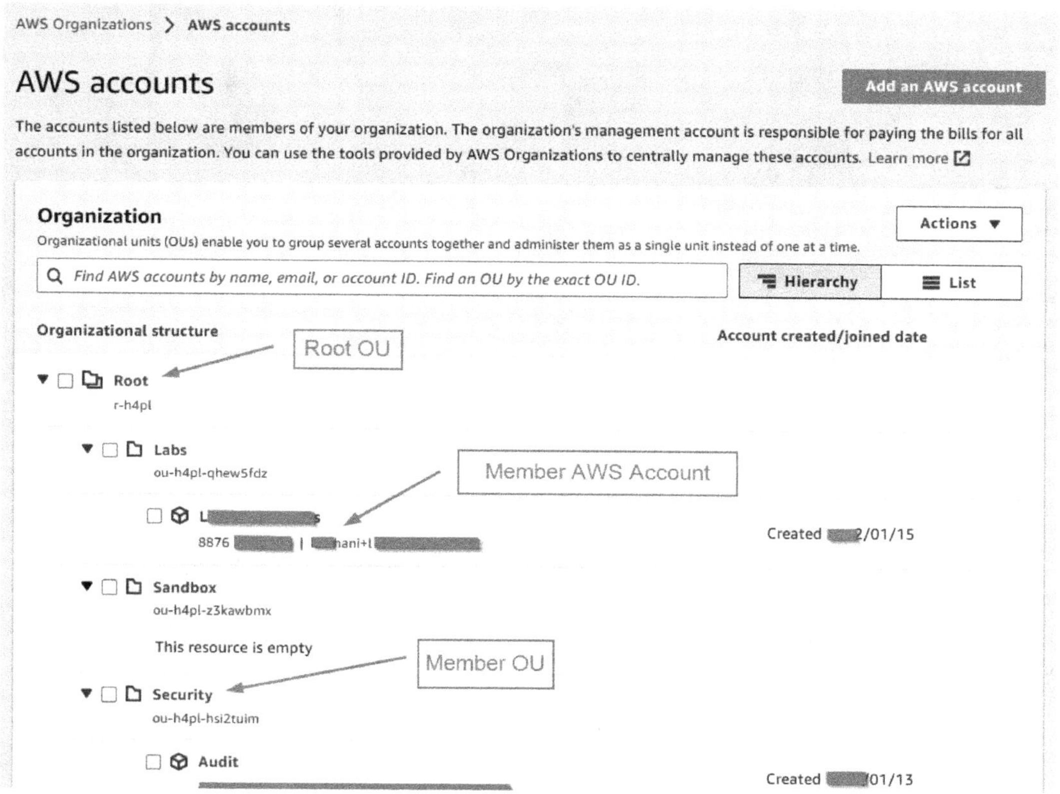

Figure 13.2 – AWS Organizations tree view in the AWS Console

The number of **organizational units** (**OU**) and the number of AWS accounts in each OU depends on the segregation requirements for applications and the hierarchical structure of the company leveraging AWS. As a part of the best practices, AWS recommends leveraging **AWS Control Tower**, which provides a pre-configured setup for a secure and compliant multi-account environment on AWS. It automates the landing zone setup, which includes the creation of a multi-account structure, IAM roles and policies, and baselining network and security configuration.

AWS recommends having a separate **Security OU** with one or multiple accounts managing security tooling and auditing/logging accounts, and an **infrastructure OU** with one or more AWS accounts managing the network, operations tooling such as automation, and so on:

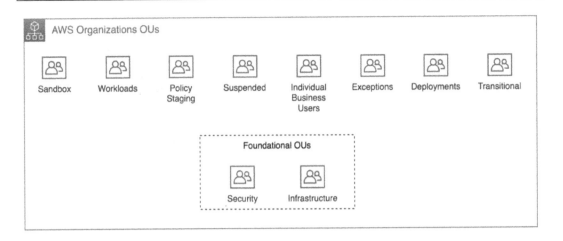

Figure 13.3 - AWS recommendation for OUs

During *re:Invent 2022*, AWS introduced a new feature called **cross-account observability**. With Amazon CloudWatch cross-account observability, you can monitor and troubleshoot applications that span multiple accounts within a Region. You can search, visualize, and analyze your metrics, logs, and traces in any of the linked accounts without account boundaries. However, the placement of this unified account, called a **monitoring account**, in an AWS Organizations structure has not been provided. But as per my understanding and experience, it should be under **Infrastructure OU**. A new AWS account should be created with the **Observability/Monitoring** account name and should be used to do cross-account observability when you would like your organization to have a bird's-eye view of all applications metrics, logs, and traces. Also, you can use the observability account/monitoring account to understand the interactions from the applications spread across multiple accounts.

Now that we've established the ideal location for the monitoring/observability account in a multi-account environment, let's explore CloudWatch's cross-account observability feature for AWS Native Services. We'll delve into how to configure it and its limitations.

Exploring CloudWatch cross-account observability

AWS CloudWatch cross-account observability is a feature that allows you to query and monitor resources across multiple AWS accounts for resources in a Region from a single AWS account. This feature simplifies the process of analyzing data across multiple accounts, allowing you to quickly identify and resolve issues.

With CloudWatch cross-account observability, you can search, analyze and visualize cross-application telemetry data, including logs, metrics, and traces, as if you are operating in a single account without any account boundaries.

The benefits of using cross-account observability are as follows:

- **Unified view**: It provides a bird's-eye view across your organization so that you can pinpoint application issues at scale and identify the affected users easily
- **Reduced mean time to resolution (MTTR)**: It will help you reduce the average time to resolve an issue as you can understand distributed tracing in not just a single application but across applications running in multiple accounts
- **Easy to set up**: It is easy to set up in a standalone AWS account scenario or when you are working in a standard AWS Organizations-based multi-account setup

It also comes with no additional cost, as you are querying the data and visualizing from a unified location.

Cross-account CloudWatch observability can be set up in two different scenarios:

- **Multiple individual AWS accounts**
- **AWS Organizations**

Even though it supports both scenarios, AWS recommends using the AWS Organizations' way of setup as part of the multi-account management best practices. As we discussed in the *Understanding multi-account and multi-Region topologies* section, it is best to create an AWS monitoring account in the infrastructure OU and configure a central monitoring account to visualize the CloudWatch data across multiple accounts and multiple Regions. You can learn more about creating organizations and multi-account setup here: https://docs.aws.amazon.com/organizations/latest/userguide/orgs_tutorials_basic.html.

How cross-account observability works

To better understand the concept of a **source account**, it's important to note that this refers to a child account where your application runs. In order to enable cross-account monitoring, each source account must provide permission to query the data from the monitoring account.

You can visualize the relationship between monitoring accounts and source accounts in *Figure 13.4*:

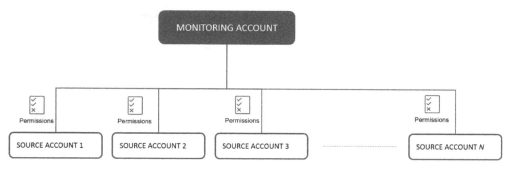

Figure 13.4 – Monitoring account and the source AWS accounts

With this configuration, all of your metrics, logs, and traces will still be stored within your own account, but the monitoring account will have the ability to query and analyze them. This allows you to maintain full control over your data while still benefiting from a comprehensive, centralized monitoring solution.

Configuring CloudWatch cross-account observability

Let's look into how to configure Observability in this monitoring account from AWS CloudWatch Console:

1.   Let's navigate to **CloudWatch | Settings**:

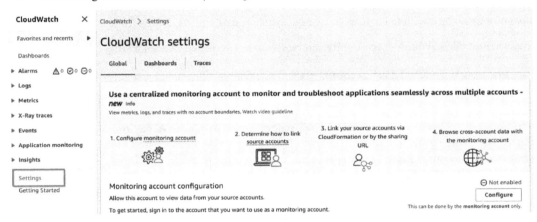

Figure 13.5 – CloudWatch Settings view in a monitoring account

2.   Navigate to the **Global | Monitoring account configuration** option and then click on **Configure**:

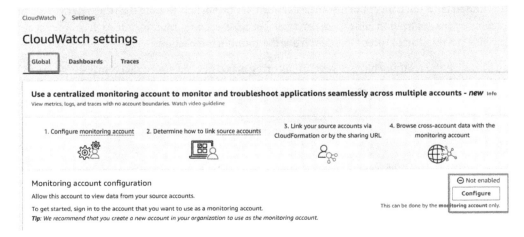

Figure 13.6 – Configuration of the monitoring account

3. Select **Logs**, **Metrics**, and **Traces** and provide the account numbers from which you would like to query the metrics, as shown in *Figure 13.7*:

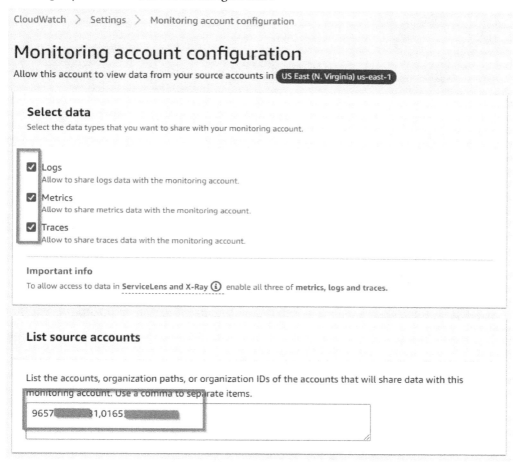

CloudWatch ＞ Settings ＞ Monitoring account configuration

Monitoring account configuration

Allow this account to view data from your source accounts in US East (N. Virginia) us-east-1

Select data
Select the data types that you want to share with your monitoring account.

☑ Logs
Allow to share logs data with the monitoring account.

☑ Metrics
Allow to share metrics data with the monitoring account.

☑ Traces
Allow to share traces data with the monitoring account.

Important info
To allow access to data in **ServiceLens and X-Ray** ⓘ enable all three of **metrics, logs and traces.**

List source accounts

List the accounts, organization paths, or organization IDs of the accounts that will share data with this monitoring account. Use a comma to separate items.

9657████████1,0165█████████

Figure 13.7 – Allow Logs, Metrics, and Traces to query from the monitoring account

We are sharing all three types of golden signals, namely logs, metrics, and traces, with the monitoring account from the **List source accounts** section. In this example, we have listed two source account numbers separated by a comma. Alternatively, you can also use the AWS organization ID.

4. To identify each account that has been added as a source account, you can use the default options for the **Account name** and **$Accountname** fields as variables, as shown in *Figure 13.8*. Then click **Configure**:

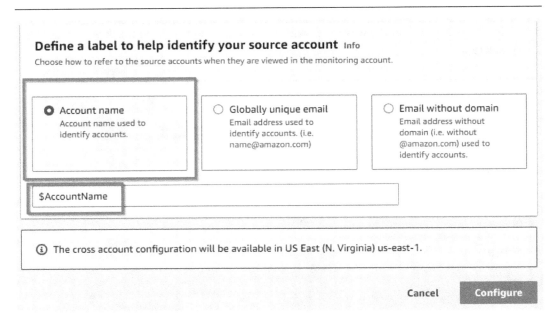

Figure 13.8 – Display name of the source accounts in the monitoring account

You should see a success message saying **You have successfully enabled the monitoring account**:

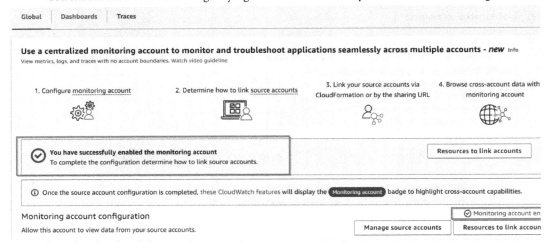

Figure 13.9 – Confirmation message that the monitoring account is successfully configured

5. Once you have enabled the monitoring account feature in your AWS account, you can download the **CloudFormation** template and deploy the same in each AWS account where you would like to share the data with the monitoring account. Do this by navigating to the **Resources to link accounts** tab, as shown in *Figure 13.10*:

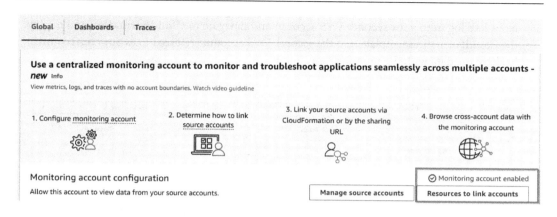

Figure 13.10 – Resources to link accounts

You will have two different options to link the source accounts to the monitoring account. Either using **AWS Organization** or **Any Account**. It is recommended to use **AWS Organizations** when you have multiple accounts used in your organization. We have discussed the placement of the AWS monitoring account in an AWS organization in the *Understanding multi-account and multi-Region topologies* section.

In this exercise, I have a multi-account organization structure, and I am choosing the AWS Organization and downloading the CloudFormation template:

```
1   AWSTemplateFormatVersion: 2010-09-09
2
3   Conditions:
4     SkipMonitoringAccount: !Not
5       - !Equals
6         - !Ref AWS::AccountId
7         - "          "
8
9   Resources:
10    Link:
11      Type: AWS::Oam::Link
12      Condition: SkipMonitoringAccount
13      Properties:
14        LabelTemplate: "$AccountName"
15        ResourceTypes:
16          - "AWS::CloudWatch::Metric"
17          - "AWS::Logs::LogGroup"
18          - "AWS::XRay::Trace"
19        SinkIdentifier: "arn:aws:oam:us-east-1:          :sink/dc6156b4-6d94-49e7-a571-6b4656882cbc"
```

Figure 13.11 – Monitoring account CloudFormation template in an AWS organization

6. Now, let's log in to your second AWS account and navigate to **CloudFormation** and deploy the CloudFormation template into the second AWS account (referred to as **SourceAccount1**) to share the data with the monitoring account:

Figure 13.12 – Deployment of the CloudFormation template in SourceAccount1

Upon successfully deploying the CloudFormation template, you should see the corresponding confirmation message, illustrated in *Figure 13.13*:

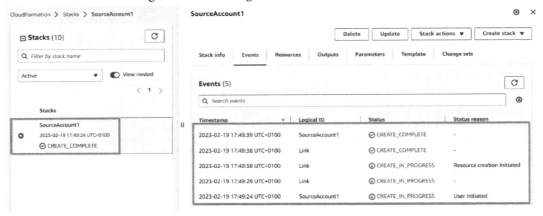

Figure 13.13 – Successful deployment of the CloudFormation template

At this point, you should be able to view and query the data of the source account from the monitoring account.

7. Switch to the AWS monitoring account to see whether the configuration is successful. Navigate to **CloudWatch | Settings | Global** and click on **Manage source accounts**:

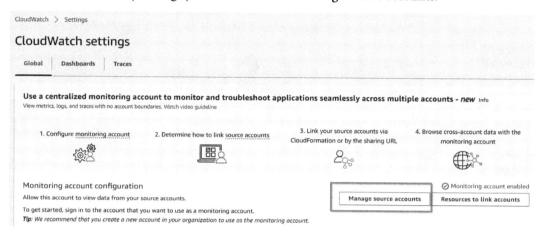

Figure 13.14 – Manage source accounts from the monitoring account

You should be able to see the linked source account in the monitoring account:

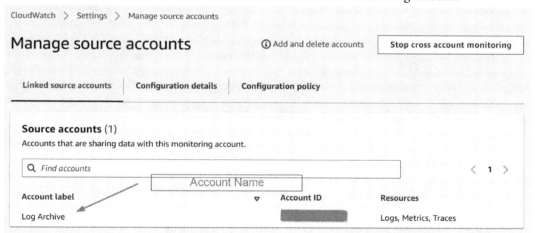

Figure 13.15 – Source account information from the monitoring account

You can repeat the same for multiple AWS accounts in your organization.

Now, let's look into the outcome of the configuration carried out till now:

1. AWS has added a new label to identify the monitoring account, which you should be able to quickly identify when you log in to the **CloudWatch** console of **Monitoring account**, as shown in *Figure 13.16*:

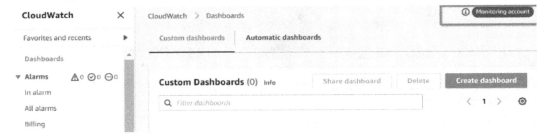

Figure 13.16 – The Monitoring account label in the CloudWatch console

2. If you go to **CloudWatch** | **Metrics** or **CloudWatch** | **Log groups**, you'll be able to view all the metrics and search logs from all the source accounts:

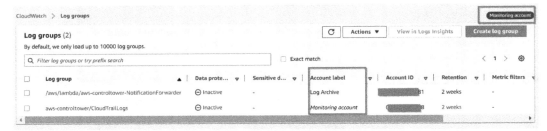

Figure 13.17 – Visibility of CloudWatch Log groups from the monitoring account

Cross-account observability is helpful in scenarios where your applications are spread across multiple accounts but are in the same Region.

It is necessary to repeat the procedure in all Regions to comprehend them all within an AWS account.

In this section, we learned about the need for AWS CloudWatch cross-account observability and how to configure it using the AWS Console in a multi-account environment.

In the next section, we will look into another AWS functionality called **CloudWatch Cross-Account Cross-Region**, which is helpful when you want to query data and build a dashboard from a unified AWS CloudWatch dashboard without switching multiple AWS accounts.

Exploring cross-account cross-Region CloudWatch

Cross-account cross-Region CloudWatch is a feature offered by AWS that enables you to monitor your resources across multiple AWS accounts and Regions from a central location. This feature allows you to access metrics and logs from multiple accounts and Regions within a single dashboard.

This is a two-step process, where you enable cross-account cross-Region functionality in your AWS Organizations master account and then add each source account to share the data with your AWS Organizations monitoring account. Let's look at how to configure this now.

Configuring AWS cross-account cross-Region in AWS Organizations

Step 1: Follow these steps to enable cross-account cross-Region functionality in your AWS Organization master account:

1. First, log in to your AWS Organizations master account.

2. Then navigate to **CloudWatch** | **Settings** | **View cross-account cross-region**, and click **Configure**:

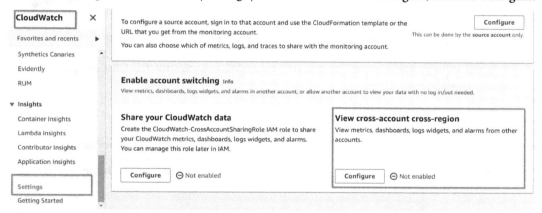

Figure 13.18 – Cross-account cross-region configuration in CloudWatch

3. Select **AWS Organization account selector**, then **Save Changes**:

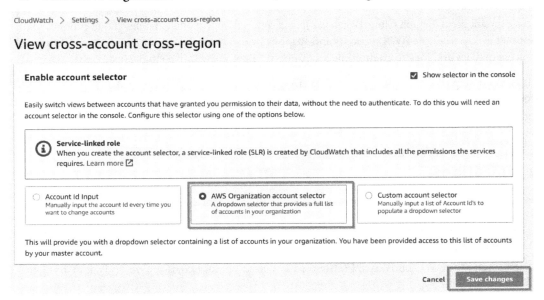

Figure 13.19 – Enabling AWS Organization to share the CloudWatch data

This will allow you to query the data across your AWS organization from the delegated AWS Monitoring account.

Step 2: Add each source account to share the data with your AWS Organizations monitoring account.

As a part of this step, you need to allow each source account to share the data with the monitoring account by deploying a CloudFormation template to allow sharing the data:

1. Log in to the AWS source account (to deploy the CloudFormation template). Select **CloudWatch | Settings** and then select **Configure** under **Share your CloudWatch data**:

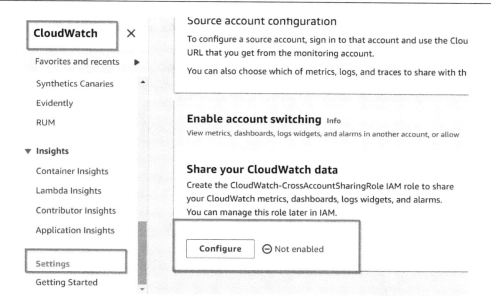

Figure 13.20 – Enabling source accounts to share the CloudWatch data with the monitoring account

2. Provide the AWS account number for the monitoring account from which you would like to query the data:

Figure 13.21 – Providing the AWS monitoring account number to share the data with

3.   Click on **Launch CloudFormation template** to deploy the CloudFormation template to create the required IAM roles:

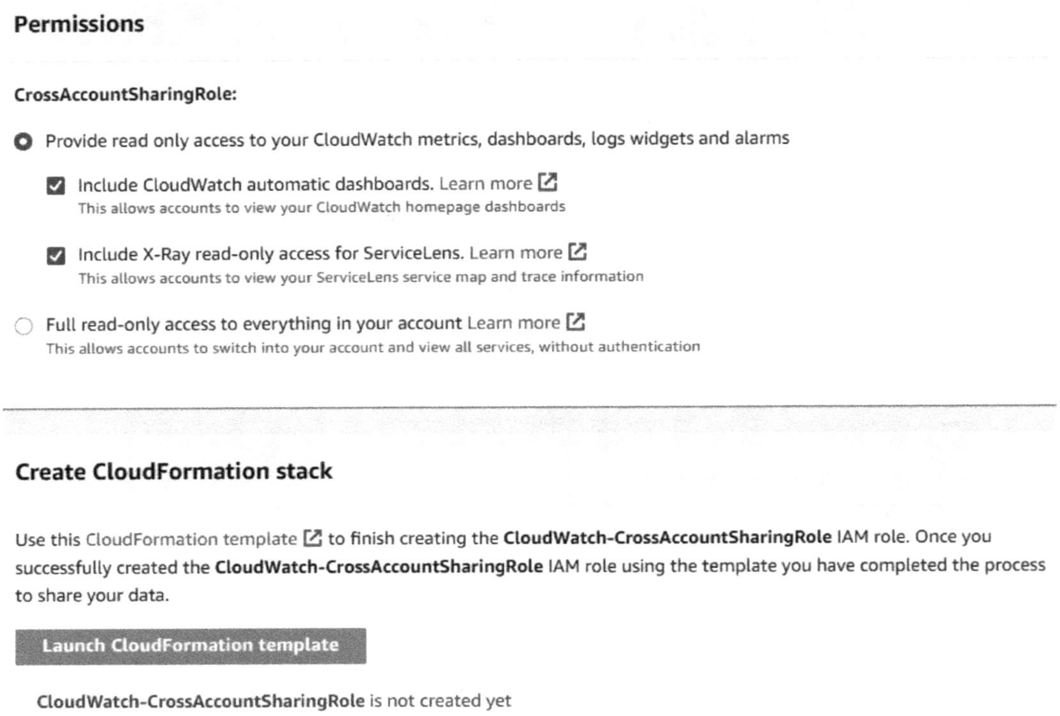

Permissions

CrossAccountSharingRole:

⊙ Provide read only access to your CloudWatch metrics, dashboards, logs widgets and alarms

    ☑ Include CloudWatch automatic dashboards. Learn more ☐
      This allows accounts to view your CloudWatch homepage dashboards

    ☑ Include X-Ray read-only access for ServiceLens. Learn more ☐
      This allows accounts to view your ServiceLens service map and trace information

○ Full read-only access to everything in your account Learn more ☐
  This allows accounts to switch into your account and view all services, without authentication

Create CloudFormation stack

Use this CloudFormation template ☐ to finish creating the **CloudWatch-CrossAccountSharingRole** IAM role. Once you successfully created the **CloudWatch-CrossAccountSharingRole** IAM role using the template you have completed the process to share your data.

[Launch CloudFormation template]

CloudWatch-CrossAccountSharingRole is not created yet

Figure 13.22 – Deploy the CloudFormation template to deploy the IAM role

The deployment of the CloudFormation template is now complete:

Figure 13.23 – Deploy the CloudFormation template to deploy the IAM role

When you switch to your monitoring account, you should be able to visualize drop-down options for account and Region to query the metrics and create dashboards.

You can see in the following screenshot that you can visualize the X-Ray traces in the **US East (Ohio)** Region in the source account from your monitoring account in the **N. Virginia** Region:

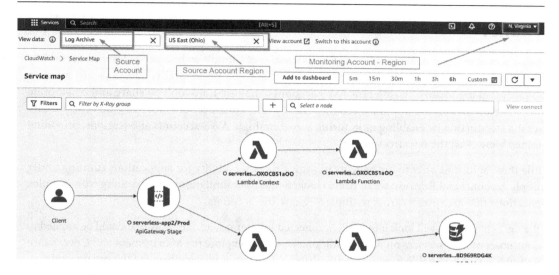

Figure 13.24 – Visualizing X-Ray traces in a cross-account cross-Region scenario

Next, let's look into the gaps in cross-account and cross-Region observability and see what additional consideration would be required for the customers.

Limitations of CloudWatch cross-account cross-Region observability

When dealing with disaster recovery scenarios for applications spread across multiple AWS Regions, the cross-account cross-Region functionality can be a valuable tool for gaining insights into the application's performance. However, it's important to note that this functionality has some limitations, particularly when it comes to end-to-end tracing.

While the cross-account cross-Region functionality can provide metrics and aggregate them across Regions, it may not be sufficient to gain a complete picture of the application's performance. To truly understand how the application functions, you may need to switch to each Region individually to visualize the application trace and query relevant logs.

This means that when using this functionality, it's important to be aware of its limitations and have a plan in place for accessing the necessary logs and traces in each Region as needed. By staying proactive and prepared, you can help ensure that you have the insights you need to effectively manage your application's performance and minimize downtime in the event of a disaster.

Summary

In this chapter, we have explored two key functionalities of CloudWatch: cross-account observability and cross-account cross-Region observability. Cross-account observability allows you to monitor resources that are located in different AWS accounts, making it possible to share and collaborate on monitoring data across multiple accounts. Meanwhile, cross-account cross-Region observability takes it a step further by enabling monitoring across multiple AWS accounts and Regions, providing a unified view of all the resources in use.

While these solutions offer many benefits in achieving observability for applications running under multiple accounts and Regions, we have also looked into their limitations when dealing with complex applications that are spread across multiple Regions and accounts.

In the next chapter, we will look into Well-Architected Framework and see how that could be applied to various observability services on AWS. Additionally, we will explore the Management and Governance Lens and how interoperable functions with observability could provide benefits the organization.

Questions

1. Describe the use of AWS Organizations.

2. What are the requirements of cross-account observability?

3. What are the use cases of cross-account observability?

4. Explain the difference between cross-account and cross-Region observability.

14

Be Well-Architected for Operational Excellence

In the previous chapter, we discussed using observability services from AWS to help you accelerate your cloud adoption journey. In this chapter, we will examine using the **Well-Architected Framework** for observability workloads, such as CloudWatch metrics, Logs, X-Ray traces, and open source managed services from AWS. We will also learn about best practices according to the AWS Well-Architected Framework pillars and about tools from AWS and the community that can help with adopting best practices for optimizing observability workloads and achieving the desired business outcomes.

Observability is one of the important parts of achieving the Well-Architected Framework for cloud workloads. Designing an observability toolset or observability solution on AWS that adheres to the best practices outlined in the the Well-Architected Framework requires consideration of multiple aspects.

In this chapter, we will understand how to apply the Well-Architected Framework for your observability solutions on AWS and look at cost optimization techniques for observability solutions. Then, we will dig deeper to understand how observability is an important part of your cloud journey and understand interoperable functions of observability through a management and governance lens.

This chapter covers the following main topics:

- AWS' Well-Architected Framework
- Applying the Well-Architected Framework and exploring automated solutions
- Understanding management and governance in the Well-Architected Framework

Technical requirements

You will need to have a working AWS account to look into the concepts being discussed and verify them as we progress in the chapter.

If you are an architect and looking to design a solution for observability, it is always a good idea to look into the AWS Well-Architected Framework, which provides you with the foundations of best practices to follow when designing a workload to run in AWS. Let's understand the foundations of the Well-Architected Framework before we go into the practice of implementing them for observability solutions on AWS.

An overview of the AWS Well-Architected Framework

The AWS Well-Architected Framework is a set of guiding tenets that will help you improve the quality of the workload. It consists of six pillars, namely **Operational Excellence**, **Security**, **Reliability**, **Performance Efficiency**, **Cost Optimization**, and **Sustainability**. You can see the six pillars in the following figure:

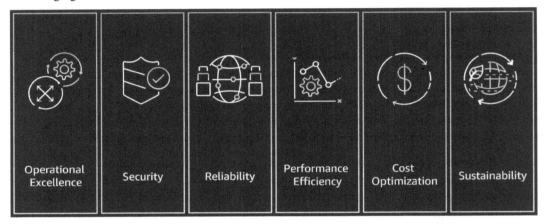

Figure 14.1 – AWS Well-Architected Framework pillars

Now, let's understand the fundamental concepts of each pillar in the AWS Well-Architected Framework:

- **Operational Excellence:** This pillar centers on optimizing operational efficiency through effective workload monitoring and the ongoing improvement of processes and procedures. To attain operational excellence, we will explore how an *operations-first* approach can be incorporated into the design of an observability framework to achieve the desired state.

- **Security**: This pillar focuses on protecting information and systems. To achieve the necessary security measures for observability workloads, we will explore account management and separation considerations for observability workloads and also discuss how to operate observability workloads such as CloudWatch Logs encryption securely as a part of achieving security goals for observability workloads.

- **Reliability**: This pillar focuses on restoring applications to full functionality in the event of failure. To ensure reliability for observability applications running on instances such as Amazon OpenSearch Service, we will examine how to prioritize reliability and achieve this goal for observability workloads.

- **Performance Efficiency**: This pillar provides guidance on efficiently scaling architecture to meet users' demands. To prioritize performance optimization, we will explore adopting various AWS-native solutions.

- **Cost Optimization**: This pillar provides guidance on optimizing workload costs. We will explore the available cost optimization techniques and tools available to optimize the cost of the observability workloads running on AWS.

- **Sustainability**: This pillar focuses on the environmental impact of the workloads. We will explore how adopting AWS observability solutions can minimize the environmental impact and investigate various scenarios available in designing observability strategies to be more sustainable.

You can read more about the Well-Architected Framework and the best practices to follow for your workloads in the documentation at `https://aws.amazon.com/architecture/well-architected/`.

Now that we have an overview of the Well-Architected framework, let's dig deeper and explore how to apply it based on *my experience* with observability on AWS!

Applying the Well-architected framework and exploring automated solutions

Let's understand how you can apply the Well-architected framework for your cloud observability solution and understand the automated solutions available to meet the business requirements, such as high availability, resiliency, and failover mechanisms.

Operational excellence

The **Operational Excellence** design principles focus on five major components: *perform operations as a code*; *make frequent, small, and reversible changes*; *refine operations procedures frequently*; *anticipate failures*; and *learn from all operational failures*. I suggest you look into the details of the Operational Excellence pillar and go through the design principles and the best practice definitions from AWS at `https://docs.aws.amazon.com/wellarchitected/latest/operational-excellence-pillar/welcome.html`.

Now let's see how we can apply these design principles and understand the best practice recommendations from AWS and how they are relevant for observability workloads.

Evaluating customer needs

As a part of the Organization Priorities best practices (OPS01-BP01, OPS01-BP02, and OPS01-BP06), evaluating both your internal and external customer requirements is one of the best practices to follow for your observability workload.

The observability toolset in AWS provides different tools for meeting your customer needs. As discussed in *Chapter 2, Overview of Observability Landscape on AWS*, you have two different options to choose from the toolset: either cloud-native services such as CloudWatch or open source managed services such as Prometheus, Grafana, and Amazon OpenSearch. It is also possible to select the best of both worlds and design your observability strategy based on that. Always evaluate the internal and external customer needs and look into the selection of the services required to meet the business requirements.

Fully automating integration and the deployment of agent workloads

OPS05-BP10 speaks about fully automating the integration and deployment of workloads. One of the important things to consider in deployment and integration is observability agent deployment for workloads running on different compute platforms such as EC2, containers, and Lambda functions and their configuration.

Designing with operations in mind requires the architecture team to prioritize automating the onboarding of different agents' installation for workloads running on AWS. Based on the toolset being adopted for observability as discussed in *Chapter 2, Overview of the Observability Landscape on AWS*, there are four different agents that are discussed in this book the: **CloudWatch agent**, **AWS X-Ray agent**, **OpenTelemetry agent**, and **FireLens agent**. Based on the workload you are looking to instrument, you could leverage CloudFormation as a method to roll out the agent as a part of the workload deployment.

It is not always possible to use CloudFormation as a method to roll out agents as a part of a workload deployment. In this case, if you are looking to automate the overall agent rollout along with application monitoring, AWS Application Insights provides a method to automate the agent installation along with the configuration of best practices for your application monitoring for the supported workloads.

If you are looking for agent rollout for EC2 instances, you could also leverage AWS **Simple System Manager** (**SSM**) automation to install the agents and configure your workloads in a standard way.

We have covered various methods to automate agent installation for different workloads running on AWS on EC2, ECS, EKS, and Lambda in *Chapter 3, Gathering Operational Data and Alerting Using Amazon CloudWatch*; *Chapter 6, Observability for Containerized Applications on AWS*; *Chapter 7, Observability for Serverless Applications on AWS*; and *Chapter 9, Collecting Metrics and Traces Using OpenTelemetry*.

If you are using a third-party tool such as Terraform to provision your infrastructure, you can include the agent deployment and configuration as a part of your Terraform templates.

For specific workloads such as serverless, if you are using a serverless application model for deployment of serverless applications, you could include observability configuration as a part of the **serverless application model (SAM)** templates.

Using a process for event, incident, and problem management

You could look into building an observability dashboard that covers the operational health of your observability workloads. You could look into prescriptive guidance on the different types of dashboards such as Customer Experience, System-Level, Cost Optimization, and so on, for enhancing your operations at `https://aws.amazon.com/builders-library/building-dashboards-for-operational-visibility/`.

Alerting when workload outcomes are at risk

As a part of OPS09-BP06 best practices, it is important to set up alerts when operations outcomes are at risk. In the case of the non-availability of CloudWatch, X-Ray, and so on, or managed AWS services such as Amazon Managed Grafana and Prometheus, you should have a mechanism when the core service itself is not available.

AWS provides a **Health Dashboard** (e) to provide the availability and operations of AWS services. You can personalize the Health Dashboard to get the service health of AWS observability services that are being used as a part of your observability solution to understand any outages that may affect your application monitoring and availability metrics.

Tracking CloudWatch alarm changes

As a part of OPS06-BP01 and OPS06-BP02, there should be a plan for unsuccessful changes along with testing and validating the changes. One of the frequent changes made at runtime is the *CloudWatch alarms.*

During the operational changes, we have difficulty changing alarm configurations and receive a flood of alerts when the alarm configuration goes wrong. AWS Config provides a method to track the current and historical configuration of your alarm and notify you via Amazon **Simple Notification Services (SNS)** when your alarm configuration changes. You can use the config rules to verify whether AWS resources are having CloudWatch alarms for the specified metrics, and create a notification when the settings are not right. You can also monitor whether you configured all the alarms with at least one action, as an alarm without an action is not much use. When you use AWS Config to monitor these settings, it will alert you when an operations noise is created because of unwanted changes and all your alarms have some sort of action associated.

You could enable the following config rules to streamline changes and the configuration of AWS CloudWatch alarms using AWS Config rules:

a. `CloudWatch-Alarm-settings-check`

b. `CloudWatch-alarm-resource-check`

c. `CloudWatch-alarm-action-enabled-check`

d. `CloudWatch-alarm-action-check`

Additional details about the AWS config rules can be found at `https://aws.amazon.com/blogs/mt/aws-config-support-for-amazon-cloudwatch-alarms/`.

Alternatively, you can leverage the AWS Config Conformance packs, *Operational Best Practices for CloudWatch*, found at `https://docs.aws.amazon.com/config/latest/developerguide/operational-best-practices-for-amazon-cloudwatch.html`, which will deploy the four AWS config rules listed previously and one additional config rule for log group encryption. This will streamline the deployment process of operational best practices for CloudWatch.

Security

The Well-Architected Framework Security pillar focuses on seven design principles: *implement a strong identity foundation*; *enable traceability*; *apply security at all layers*; *automate security best practices*; *protect data in transit and at rest*; *keep people away from data*; and *prepare for security events*. I suggest you look into the details of the Security pillar and go through the design principles and the best practice definitions from AWS at `https://docs.aws.amazon.com/wellarchitected/latest/security-pillar/welcome.html`. Now, let's look into applying some of the best practices for your observability workloads on AWS.

Identity management

For delegating permissions on AWS-native observability solutions such as CloudWatch and X-Ray, you could integrate your preferred identity provider with **IAM Identity Center** in AWS and delegate the permissions as required by the creation and segregation of roles.

If you have a multi-account and multi-region environment, there are two different ways to streamline or query the AWS-native observability setup.

Creating a monitoring account

As a part of the SEC01-BP01 (separate workloads using accounts) best practice, AWS recommends having separate monitoring and logging accounts. With AWS CloudWatch adding cross-account and cross-region functionality, it is recommended to have a separate monitoring account, which will help you in querying the metrics, logs, and traces from a single place.

AWS provides a method to query metrics, logs, traces, and dashboards for multiple account scenarios by leveraging cross-account, cross-region functionality. In a multi-account scenario, you could look into creating a dedicated monitoring account and query metrics, logs, and traces from multiple account scenarios and multiple region scenarios. You can find more about this feature at `https://docs.aws.amazon.com/AmazonCloudWatch/latest/monitoring/CloudWatch-Unified-Cross-Account-Setup.html`. This centralizes the permissions management access control for all your observability data from the central place. We discussed this setup in *Chapter 13, Observability Best Practices at Scale*, to understand how this will improve operational visibility and support in the best practice in a complex enterprise organization.

Centralized or distributed accounts for CloudWatch

If you are not looking to adopt the first option (creating a separate monitoring account) and you already have multiple accounts and multiple regions, AWS provides prescriptive guidance on the optimized way to delegate permissions for metrics and logs. It is recommended to keep the metrics and logs in the workload account where your applications are running and to centralize the logs into a separate logging account to query and search the logs from a unified location by leveraging Amazon OpenSearch Service. You can learn about additional considerations for metrics and logging at `https://docs.aws.amazon.com/prescriptive-guidance/latest/implementing-logging-monitoring-cloudwatch/cloudwatch-centralized-distributed-accounts.html`.

AWS Managed Services for observability in a multi-account setup

If you are using Amazon-Managed Prometheus, Amazon-Managed Grafana, and Amazon OpenSearch Service, you could leverage the central monitoring account methodology to ingest the metrics, logs, and traces from multiple AWS accounts into an **Amazon Managed Prometheus (AMP)** workspace being run in the centralized monitoring account. This approach will allow you to delegate permission on the AMP workspace or Amazon Managed Grafana from a central place and track changes to the security delegation from a central place. You can find more information on how to set up cross-account ingestion on the AWS blog for Amazon-Managed Prometheus at `https://aws.amazon.com/blogs/opensource/setting-up-cross-account-ingestion-into-amazon-managed-service-for-prometheus/` and Amazon-Managed Grafana at `https://aws.amazon.com/blogs/opensource/setting-up-amazon-managed-grafana-cross-account-data-source-using-customer-managed-iam-roles/`.

Protecting data in transit and at rest

SEC08-BP03 requires enforcing encryption at rest for sensitive data in your cloud environment. Fortunately, CloudWatch provides robust encryption options to ensure that your data is secure, both in transit and at rest.

CloudWatch log groups can be encrypted using AWS KMS keys, providing you with complete control over the encryption keys used to protect your data. This enables you to manage and rotate your encryption keys as per your security policies.

If you are using Amazon-Managed Grafana, Amazon-Managed Prometheus, or Amazon OpenSearch Service, AWS follows the shared responsibility model for data protection. In this model, AWS is responsible for securing the underlying infrastructure of these services, while you are responsible for securing your data and any custom code you may use with these services. In terms of data encryption, these services automatically encrypt data in transit using SSL, ensuring that your data is protected during transmission. You can also encrypt your data at rest using AWS KMS, which provides robust encryption options and makes it easy to manage encryption keys.

Compliance

As per SEC07-BP02 (define data identification and classification), it is good to enforce protection for log data for any personalized information. To safeguard sensitive application information and **Personally Identifiable Information** (**PII**), you can enable data protection for CloudWatch log groups. You can protect the data in CloudWatch Logs by applying data protection policies, which can help you discover the sensitive data logged by systems and applications and protect them as per the data protection policies configured. This will help in potentially reducing the exposure of sensitive data due to application security vulnerabilities. If you would like to look into the details of protecting the data, you can refer to `https://aws.amazon.com/blogs/aws/protect-sensitive-data-with-amazon-cloudwatch-logs/`.

The following is the configuration capability of protecting sensitive data from CloudWatch log groups based on the type of data being exposed:

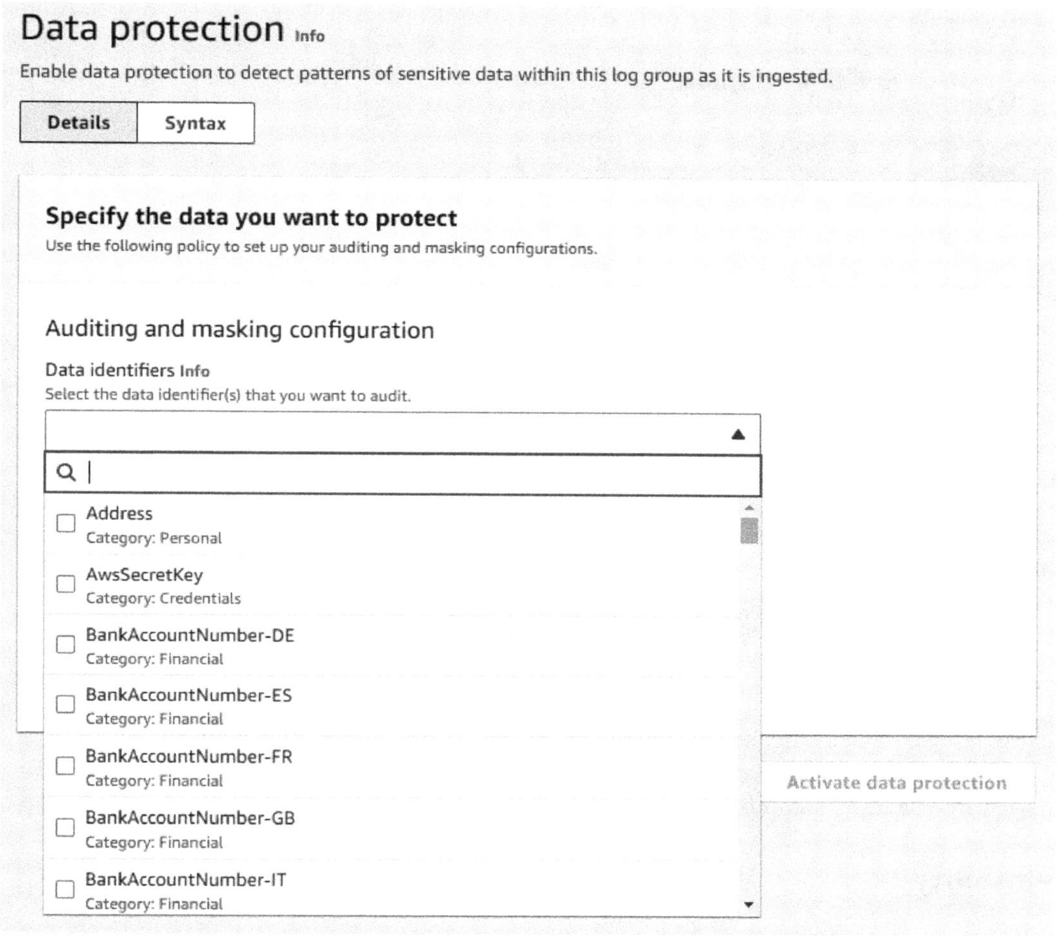

Data protection Info

Enable data protection to detect patterns of sensitive data within this log group as it is ingested.

| Details | Syntax |

Specify the data you want to protect

Use the following policy to set up your auditing and masking configurations.

Auditing and masking configuration

Data identifiers Info
Select the data identifier(s) that you want to audit.

▲

Q |

☐ Address
Category: Personal

☐ AwsSecretKey
Category: Credentials

☐ BankAccountNumber-DE
Category: Financial

☐ BankAccountNumber-ES
Category: Financial

☐ BankAccountNumber-FR
Category: Financial

Activate data protection

☐ BankAccountNumber-GB
Category: Financial

☐ BankAccountNumber-IT
Category: Financial

▼

Figure 14.2 – CloudWatch log group data protection

If you are looking for an industry-standard compliance mechanism for CloudWatch, Amazon CloudWatch is FedRAMP- and PCI-compliant.

For Amazon OpenSearch Service, you can leverage AWS Config to measure the compliance of security best practices by deploying the *Security Best Practices for Amazon OpenSearch Service* conformance pack by region. This automates the measurement of compliance as per the best practices from AWS for Amazon OpenSearch Service. You can find the sample template, as shown in the following figure, in AWS Config:

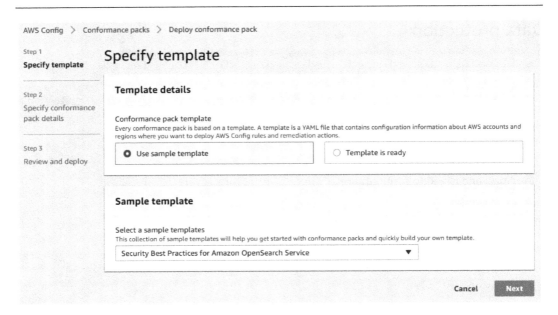

AWS Config > Conformance packs > Deploy conformance pack

Step 1
Specify template

Specify template

Step 2
Specify conformance
pack details

Template details

Conformance pack template
Every conformance pack is based on a template. A template is a YAML file that contains configuration information about AWS accounts and regions where you want to deploy AWS Config rules and remediation actions.

Step 3
Review and deploy

● Use sample template   ○ Template is ready

Sample template

Select a sample templates
This collection of sample templates will help you get started with conformance packs and quickly build your own template.

Security Best Practices for Amazon OpenSearch Service   ▼

Cancel   Next

Figure 14.3 – AWS Config conformance pack

Enabling traceability

You can integrate and configure automation for CloudTrial logging by setting up the destination for CloudTrial management logs to CloudWatch, which will help you query and set up automation for any security-related incidents based on the CloudTrail logging information.

Reliability

The Well-Architected Framework Reliability pillar focuses on five design principles: *automatically recover from failure; test recovery procedures; scale horizontally to increase aggregate workload availability; stop guessing capacity;* and *manage change through automation*. I suggest you look into the details of the Reliability pillar and go through the design principles and the best practice definitions from AWS at

`https://docs.aws.amazon.com/wellarchitected/latest/reliability-pillar/welcome.html`. Now, let's look into applying some of the best practices for your observability workloads on AWS.

Managing service quotas and constraints

As per REL01-BP01 (aware of service quotas and constraints), REL01-BP02 (manage service quotas across accounts and regions), and REL01-BP05 (automate quota management), it is best practice to manage service quotas for services and automate where possible. To manage the service quotas for CloudWatch, you can create a CloudWatch dashboard using the metric math functionality. If you would like to understand more about metric math, please refer to *Chapter 5, Insights into Operational Data with CloudWatch*. A helpful document on how to configure the service quotas for CloudWatch using metric math can be found at `https://docs.aws.amazon.com/AmazonCloudWatch/latest/monitoring/CloudWatch-Quotas-Visualize-Alarms.html`.

If you are looking for a more comprehensive solution for all the supported services, there is a ready-made solution implementation from AWS called **Quota Monitor**. This can be deployed to monitor actions on the service alerts and integrate with your favorite notification mechanism as required. It can be found at `https://aws.amazon.com/solutions/implementations/quota-monitor/`.

Failure management

As per REL10-BP01 (deployment of workloads to multiple locations), when you have a publicly accessible application in different regions and would like to set up observability across regions, you could use the deep application observability solution to understand the failover and deployment of best practices for application observability from the solution at `https://aws.amazon.com/solutions/guidance/deep-application-observability-on-aws/`.

Implementing change

As per REL08-BP05 (deploy changes with automation), one of the changes you may require to do is the CloudWatch agent configuration. If you are using the CloudWatch unified agent, you could look into managing the agent configuration in your S3 bucket and push the configuration using SSM automation. Alternatively, you could also leverage the Git repository to publish your agent configuration and push the configuration to the CloudWatch agents as a part of change management. We have discussed this configuration in *Chapter 3, Gathering Operational Data and Alerting Using Amazon CloudWatch*, on how to automate the CloudWatch agent using SSM automation.

Performance efficiency

The Well-Architected Framework Performance Efficiency pillar focuses on five design principles, namely *democratize advanced technologies*; *go global in minutes*; *use serverless architectures*; *experiment more often*; and *consider mechanical sympathy*. You could find details about this definition of the design principles and best practices in the documentation at `https://docs.aws.amazon.com/wellarchitected/latest/performance-efficiency-pillar/welcome.html`.

Let's look at some of the best practices we could adopt for the observability workloads from the Performance Efficiency design principle point of view.

Performance architecture selection

PERF01-BP01 describes the available services and resources. We have discussed the overall observability solutions available on AWS as a part of *Chapter 2, Overview of the Observability Landscape on AWS*. You should look into the available services and select the services that are best for your applications.

Architecture considerations

As per PERF02-BP1, to evaluate the available compute options, CloudWatch is a managed service from AWS. If you are looking to adopt any of the open source observability solutions, it is always good to consider using Amazon-Managed Grafana or Amazon-Managed Prometheus as a first choice before deploying and managing them on your own on an EC2 instance. If you are looking to adopt OpenSearch Service, consider looking into OpenSearch Serverless (`https://aws.amazon.com/opensearch-service/features/serverless/`) instead of deploying OpenSearch on a managed EC2 cluster from AWS. This not only improves the scalability but also helps you with performance optimization based on the number of queries being executed.

Cost optimization

The Cost Optimization pillar in the Well-Architected Framework provides five design principles for optimizing the cost, namely *implement Cloud Financial Management (CFM)*; *adopt a consumption model*; *measure overall efficiency*; *stop spending money on undifferentiated heavy lifting*; and *analyze and attribute expenditure*. I suggest you look into the details of the Cost Optimization pillar and go through the design principles and the best practice definitions from AWS at `https://docs.aws.amazon.com/wellarchitected/latest/cost-optimization-pillar/welcome.html.html`.

Let's look into implementing these best practices for the observability workloads on AWS.

Practicing CFM

It is recommended to implement CFM as an organization in the overall cloud governance function. AWS provides various tools for understanding the optimize the cost model. AWS Cost Explorer is a tool that provides dashboards and reports for your overall cloud spend. You could filter the usage of the Cost Explorer only by the service, CloudWatch, and understand the top API operations, which are the high-cost consumers. A high-level summary report is shown in the following figure. If you look into the details, the high cost is caused by CloudWatch Logs, which are used for the embedded metric format and PutLogEvents API.

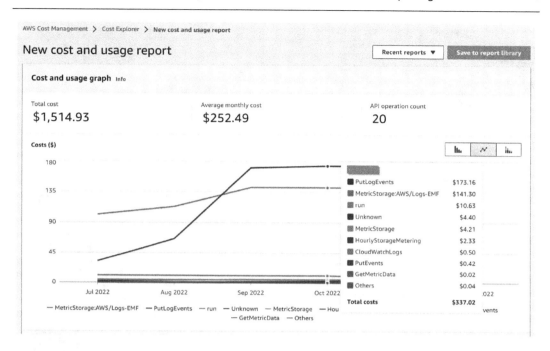

AWS Cost Management > Cost Explorer > New cost and usage report

New cost and usage report

Figure 14.4 – AWS Cost Explorer

Expenditure and usage awareness

As per the best practice COST03-BP02, it is recommended to identify cost attribution categories and COST03-BP04 (configure billing and cost management) tools. Also, as a part of COST01-BP06, we should have tools to understand the overall cost and categorize them by workload. To address this, AWS has a solution called **Cloud Intelligence Dashboards** (https://github.com/aws-samples/aws-cudos-framework-deployment). As a part of Cloud Intelligence Dashboards, you have a tab for **Monitoring and Observability**, which will help you in understanding the cost along with detailed information about the top 10 CloudWatch resources that are generating the cost. This will help you in fine-tuning the cost. You can look into the sample dashboard provided by the tool in the following figure:

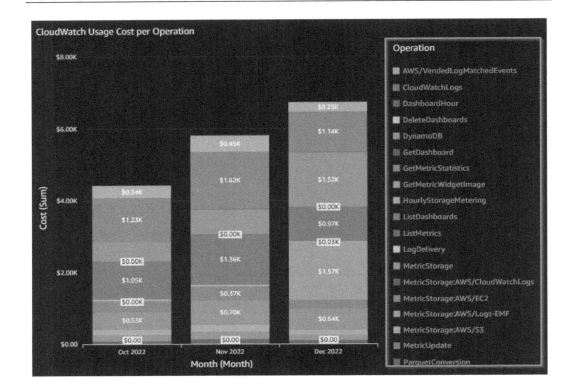

Figure 14.5 – Monitoring and Observability in the Cloud Intelligence Dashboards

You can also find the GitHub repo for the Cloud Intelligence Dashboards from AWS in the GitHub repository at https://github.com/aws-samples/aws-cudos-framework-deployment. AWS also provides guidance on reducing the CloudWatch cost based on the Cloud Intelligence Dashboards at https://aws.amazon.com/premiumsupport/knowledge-center/cloudwatch-understand-and-reduce-charges/?nc1=h_ls and https://aws.amazon.com/premiumsupport/knowledge-center/cloudwatch-logs-bill-increase/.

Optimizing over time

As per COST10-BP01, you should have a process to review your workload and optimize it in terms of cost. Let's look into one of the examples of optimizing the CloudWatch cost for the **Elastic Kubernetes Service (EKS)** workload. When leveraging Container Insights along with **AWS Distro for OpenTelemetry (ADOT)**, which was discussed as a part of *Chapter 10, Deploying and Configuring an Amazon Managed Service for Prometheus*, you could look at reducing the cost of Container Insights. AWS has published a blog on how the cost could be optimized when ADOT along with Container Insights in two different ways, namely *filtering metrics using processors* and *customizing metrics and* *dimensions* (`https://aws.amazon.com/blogs/containers/cost-savings-by-customizing-metrics-sent-by-container-insights-in-amazon-eks/`). You can see that overall cost-reducing is close to 60%.

Metric Math

You can use CloudWatch metric math, which was discussed in *Chapter 5, Insights into Operational Data with CloudWatch*, to derive new metrics from existing metrics rather than ingest a new metric, which will increase the cost of CloudWatch. For example, if you are looking to derive the sum of overall sales, you can use the metric math functions to sum the individual sales. This can reduce the volume of data sent to CloudWatch and help reduce the cost of using the service.

CloudWatch log group cost optimization

To manage the cost of your CloudWatch logs effectively, it's advisable to establish a retention period that aligns with your business needs. By doing so, you can avoid retaining logs for longer than necessary, which can result in higher storage costs.

It's also important to consider any compliance requirements that apply to your logs. Depending on your industry or regulatory environment, you may need to retain logs for a specific duration or ensure that they are stored securely to meet compliance obligations.

Based on the type of data being stored in each CloudWatch log group, set the retention timelines. You can edit the log group retention setting and set up the retention timeline as shown in the following figure:

Figure 14.6 – Configure CloudWatch log retention

Managed services

Always prefer adopting managed services or serverless options over building your own solution. This approach offers several benefits including reduced maintenance requirements, a lower total cost of ownership, and faster time to market. For example, leverage Amazon Managed Grafana over installing your own Grafana on **Elastic Compute Cloud** (**EC2**). Leverage the serverless Amazon OpenSearch Service over the EC2-based OpenSearch Service. This not only optimizes the cost but also helps you achieve your sustainability goals.

Leveraging logs for sending metric data

If you are looking to optimize the cost of storing metrics in the CloudWatch namespace, you could consider sending the metric data as a log stream into CloudWatch Logs, using metric filters to create metrics for the required dimension, and also querying the data on demand as a metric. This will help in optimizing the cost when you are looking to publish the metric in multiple dimensions.

Sustainability

The Sustainability pillar's design principles are *understand your impact*; *establish sustainability goals*; *maximize utilization*; *anticipate and adopt new, more efficient hardware and software offerings*; and *use managed services and reduce the downstream impact of your cloud workloads*. You can find more details about these design principles and best practices for sustainability for your workload at `https://docs.aws.amazon.com/wellarchitected/latest/sustainability-pillar/sustainability-pillar.html`.

Scaling infrastructure with user load

As a part of SUS-BP01 (scale infrastructure with user load), it is advisable to choose serverless or managed services over EC2 instances and building your own service. For observability workloads, use CloudWatch-native services as relevant and fall back on AWS Managed Services as required for additional functionalities, or use a hybrid approach based on the workload demands.

Optimizing logging strategies

Send only relevant information to CloudWatch Logs. You could use the filtering option available in the CloudWatch agent to limit the amount of data being sent for retention in the CloudWatch log groups. This will reduce the amount of data being stored and support sustainability goals for your organization.

In this section, we understood the pillars of the AWS Well-Architected framework and discussed how to implement the best practices for your observability solutions on AWS. In the next section, we will look through the management and governance lens in the Well-Architected framework.

Understanding management and governance in the Well-Architected Framework

The AWS Well-Architected Framework provides the best practices and guidelines to follow when you are designing your workload to run on AWS. To strengthen the framework, the management and governance lens especially focuses on applying these principles to ensure that you have the necessary processes, tools, and practices to build cloud-ready environments. The goal of the **management and governance** (**M&G**) lens focuses on four key parts:

- **Use what you know**: Allows for quick implementation of scalable and secure cloud management, removing the need for re-investment by using familiar and existing tools

- **Increase speed to value**: Accelerate cloud migrations by following the best practices learned from thousands of successful migrations and activations

- **Increase efficiency with interoperability**: Capabilities should interoperate and inform each other to achieve greater scale than possible if they are considered in isolation

- **Balance agility and governance**: The M&G lens provides prescriptive guidance on how to reduce the friction between the ability to govern and the ability to be agile while reducing friction in operational capabilities

When you are on a migration journey to the cloud, and you use different migration approaches such as migrating/lifting and shifting, modernizing or re-architecting applications, or building new, you need a cloud-ready environment that takes into consideration all the M&G functions. The M&G lens provides prescriptive best practices that help you define how you can manage and govern across the life cycle of the cloud. There are eight functions that are most pertinent for successful customers in cloud-ready environments. Let's look at these eight functions:

- **Controls and Guardrails**: This function covers the policies, procedures, and technical controls that help ensure compliance with relevant regulations and standards, as well as the overall integrity and security of your cloud environment

- **Network Connectivity**: This function covers the design and management of your network infrastructure, including the connectivity between your on-premises environment and the cloud

- **Identity Management**: This function covers the management of user identities, authentication, and access control in your cloud environment

- **Security Management**: This function covers the policies, procedures, and technical controls that help protect your cloud environment from security threats

- **IT Service Management (ITSM)**: This function covers the management of IT services, including the planning, design, delivery, and support of those services

- **Observability**: This function covers the monitoring and analysis of your cloud environment to help identify issues, optimize performance, and improve reliability

- **CFM**: This function covers the management of costs and billing in your cloud environment, including the optimization of resource utilization and the tracking of spending

- **Sourcing and Distribution**: This function covers the management of resources in your cloud environment, including the procurement of hardware and software, as well as the distribution of those resources to meet the needs of your organization

If you are looking to understand what each function is all about, you can find the reference and prescriptive guidance on adopting M&G at `https://docs.aws.amazon.com/wellarchitected/latest/management-and-governance-guide/management-and-governance-cloud-environment-guide.html`.

How those functions interoperate is a critical factor in their success. What I mean by interoperate is the interaction between functions. The output of one function should feed input to another. As an example, the output of the Security Management function (logs of API calls and events) is used as input for the Observability function (monitoring and analyzing system behavior). The output of the Observability function (resource utilization and performance data) is used as input for the CFM function (analyzing and optimizing costs). By interoperating in this way, the M&G functions can work together to create a more effective and efficient cloud environment.

Here is the snapshot of the AWS services that are supporting these functions:

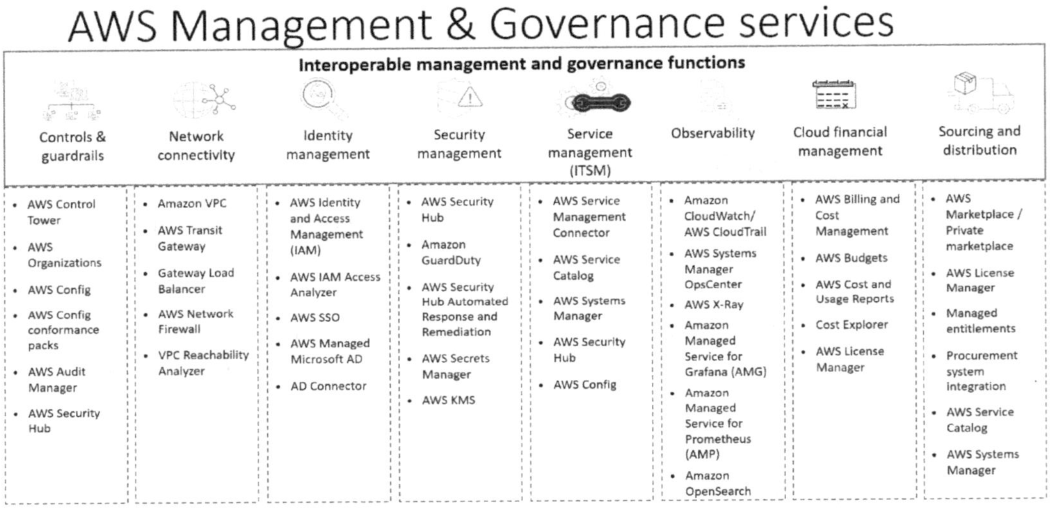

AWS Management & Governance services

Interoperable management and governance functions

Controls & guardrails	Network connectivity	Identity management	Security management	Service management (ITSM)	Observability	Cloud financial management	Sourcing and distribution
• AWS Control Tower • AWS Organizations • AWS Config • AWS Config conformance packs • AWS Audit Manager • AWS Security Hub	• Amazon VPC • AWS Transit Gateway • Gateway Load Balancer • AWS Network Firewall • VPC Reachability Analyzer	• AWS Identity and Access Management (IAM) • AWS IAM Access Analyzer • AWS SSO • AWS Managed Microsoft AD • AD Connector	• AWS Security Hub • Amazon GuardDuty • AWS Security Hub Automated Response and Remediation • AWS Secrets Manager • AWS KMS	• AWS Service Management Connector • AWS Service Catalog • AWS Systems Manager • AWS Security Hub • AWS Config	• Amazon CloudWatch/ AWS CloudTrail • AWS Systems Manager OpsCenter • AWS X-Ray • Amazon Managed Service for Grafana (AMG) • Amazon Managed Service for Prometheus (AMP) • Amazon OpenSearch	• AWS Billing and Cost Management • AWS Budgets • AWS Cost and Usage Reports • Cost Explorer • AWS License Manager	• AWS Marketplace / Private marketplace • AWS License Manager • Managed entitlements • Procurement system integration • AWS Service Catalog • AWS Systems Manager

Figure 14.7 – M&G with AWS services

Now let's look at the Observability functionality in the M&G lens.

Observability is essential for all teams who operate and manage cloud applications and services. It helps teams understand the behavior and performance of their systems by providing visibility into workload events and operations metrics. This information can then be used to take appropriate action and improve the reliability and efficiency of the cloud environment. The Observability function of M&G provides guidance on how to prioritize the implementation of observability in your cloud environment. Implementation priorities as defined for Observability in M&G are as follows:

- **Collect, aggregate, and protect event and log data**:

  Logs, metrics, and traces should be collected across the different observability categories. You should look into gathering, compiling, and securing event and log information by utilizing various observability tools.

Utilize AWS CloudTrail logging for control plane observability, and Amazon VPC Flow Logs, VPC Reachability Analyzer, and Amazon Inspector Network reachability services for network observability, including the monitoring of network firewalls, intrusion detection and prevention, load balancers, WAF, and proxy tools. Also, implement distributed tracing within your application for observability of serverless, container, storage, and database workloads.

- **Build capabilities to analyze and visualize logs events and traces**:

As your organization grows on AWS, it becomes increasingly important to have the ability to organize, present, and monitor your log data and metrics. By utilizing the correlation of logs and performance metrics from multiple sources, you can gain valuable insights and an understanding of your systems. To ensure the security of your organization, it is essential to implement rules that quickly respond to any identified security incidents or patterns in your logs. A consistent monitoring plan will allow you to adapt and evolve your observability capabilities as your organization continues to migrate and develop solutions on AWS.

- **Add detection and alerts for anomalous patterns across environments**:

Take a proactive approach to identify known vulnerabilities and detect abnormal patterns of events and activities. You should begin with identifying patterns or signs of unauthorized account access or privileges, such as login activity to cloud management consoles, modifications or attempted modifications to crucial cloud objects and data, and the creation, deletion, or alteration of credentials or cryptographic keys. Additionally, detect incidents and patterns of denied access, unidentified network traffic, unusual increases in cloud service costs, and uncommon application traffic behavior.

- **Define, automate, and measure response and remediation**:

Define acceptable behavior limits in conjunction with business metrics to comprehend key performance indicators for workloads and environments. Identify appropriate incident and response actions to take. Utilize **security information and event management (SIEM)** solutions to observe workloads in real time, discover security issues, and hasten root-cause analysis. Leverage **security orchestration, automation, and response (SOAR)** platforms along with responses generated from recorded events using tools such as AWS Lambda. Implement a procedure to continuously enhance the **mean time to identify (MTTI)** the root cause and **mean time to respond (MTTR)** to problems.

The M&G lens provides guidance on the essential eight capabilities for creating a cloud-ready environment that is prepared for migration, able to scale, and operating efficiently. When considering these capabilities, it is important to consider how they interoperate, what priority should be given to their implementation, and which native AWS and partner services can support them. By designing the functions to work together effectively, you can create a more effective and efficient cloud environment. This can be done using a combination of manual, tightly integrated, loosely coupled, or automated mechanisms.

The interoperability of these functions can be understood from an observability point of view to gain visibility into the operation of your cloud environment. It is important to define, capture, and analyze operations metrics. The Controls and Guardrails function can help observe changes and highlight them in observability tools, while the Network function can capture network flow logs and send them to central infrastructure log archives and aggregation tools. The Identity Management function can record changes with observability tools and set up automated alerting. By tuning observability and monitoring with inputs from the Security Management function, you can reduce static and improve the reliability of your systems. The Service Management function can integrate observability with operational tooling, and the CFM function can include observability measures to alert for changes in incurred and forecasted costs. The Sourcing and Distribution function can ensure that customers and purchased solutions include instrumentation to support observability and monitoring capabilities. By using these interdependent management and governance functions, you can create a more effective and efficient cloud environment.

Summary

In this chapter, we looked at an overview of the AWS Well-Architected framework and looked into the six pillars of the framework. Further, we have gone through how to apply the Well-Architected framework best practices to observability workloads running on AWS and explored various tools and solutions available to optimize the workloads as per the best practice requirements. Further, we talked about the M&G lens and the role of observability in the M&G lens. The M&G lens provides guidance beyond observability and throws light on how different functions should interoperate to derive the maximum benefits when running workloads on AWS. Then, we looked into how to set up an interoperable environment for the M&G pillars with respect to a observability solution and how they could be reused in other pillars to prepare your cloud-ready environment.

In the next chapter, we'll delve into the Cloud Adoption Framework and explore the critical factors to consider when establishing an effective observability strategy for your organization. We'll also examine how observability can play a vital role in accelerating the adoption of cloud services and discuss other essential considerations beyond observability from a cloud operations perspective. By understanding these key concepts, you can develop a comprehensive approach to managing and optimizing your cloud environment to meet your business needs.

Questions

1. What are the six pillars of the AWS Well-Architected Framework?
2. What is the significance of the M&G lens in the Well-Architected Framework?
3. How will interoperability between the M&G lens pillars help you achieve success?
4. What are the best practices to adopt from the Security pillar for your observability workloads?

15
The Role of Observability in the Cloud Adoption Framework

The **Cloud Adoption Framework** (**CAF**) provides customers with a set of guidelines and best practices to assist them in digitally transforming business outcomes by innovatively using AWS. The CAF has undergone several updates, with versions 1.0 and 2.0, and the current version, 3.0. In version 1.0, the framework consists of seven perspectives, each containing a group of components and activities necessary for successful cloud adoption. In version 2.0, the perspectives were consolidated to six and a new governance perspective was added. This version focuses on the capabilities that will be affected in an organization during the move to the cloud. With version 3.0, the focus has shifted from cloud adoption to accelerating digital transformation and achieving desired business outcomes through the adoption of the cloud. The high-level changes in the CAF across its versions can be understood from the following figure:

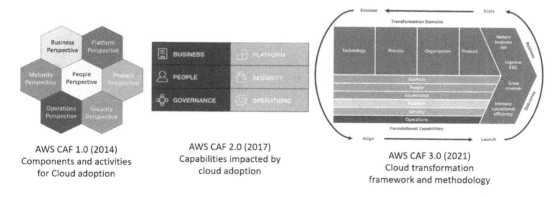

AWS CAF 1.0 (2014)
Components and activities
for Cloud adoption

AWS CAF 2.0 (2017)
Capabilities impacted by
cloud adoption

AWS CAF 3.0 (2021)
Cloud transformation
framework and methodology

Figure 15.1 – History of the CAF

In this chapter, we will look at the following main topics:

- Overview of Cloud Adoption Framework 3.0
- Developing an observability strategy for an organization
- Role of observability in the CAF and the best practices for quicker adoption of cloud

Let's navigate into CAF 3.0 and understand the different aspects that need to be considered to achieve the desired business outcome, how observability could support achieving those outcomes, and look into the observability strategy.

Overview of Cloud Adoption Framework 3.0

Let's understand the various components of CAF 3.0. The CAF discusses and provides guidance on multiple components:

- Phases of the cloud transformation journey
- Transformation domains
- Foundational capabilities
- Business outcomes

You could visualize the same in the AWS CAF 3.0 value chain figure here:

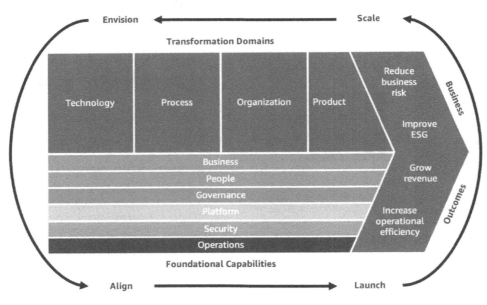

Figure 15.2 – AWS CAF value chain

Let's deep dive into each of these components.

Cloud transformation journey

The cloud transformation journey discusses the agile approach for organizations on realizing the business value of the cloud. If you look at the outer circle in the CAF, there are four different phases, namely **Envision**, **Align**, **Launch**, and **Scale**.

Let's dive deeper into them and look at the focus of the phases in the cloud transformation journey:

- **Envision:** The envision phase focuses on how customers identify and prioritize transformation opportunities in line with their strategic objectives. So it is always working backward from customer strategic business objectives to identify transformation opportunities in an organization. As a part of this exercise, you would look into identifying key stakeholders for driving business decisions, which is one of the critical aspects of the envision phase.

- **Align:** In the align phase, the focus is on identifying the broader set of stakeholders across the organization and also focus on cross-organizational dependencies and capability gaps. So, this phase will help customers create strategies for improving their cloud readiness as well as ensure stakeholder alignment and facilitate relevant organizational change in management activities.

- **Launch:** In the launch phase, the focus is on delivering pilots in production relatively rapidly. So, within 60-90 days, look at providing incremental business value quickly. These pilots should be highly impactful and, when successful, they should influence the future direction. Learning from pilots can help customers adjust their approach before scaling to full production.

- **Scale:** Upon successful pilot evaluations, expand service across the organization for maximum business impact and scale the service across the organization. Ensure cloud investments yield sustained benefits over the long term.

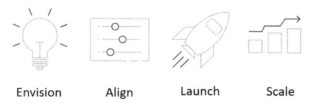

Envision   Align   Launch   Scale

Figure 15.3 – Phases of the cloud journey in the CAF

Next is to look into the transformation domains, which are spread across technology, process, organization, and product.

Transformation domains

The cloud transformation value chain focuses on four transformation domains that will help you realize the full value and potential of the cloud – namely technological transformation, process transformation, organization transformation, and product transformation.

Let's see what they focus on:

- **Technological transformation**: Technological transformation focuses on using the cloud to migrate and modernize your legacy infrastructure, applications, data, and analytics platforms by utilizing the cloud.

- **Process transformation**: Process transformation aims to digitize, automate, and optimize your business operations. This could involve using new data and machine learning to enhance customer service, employee productivity and decision-making, business, and so on. The outcome of this transformation will result in improved operational efficiency, reduced operating costs, and a better experience for employees and customers.

- **Organization transformation**: Organization transformation centers on rethinking your operating model, which outlines how your business and technology teams work together to deliver customer value and achieve strategic goals.

- **Product transformation**: Product transformation involves rethinking your business model by developing innovative value propositions and revenue streams, enabling you to target new customers and penetrate new markets.

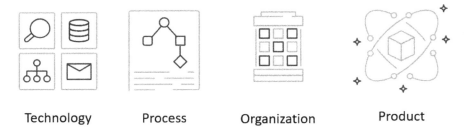

Technology   Process   Organization   Product

Figure 15.4 – Transformation domains in the CAF

Foundational capabilities

Next comes the foundational capabilities, which are grouped into six different perspectives. A **capability** is an organizational ability to leverage processes to deploy resources to achieve a particular outcome. The AWS CAF identifies 47 foundational capabilities that will help you with successful cloud transformations. These capabilities will provide you with the best practice guidance that helps you improve your cloud readiness and support your digital transformation goals.

The AWS CAF categorizes these 47 capabilities into 6 perspectives, namely **Business, People, Governance, Platform, Security**, and **Operations**. Each perspective encompasses a set of related capabilities that are owned or managed by specific stakeholders during the cloud transformation journey:

- The **Business** perspective helps ensure that your cloud investments speed up your digital transformation ambitions and business outcomes

- The **People** perspective serves as a bridge between technology and business in accelerating your cloud journey to help organizations grow to a culture of continuous growth and learning, where change will be treated as business-as-normal, with a focus on culture, organizational structure, leadership, and workforce

- The **Governance** perspective helps you orchestrate your cloud initiatives while maximizing organizational benefits and minimizing cloud transformation-related risks

- The **Platform** perspective helps you build an enterprise-grade, scalable, hybrid cloud platform, modernize your existing workloads, and implement cloud-native solutions

- The **Security** perspective focuses on ensuring the confidentiality, integrity, and availability of your data and cloud workloads

- The **Operations** perspective helps you ensure that your cloud services are delivered at a level that meets the needs of your business

Figure 15.5 – Perspective and capabilities in the CAF

Business outcomes

When you look at the outcomes of CAF 3.0, the benefits of adopting the CAF help businesses reduce the risk profile by improving reliability, business continuity, increasing performance, and enhancing security. The other benefits are improvement in **environmental, social, and governance** (**ESG**) performance by providing insights into sustainability. The CAF also helps a business to grow new revenue streams through rapid innovation in their products and services by cloud transformation and increases operational efficiency by reducing operating costs, improving productivity, and enhancing customer experience.

Now, let's see how we can adopt the CAF from the point of observability solutions and look into how it will support accelerating your cloud transformation journey.

Developing an observability strategy for your organization

An **observability strategy** is essential for any organization looking to gain a deeper understanding of its IT operations and systems. It addresses key questions such as the following:

- How do we effectively monitor our systems in the cloud?

- How can we gain visibility into serverless workloads?

- Why do certain workloads experience issues, and how can we prevent them from recurring?

- Which monitoring tools are best suited for our specific needs?

If you are facing any of these challenges, it is time to consider developing an observability strategy for your organization.

Having an effective observability strategy is crucial for gaining a thorough understanding of IT operations, workloads, and their effect on business outcomes and risks. It enables organizations to reduce downtime, enhance customer satisfaction, enhance operational control, and increase efficiency. The observability strategy is based on the **golden triangle of observability**, which includes metrics, logs, and traces, as discussed in *Chapter 1, Observability 101*. Defining an observability strategy is beneficial as it helps organizations realize its value and its impact on the overall success of the organization.

Benefits of defining an observability strategy

We could summarize the benefits of defining the observability strategy using the observability value curve. The observability value curve provides organizations with a clear understanding of the benefits they can expect to see by implementing the observability strategy. Let's look at the observability value curve:

- **Understanding operational health**: This is a crucial aspect of ensuring the smooth functioning of any system. It involves revealing the health of both primary and supporting resources to identify potential issues and address them proactively. Observing critical signals and metrics is key in improving the **mean time to detect** (**MTTD**) failures, which is the average time to identify a problem in a system. By focusing on operational health and reducing MTTD, organizations can minimize downtime and ensure the availability of their systems, which is critical to delivering high-quality services to their customers.

- **Reducing mean time to recover** (**MTTR**): This aspect helps to quickly identify the root cause of issues and minimize the time needed to diagnose and resolve problems, restoring systems from failure more efficiently through targeted troubleshooting.

- **Identification of strategic improvement opportunities**: By identifying the weak links in your distributed application environment, you can uncover opportunities for strategic improvements that will enhance the overall performance and reliability of your systems. This process involves evaluating the various components of your application infrastructure and determining where improvements can be made to optimize performance, reduce downtime, and improve overall customer satisfaction. The identification of strategic improvement opportunities can help organizations stay ahead of potential issues and prevent system failures, leading to a more resilient and efficient application environment.

- **Enabling data-driven decisions**: By leveraging observability data, organizations can enhance their decision-making capabilities and move away from relying on symptoms or intuition. Observability data provides organizations with a comprehensive view of their systems and the underlying data, enabling them to make informed decisions that are based on facts and evidence. This data-driven approach can help organizations avoid deciding a course of action based on assumptions or guesses, which can lead to unintended consequences and suboptimal outcomes.

- **Improving business and technical agility**: By utilizing observability to monitor systems, organizations can speed up developer velocity and help developers release code with confidence. Observability provides organizations with a comprehensive view of their systems, enabling developers to identify potential issues and resolve them more quickly. This proactive approach helps to reduce the time to market for new features and services. Observability not only helps organizations improve the quality and speed of their software development process but it also provides them with greater operational visibility and control, leading to improved business and technical agility.

The observability value curve can be summarized as shown in the following figure:

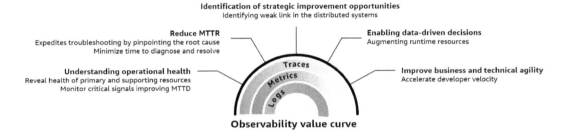

Figure 15.6 – Observability value curve

The output of the observability strategy

Once you have completed your observability strategy, you should look at achieving the following output:

- Develop a clear plan for implementing observability tools and techniques to address areas of uncertainty and gain deeper insights into the IT environment

- By having a well-defined observability plan, you can clearly identify what metrics are important to track and how to measure them to support decision-making and drive continuous improvement

Now that we have understood the benefits, outcomes, and output of defining the observability strategy, let's look into how to approach the observability strategy and the techniques that could be leveraged to derive the same.

Applying an observability strategy

The overall circle of observability strategy planning for an organization can be understood from the following figure:

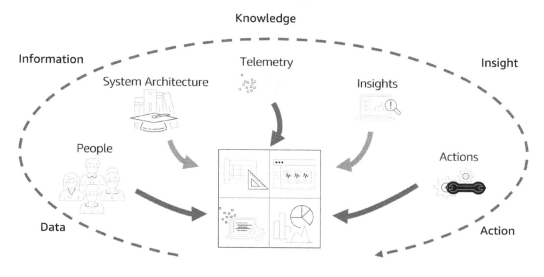

Figure 15.7 – Overview of an observability strategy

An observability strategy in an organization involves establishing a people-focused culture, gaining insight into the system architecture, defining telemetry needs based on that architecture, selecting the appropriate observability tools to meet those needs, generating operational insights, and taking action based on those insights. This continuous cycle transforms operational data into actionable information, leading to improved understanding and knowledge, and resulting in the ability to swiftly address any issues and maintain a stable system. Let's look into the first aspect of **people culture**.

People culture in an observability strategy

One of the important transformations required is the **people** transformation.

Starting the observability culture

When considering people's transformation in your organization, think of observability as a service or product that you can offer to internal teams. To effectively implement observability, it's important to establish an observability team that can provide a holistic approach, standardize tooling, and educate teams on how to effectively use observability solutions. However, it's important to be aware of potential downsides, such as reliance on the observability team, a focus on cause-based alerting, lack of design, and the potential for multiple tools and a high workload.

An observability team in a **Cloud Center of Excellence (CCOE)** will help you to do the following:

- Ensure full-stack observability for all applications
- Efficiently select tools for log, trace, and metric collection
- Enforce standards for consistent logging across all applications
- Put in place a strategy for centralized log collection, storage, and analysis
- Establish KPIs to measure workload performance
- Standardize a set of metrics to measure the achievement of KPIs
- Put in place a plan to remediate performance-related issues for observability service

When establishing an observability culture within an organization, there are three key areas to consider: product offering, commonality, and training/education. Each of these aspects plays a crucial role in ensuring the success of your observability strategy. Let's dive into the details of these:

- **Product offering**: To implement observability in an organization, start by considering it as a product offering and determining the user experience you want to provide for your teams. Adhere to the 80:20 rule, where you focus on standardizing common use cases and making them easy to adopt, while also providing flexibility for building more complex use cases.

- **Commonality**: To successfully achieve the adoption of observability within your organization, it's important to focus on commonality and ease of use. Use out-of-the-box functionality to make it simpler for teams to adopt. Integrate observability into your application stack, use shared code, and establish common alerting mechanisms and dashboards for commonly used services on AWS in your organization. This will make it easy for teams to use observability without spending time on standardization. Standardize the structure and format of the dashboards across different teams and encourage the use of common observability patterns, such as log formatting, exception handling, and metric conventions.

- **Promote education and training**: To effectively implement observability within your organization, it's important to educate internal teams on how to use it. Create bootcamp-style training sessions for teams transitioning to the cloud or for new employees joining the organization. These sessions can cover topics such as how to use observability tools, foundational use cases, and leveraging the same in adopting different types of applications. Establish a clear process for requesting help and support when teams need help with adopting observability for their applications. Through education and training, teams will be better equipped to quickly adopt and effectively use observability services.

Observability culture can be summarized as shown in the following figure:

Figure 15.8 – Adoption for the people culture

Maintaining the observability culture

Once the observability culture has been established, the next step is to ensure its ongoing success. It is crucial to maintain the culture and continuously improve the observability offering within the organization:

- **Help**: To sustain the culture, provide clear channels of communication and support. Standardize the process for requesting help and support, establish chat channels for discussing relevant topics, and create a mechanism for raising exceptions when the standard process is not suitable. Schedule regular meetings with teams to gather feedback and improve the roadmap.

- **Share**: Regularly communicate new developments and improvements to the teams, share customer success stories, and continually roll out new features and enhancements. Organize demos and gather feedback during these sessions, and promote testimonials from satisfied customers.

- **Practice**: Remember that humans learn best by doing. Expertise is not evenly distributed, so consider organizing events such as game days, chaos management, and testing opportunities to improve observability practice.

- **Feedback**: Take feedback seriously, ask for improvements, and gather feedback through survey forms, embedded forms in alerts, and dashboards. Pay attention during the incidents related to observability and learn from the failures to improvise the observability system as a part of agile methodology.

Maintaining observability culture can be summarized as follows:

Figure 15.9 – Maintaining the observability culture

System architecture in an observability strategy

The system architecture is a crucial component of an observability strategy in an organization. Understanding the architecture of the systems being monitored is essential for defining the telemetry requirements and selecting the appropriate tools. A clear understanding of the architecture helps to identify the areas that need visibility and where telemetry data should be captured, stored, and analyzed. This, in turn, enables the creation of an effective observability strategy that provides the necessary insights into system performance and behavior.

When examining the system architecture in the cloud, it's important to keep in mind that it differs from traditional on-premises architecture. There are several cloud architecture patterns to choose from, including multi-tier architecture, microservices architecture, serverless architecture, event-driven architecture, edge computing architecture, or hybrid cloud architecture. Each pattern has its own unique benefits and considerations. It is crucial to consider and map the data flows, identify dependencies and interconnections, and determine the critical components that need to be monitored. Once the architecture is understood, it is possible to determine the telemetry requirements, such as the types of data that need to be captured, the frequency at which data needs to be collected, and the sources of that data.

The architecture plays a key role in determining the appropriate toolsets required for effective observability. This includes selecting tools that can integrate with the existing architecture, handle the volume and variety of data, and provide the required functionality for analysis and reporting.

In summary, the system architecture is an essential aspect of an observability strategy, as it provides the foundation for defining telemetry requirements, selecting tools, and generating operational insights. It is important to take the time to fully understand the architecture to ensure a successful observability strategy. In *Chapter 3, Gathering Operational Data and Alerting Using Amazon CloudWatch*, to *Chapter 7, Observability for Serverless Application on AWS*, we covered observability services available from AWS for multi-tier architecture, microservices architecture, and serverless architecture. We also covered the open source managed services available on AWS in *Chapters 9, Collecting Metrics and Traces Using OpenTelemetry, Chapter 10, Deploying and Configuring an Amazon Managed Service for Prometheus*, and *Chapter 11, Deploying the Elasticsearch, Logstash, and Kibana Stack Using Amazon OpenSearch Service*.

Telemetry in an observability strategy

When building an observability strategy, it's important to start with a clear understanding of your business outcomes and goals. This will ensure that the strategy aligns with the needs of the organization and supports decision-making. A five-step approach can be used to define the observability strategy as telemetry and the outcome is outlined in the following figure:

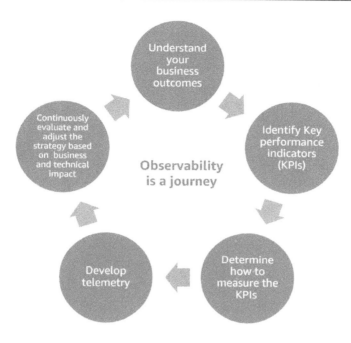

Figure 15.10 – Approach to observability strategy

Let's look into understanding each of the processes in the approach to observability strategy:

- **Understand your business outcomes**: Work backward based on your desired business outcomes by consulting with relevant stakeholders and understanding their priorities. Document these outcomes for reference. We are looking to start with the end in mind as a part of this process.

- **Identify KPIs**: Develop KPIs that align with the business outcomes and can be used to track progress. Keep in mind that monitoring and measuring these KPIs is an ongoing effort. This will help you set up clear goals for observability based on business outcomes.

- **Determine how to measure KPIs**: KPIs are definitions of what is required to evaluate the success of business outcomes, but they are not specific metrics. Translate the KPIs into specific metrics that can be tracked and measured. This will involve collaboration between business and IT teams to identify the relevant workloads and components that need to be observed. Consider the level of granularity that will be necessary for accurate measurement.

- **Develop telemetry**: Use the identified metrics and measurement criteria to develop the telemetry infrastructure to support data collection and analysis.

- **Continuously evaluate the business and technical impact**: Regularly assess the effectiveness of the KPIs and technical measurements in achieving the desired business outcomes. Make adjustments as necessary to ensure they remain aligned with the overall objectives.

Capturing the observability discussions

When you are conducting the workshop for identifying the business outcomes, you could look into capturing the input, output, and outcomes as a part of the simple template that will help you further guide in defining the technical requirements based on the business outcomes. We show a simple template here that could help you capture end-to-end tracking of your observability strategy:

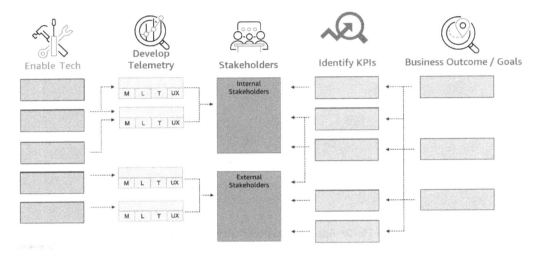

Figure 15.11 – Observability strategy template

Now that you clearly understand the need for an observability strategy, as well as the approach and method for conducting a workshop, let's dive into identifying the specific details required to fulfill the template:

1. First, identify all internal and external stakeholders. Internal stakeholders typically include business owners, product owners, operations teams, developers, and security and compliance teams. Typically, you could group them into operations, developers/engineers, and management. External stakeholders include end users, partners, vendors, suppliers, and providers. Make sure to also consider any compliance requirements that need to be met, such as those from government regulators, accreditation bodies, or industry organizations. Once you have identified all stakeholders, create personas for each, such as a chief operating officer, site reliability engineer, or development team, and map out the required insights for each persona.

2. Next, focus on identifying the business outcomes required by each stakeholder. Examples of typical business outcomes include improving the quality of the user experience to drive increased revenue and compliance reporting to meet regulatory requirements. Derive KPIs from these business outcomes. For example, to improve the user experience for a web application, you should focus **service-level objectives** (**SLOs**) on the site availability and responsiveness, and for compliance reporting, you should focus on the time it takes to identify and mitigate vulnerabilities.

3. The next step is to determine the telemetry needed based on the system architecture to measure these KPIs. Look at **service-level indicators** (**SLIs**) such as sentiment customer scores, web page response times, and other metrics to understand the user experience and measure against the KPIs. Additionally, consider system-level metrics to assess the operational effectiveness of the website. Ensure that all telemetry requirements are **specific, measurable, achievable, relevant, and time-bound** (**SMART**). On a high level, the telemetry related to metrics, logs, traces, and end user experience supports mapping against all the KPIs being identified as a part of the exercise.

4. The next step is to map the technology or tools that will help you in measuring the telemetry defined. You can adopt one of the following approaches based on the system architecture and telemetry requirements to achieve the desired business outcomes. There could be multiple approaches to finalizing the tooling requirements, but we listed a few here:

   - **Cloud-Native**: AWS offers a wide range of cloud-native observability solutions, which we have discussed in *Chapter 2, Overview of the Observability Landscape on AWS*. We also covered details about them from *Chapter 3, Gathering Operational Data and Alerting Using Amazon CloudWatch,* through *Chapter 8, End User Experience Monitoring on AWS*, and in *Chapter 12, Augmenting the Human Operator with Amazon DevOps Guru.*

   - **Open Source Observability**: AWS offers a range of open source managed observability solutions for a variety of workloads running either on the cloud or on-premises. We have covered details about them from *Chapter 9, Collecting Metrics and Traces Using OpenTelemetry,* through *Chapter 11, Deploying the Elasticsearch, Logstash, and Kibana Stack Using Amazon OpenSearch Service.*

   - **Cloud Native and Open Source Managed**: Observability requirements for an organization at times would be complex and we may need to adopt a combination of AWS native services and also open source managed services for our requirements.

   - **Partner Tools**: AWS offers a wide range of Partner tools from the AWS marketplace. Based on the complexity of the requirements and the metrics you are looking to measure against the KPIs, you could adopt one or more of the tools for your cloud observability requirements.

Insights into the observability strategy

We have previously discussed how people, system architecture, and telemetry play a crucial role in determining an observability strategy. However, without proper insights and actions, the strategy will not be fully effective. We can group insights from the information into six different categories. Let's delve deeper into each of these insights:

- **Fault management**: Monitoring your applications to prevent or respond to incidents.

- **Configuration management**: Monitoring and tracking configurations and changes made to a cloud infrastructure supporting your application components.

- **Accounting management**: Monitoring the usage of cloud resources and allocating costs to specific users, departments, or projects based on resource utilization. This helps organizations track and control their cloud expenses, promote cost-effectiveness, and optimize resource utilization.

- **Performance management**: A critical aspect of cloud computing that involves monitoring the performance of cloud components to ensure they are functioning as expected. This includes monitoring for any constraints or limitations in cloud resources. This helps organizations identify and resolve performance issues in a timely manner, preventing potential downtime and ensuring the smooth operation of cloud-based systems and services.

- **Security management**: Monitoring access and security controls, and identifying inappropriate or malicious activities.

- **User Experience (UX)**: Understanding how users interact with your system and what is important for them to navigate across your application.

By understanding these insights, organizations will make data-driven decisions, detect issues before they become critical, improve system performance, and ensure compliance and security. Insights generated from observability can be presented in various formats, such as dashboards, knowledge bases for known issues, and determining corrective actions. You could look at the overall observability plan as shown in the following figure:

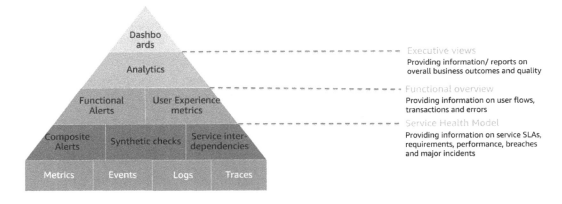

Figure 15.12 – Observability model

Actions in an observability strategy

Organizations can take it a step further by automating the response process. By building run books for corrective actions and triggering them automatically based on identified issues, organizations can improve their incident response time and minimize the impact of incidents on the system and users. This can also help in identifying patterns of issues and creating preventative measures.

Automation can also help reduce the workload of operations and development teams, allowing them to focus on other important tasks. By automating the incident response process, organizations can ensure that someone consistently takes the right actions in a timely manner, eliminating human errors, and improving overall system stability and availability.

Till now, we have discussed strategy management from the business perspective and looked at applying the same to observability. Now, let's look at the other dimensions relevant to the CAF from the adoption of the observability strategy in an organization.

Operations perspective in the CAF

In the operations perspective, there is an **Observability and Event Management (AIOps)** capability that should be looked into when you are looking into the tooling strategy for observability on AWS. We have covered general guidance on observability tooling in this chapter in the *Telemetry in an observability strategy* section. We will cover best practices when applying the Well-Architected Framework for your observability workloads as a part of *Chapter 14, Be Well-Architected for Operational Excellence*. Let's look into the observability maturity model from an operations perspective for different workloads.

Observability maturity model

An organization could look into the observability maturity model to determine the level of adoption of observability and make sure that they could leverage full-stack observability options available on AWS to determine their roadmap to maturity. We have summarized a roadmap to maturity in the following figure as a guide. You should look into tweaking the maturity model to suit your organization's requirements based on the observability strategy output. Based on the level of criticality of the application from the business perspective, you could leverage a maturity model.

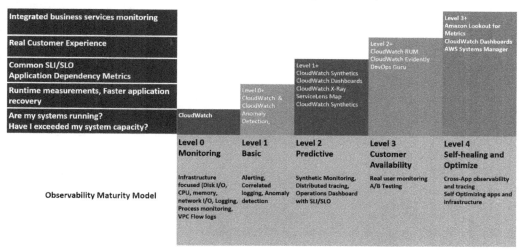

Figure 15.13 – Indicative observability maturity model

When adopting the maturity model, there are still fundamental best practices in metrics, logging, and tracing that will help you in getting the full potential of the adoption of the model.

Let's look into some of the best practices in metrics, logging, and tracing.

Best practices for metrics collection

Best practices for metrics collection include identifying KPIs, defining workload metrics, collecting and analyzing workload metrics, establishing baselines for workload metrics, learning expected patterns of activity for workload, using monitoring to generate alarm-based notifications, and reviewing metrics at regular intervals.

First, it's important to identify KPIs that apply to the organization's goals and objectives. These KPIs will serve as a guide for the metrics that need to be collected. Once the KPIs are identified, it's important to define specific workload metrics that will measure the performance of the systems and applications.

Next, it's important to collect and analyze the workload metrics in order to gain insights into the performance and health of the systems and applications. This analysis can establish baselines for the workload metrics, which can be used as a reference point for future analysis. By learning the expected patterns of activity for the workload, organizations can identify anomalies and potential issues.

Monitoring tools should generate alarm-based notifications when metrics deviate from the established baselines or expected patterns of activity. This can help organizations quickly identify and address issues before they become critical problems.

Finally, it's important to review metrics at regular intervals to ensure that they are still relevant and to identify any changes in the systems or applications that may require adjustments to the metrics being collected. This will help organizations to maintain an accurate understanding of the performance and health of their systems and applications, which is essential for improving reliability and availability.

Best practices for logging

Best practices for logging include instrumenting applications to write logs to the STDOUT (short for **Standard Output**) and STDERR (short for **Standard Error**) streams, avoiding writing any logs to the container filesystem, keeping log formats consistent, setting resource limits on log collection daemons, and using managed services such as Amazon Macie to discover sensitive data.

First, it's important to instrument your applications to write logs to the STDOUT and STDERR streams. This allows logs to be easily collected and centralized by log collection daemons such as Fluentd. This also allows logs to be easily viewed and analyzed using log management tools, such as OpenSearch and CloudWatch Logs.

To make logs more easily searchable and analyzable, it's important to keep log formats consistent. This will help to ensure that logs are easy to search, filter, and aggregate, making it easier to identify and troubleshoot issues.

It's important to set resource limits on log collection daemons in containers that will help you avoid any issues with system performance. One option to receive information based on requirements is to consider switching between different log levels. This will ensure that log collection daemons don't consume too many resources and cause issues with other system processes.

Finally, it's important to use managed services such as Amazon Macie to discover sensitive data in logs. This can help organizations to detect and address data breaches or unauthorized access to data in a timely manner, improving security and compliance.

Best practices for tracing

Best practices for tracing include ensuring that all services in the application stack emit trace data, tracing calls to AWS-managed services from your application, tracing calls to other microservices or public HTTP APIs, and creating subsegments around critical subsections of your application code. Overall, by using these best practices, organizations can gain a more complete understanding of the performance and behavior of their applications, which can help to improve reliability and troubleshoot issues more quickly.

AIOpS

AWS provides different services such as DevOps Guru, CodeGuru, AWS Systems Manager OpsCenter, CloudWatch Metric-based anomaly detection, and Amazon Lookout for metrics to leverage machine learning for your operational effectiveness without configuring manual rule-based thresholds. It should be always a good practice to leverage new features and technologies to increase operational effectiveness.

Best practices for faster observability maturity

The following are foundational practices that will help you in reaching observability maturity faster:

- **Instrument by default**: Whether you are using EC2, containers, or serverless infrastructure, the observability sources should be captured, aggregated, and evaluated.

- **Track performance metrics**: Utilize a monitoring and visibility service to track key performance metrics, such as database transactions, slow queries, I/O latency, HTTP request rate, service latency, and others.

- **Investigate metrics during an incident**: During an event or incident, utilize monitoring dashboards or reports to identify and diagnose the effects. These views give a clear understanding of the workload components that are not functioning optimally.

- **Define KPIs to gauge workload performance**: Determine the KPIs that indicate whether the workload is performing as desired. For instance, the response latency for an API-based workload can serve as a performance indicator, while the number of purchases can be the KPI for an e-commerce site.

- **Create automated alerts based on observability**: Establish KPIs to gauge the performance of your workload. Then, set up a monitoring system that sends out alarms automatically when the KPIs fall outside of their predetermined bounds using anomaly detection if available, or else using a threshold-based alerting mechanism.

- **Periodically review collected metrics**: During regular maintenance or in response to events/incidents, evaluate the metrics collected. These evaluations can help identify which metrics were crucial in resolving issues and which additional metrics, if tracked, could assist in identifying, resolving, or avoiding future issues.

- **Be proactive with observability and alarms**: By combining KPIs with observability and alerting systems, you can proactively address performance issues. Use alarms to initiate automated responses to resolve problems whenever possible. If an automatic solution is not feasible, escalate the alarm to those who are able to respond.

In this section, we discussed the importance of an observability strategy for an organization and the benefits that it brings. We then outlined the key components of an observability strategy, including its definition and potential outcomes. Additionally, we examined the process of implementing an observability strategy and the five crucial considerations involved. Moving on, we explored the operational aspect of observability in the context of cloud adoption. Finally, we outlined various observability maturity models and highlighted best practices for swiftly achieving a mature observability strategy.

In the upcoming section, we will examine the advantages of incorporating observability best practices to expedite your team's cloud adoption process.

Role of observability in the CAF and the best practices for quicker adoption of the cloud

As you are migrating to the cloud, observability forms a core pillar that will provide confidence to your **site reliability engineers** (**SREs**), DevOps engineers, and cloud operations team in faster adoption when they know that metrics are available for them to track the efficiency of their work. Some outcomes that would help increase the team's confidence because of observability are as follows:

- **Real-time visibility**: AWS observability tools provide real-time visibility into the performance and health of systems and applications. You can additionally create business KPI dashboards, as discussed in *Chapter 5, Insights into Operational Data with CloudWatch*, which will allow organizations to quickly identify and troubleshoot issues.

- **Root cause analysis**: DevOps Guru provides detailed insights into the root cause analysis and provides context and the resolution that can help organizations identify the root cause of issues, rather than just symptoms

- **Cost optimization**: You could leverage the performance data from CloudWatch and AWS Trusted Advisor to right-size the workload and achieve cost optimization by identifying underutilized resources or inefficient configurations

- **Performance optimization**: CloudWatch could help you identify and optimize bottlenecks and performance issues, leading to improved application performance

- **Security**: DevOps Guru anomaly detection integrates with CloudTrail to understand the API calls and any changes that happened at the application deployment to detect and investigate security issues, such as a data breach or unauthorized access to data

- **Automated monitoring and troubleshooting**: AWS Application Insights-based automated onboarding and ML-based monitoring could help organizations to onboard quickly and identify and resolve issues in their systems and applications, reducing downtime and improving availability

- **Predictive maintenance**: DevOps Guru Proactive Insights will help you with AI-driven predictive maintenance opportunities that can help organizations proactively identify and address potential issues before they become critical problems, reducing the need for reactive maintenance and improving system uptime

- **Self-healing systems**: DevOps Guru Insights could be integrated as AWS Systems Manager OpsItems and could be associated with run books to provide self-healing systems that automatically detect and resolve issues without human intervention, which can help organizations to improve their ability to manage and operate their systems in the cloud

- **Error tracking**: When leveraging AWS DevOps tools to deploy your applications, you could use AWS Evidently along with AWS DevOps Guru for error tracking, and CloudTrail can collect and aggregate error logs and events, making it easier to identify, diagnose and fix errors

Overall, observability and AIOps can help organizations to improve the availability, performance, and security of their systems and applications in the cloud, which will speed up cloud adoption.

In the next section, we will look into the set of practices and techniques that will help you go beyond traditional monitoring and alerting and help organizations better manage cloud infrastructure.

Beyond observability

We discussed various observability best practices in *Chapter 14, Be Well-Architected for Operational Excellence*. We discussed interoperability in *Chapter 14* while discussing the management and governance lens. From the perspective of cloud operations, we can consolidate the necessary requirements for observability into six subsections.

Observability

Throughout *Chapter 1, Observability 101*, to *Chapter 12, Augmenting the Human Operator with Amazon DevOps Guru*, we explored the multitude of options available in CloudWatch observability.

When it comes to selecting the right observability services, it's important to carefully consider the specific requirements of your organization. This includes generating metrics, logs, traces, and events that align with your unique needs and priorities.

By leveraging CloudWatch's powerful capabilities, you can gain a deep understanding of your end user experience and identify any areas that may require improvement. This can help you optimize your applications and services and, ultimately, drive better business outcomes.

However, to truly maximize the value of CloudWatch observability, it's crucial to select the right combination of services and tools for your specific use case. By carefully considering your requirements and selecting the most appropriate services, you can gain valuable insights and make data-driven decisions that enhance the performance and reliability of your applications and infrastructure.

AIOps-based operations

AIOps (short for **Artificial Intelligence for IT Operations**) is a method that leverages machine learning algorithms and other advanced technologies to optimize IT operations. Anomaly detection, correlation, dynamic thresholding, and alert rules are all key components of an AIOps system.

Amazon DevOps Guru is a fully managed AIOps service that is specifically designed to help AWS customers enhance the availability and reliability of their applications. By leveraging a range of advanced technologies, including machine learning algorithms and anomaly detection, DevOps Guru can help organizations quickly identify operational issues and receive recommendations for remediation.

Besides its sophisticated anomaly detection capabilities, DevOps Guru also correlates data from multiple services to identify patterns or relationships that may contribute to issues. By taking a holistic approach to data analysis, DevOps Guru can provide customers with valuable insights into application performance and identify the root causes of issues.

DevOps Guru can automatically adjust its recommendations based on current conditions, helping IT teams respond more effectively to issues as they arise. This automated root cause analysis can help reduce the time and effort required to identify and address issues, enabling organizations to enhance their operational efficiency and focus on driving better business outcomes.

Event management

Event management is a critical component of IT operations, and AWS Systems Manager OpsCenter offers powerful capabilities for managing events across a range of systems and applications. To optimize event management, OpsCenter leverages advanced technologies such as event normalization, event enrichment, and event deduplication:

- **Event normalization** involves transforming raw event data into a standard format that can be easily analyzed and understood. This allows IT teams to quickly identify relevant data points and gain insights into the root causes of issues. By normalizing events, OpsCenter can help reduce the time and effort required to analyze and respond to issues.

- **Event enrichment** involves enhancing event data with additional context and information. This can include data such as server or application metadata, user information, or system logs. By enriching event data, OpsCenter can provide IT teams with more comprehensive insights into issues and help them better understand the impact on their systems and applications.

- **Event deduplication** involves identifying and removing duplicate events from event streams. This can help reduce noise and improve the accuracy of event analysis. By deduplicating events, OpsCenter can help IT teams focus on the most relevant data and avoid wasting time and resources on redundant information.

With these advanced event management capabilities, OpsCenter can help IT teams streamline their operations and respond more effectively to issues as they arise. By providing automated event analysis, insights, and recommendations, OpsCenter can help organizations optimize their IT operations and enhance the performance and reliability of their applications and systems.

Service management

In today's digital landscape, companies are constantly striving to improve their service management capabilities. While many organizations rely on a standard service management system to manage an incident, change, release, knowledge, and problem management, there are additional tools and technologies that can be leveraged to enhance these processes.

For example, cloud resources generate a significant amount of events that can be used to create automated incidents using AWS Systems Manager OpsCenter. By integrating this tool with your existing service management system, you can streamline incident resolution and reduce downtime for your users.

Furthermore, it's important to understand the impact of releases on your observability and change windows. By tracking major changes and monitoring their impact on application performance, you can proactively identify and address potential issues before they impact your users.

In addition to these tools and techniques, it's also important to leverage knowledge bases to improve incident resolution times. Amazon DevOps Guru provides valuable recommendations that can be integrated into your organization's specific knowledge base for quicker incident resolution.

Overall, effective service management requires a comprehensive approach that integrates the latest tools and technologies with existing processes. By continuously improving your service management capabilities, you can enhance your organization's ability to deliver high-quality services that meet the needs of your users.

Automated resolution

Traditionally, **incident resolution** has been a manual process that relies heavily on the organization's knowledge base. However, in today's fast-paced digital landscape, it's crucial to optimize incident resolution and reduce downtime for users. AWS Systems Manager provides an automated approach to resolving issues, which can help organizations streamline their incident management processes.

By creating Systems Manager documents and mapping them to related AWS Systems Manager OpsItems, you can automate the resolution of known issues. This approach allows you to look at automating the resolution of issues via Systems Manager automation, rather than executing steps manually. By automating incident resolution, you can optimize your workforce management, reduce the risk of human error, and improve overall business KPIs.

In addition, AWS Systems Manager provides a centralized location for managing resources and automating tasks, which can help organizations improve operational efficiency and reduce costs. By leveraging Systems Manager automation, you can free up valuable resources to focus on more strategic initiatives that drive business growth.

Overall, automated incident resolution using AWS Systems Manager provides a modern, streamlined approach to incident management that can help organizations optimize their workforce, reduce downtime, and improve overall business performance.

Dashboards

In today's complex digital landscape, it's crucial to have end-to-end visibility into your applications and business metrics. One way to achieve this is by leveraging Amazon CloudWatch and Amazon QuickSight to create different types of dashboards that provide valuable insights into your data.

By creating an application dashboard, you can get a comprehensive view of your application's performance, including metrics such as latency, errors, and availability. This view can help you quickly identify and address issues that may impact user experience.

In addition, a business dashboard can provide insights into the impact of issues on your business metrics, such as revenue, customer retention, and engagement. This view can help you make informed decisions about how to prioritize and address issues that may impact your business goals.

A KPI dashboard is another important tool that can help you track your **service-level agreements** (**SLAs**) and understand the impact of issues on your operational performance. This view can help you proactively identify and address issues before they impact your SLAs.

Finally, a dashboard that provides an overview of the customer experience can help you understand how users are interacting with your applications and identify areas for improvement. This view can help you deliver a better user experience and drive customer satisfaction.

Overall, by leveraging different types of dashboards in Amazon CloudWatch and Amazon QuickSight, you can simplify your operations, improve observability data, and make informed decisions that drive business growth.

Things beyond observability can be visualized in *Figure 15.14*:

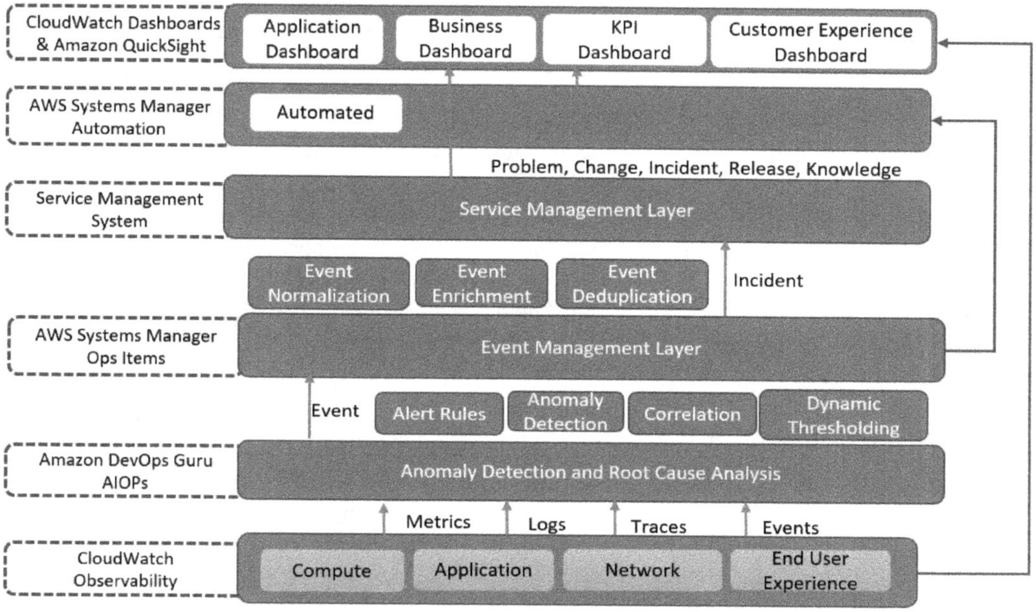

Figure 15.14 – Beyond observability for operational excellence

In this section, we understood the need to look into the holistic approach for cloud operations by looking beyond observability metrics. Considering the larger picture and analyzing factors such as AIOps, event management, automation, dashboards, and service management can optimize your cloud resources, improve operational efficiency, and enhance user experience.

Summary

In this chapter, we have looked into the transformation of the Cloud Adoption Framework over the years and looked into different aspects of the CAF V3.0. Further, we have emphasized the importance of an observability strategy for an organization and looked into the tools and techniques on how to effectively create an observability strategy, and looked at sustaining it. Further, we have looked into building an observability maturity model and best practices for quick attainment. Finally, we have explored how important observability is for organizations embarking on the journey to the cloud.

When an organization is embarking on a new journey, it is always important to track progress and share metrics for improved visibility. Observability is a key component of your cloud adoption, which provides insights into the success and areas of improvement. Observability should be considered an essential part of your cloud adoption journey.

Moreover, we have delved into how operational excellence can be achieved beyond observability, through the implementation of additional layers that support cloud operations. For instance, by adopting a comprehensive incident management system and utilizing automation tools such as AWS Systems Manager, organizations can enhance their incident response capabilities and optimize workforce management.

To conclude, achieving operational excellence in cloud operations requires a multifaceted approach that considers observability metrics, incident management, automation, event management, and dashboards. By leveraging the right tools and strategies, organizations can ensure high performance, scalability, and cost efficiency, and drive business growth in today's rapidly evolving digital landscape.

Congratulations on completing this book and gaining a comprehensive understanding of the role of observability in cloud operations, as well as the various AWS services available in this area. I hope you found this information useful and informative and that it helps you adopt best practices in your cloud operations.

Whether you are a solutions architect, DevOps engineer, or cloud engineer, the knowledge gained from this book will undoubtedly prove valuable in your professional endeavors. By leveraging the power of observability in your cloud operations, you can enhance the reliability, scalability, and performance of your applications and infrastructure.

I hope that the insights gained from this book will help you achieve even greater success in your career.

Questions

1. Explain the importance of the CAF.

2. What are the different components of the CAF?

3. What are the benefits of defining an observability strategy in an organization?

4. What are the important aspects of a *people* strategy in an observability strategy?

5. What is the importance of observability in the CAF?

Index

Other Books You May Enjoy

If you enjoyed this book, you may be interested in these other books by Packt:

AWS FinOps Simplified

Peter Chung

ISBN: 9781803247236

- Use AWS services to monitor and govern your cost, usage, and spend
- Implement automation to streamline cost optimization operations
- Design the best architecture that fits your workload and optimizes on data transfer
- Optimize costs by maximizing efficiency with elasticity strategies
- Implement cost optimization levers to save on compute and storage costs
- Bring value to your organization by identifying strategies to create and govern cost metrics

Cloud-Native Observability with OpenTelemetry

Alex Boten

ISBN: 9781801077705

- Understand the core concepts of OpenTelemetry
- Explore concepts in distributed tracing, metrics, and logging
- Discover the APIs and SDKs necessary to instrument an application using OpenTelemetry
- Explore what auto-instrumentation is and how it can help accelerate application instrumentation
- Configure and deploy the OpenTelemetry Collector
- Get to grips with how different open-source backends can be used to analyze telemetry data
- Understand how to correlate telemetry in common scenarios to get to the root cause of a problem

Packt is searching for authors like you

If you're interested in becoming an author for Packt, please visit authors.packtpub.com and apply today. We have worked with thousands of developers and tech professionals, just like you, to help them share their insight with the global tech community. You can make a general application, apply for a specific hot topic that we are recruiting an author for, or submit your own idea.

Share Your Thoughts

Now you've finished *AWS Observability Handbook*, we'd love to hear your thoughts! Scan the QR code below to go straight to the Amazon review page for this book and share your feedback or leave a review on the site that you purchased it from.

https://packt.link/r/1804616710

Your review is important to us and the tech community and will help us make sure we're delivering excellent quality content.

Download a free PDF copy of this book

Thanks for purchasing this book!

Do you like to read on the go but are unable to carry your print books everywhere? Is your eBook purchase not compatible with the device of your choice?

Don't worry, now with every Packt book you get a DRM-free PDF version of that book at no cost.

Read anywhere, any place, on any device. Search, copy, and paste code from your favorite technical books directly into your application.

The perks don't stop there, you can get exclusive access to discounts, newsletters, and great free content in your inbox daily

Follow these simple steps to get the benefits:

1. Scan the QR code or visit the link below

https://packt.link/free-ebook/9781804616710

2. Submit your proof of purchase

3. That's it! We'll send your free PDF and other benefits to your email directly

www.ingramcontent.com/pod-product-compliance
Lightning Source LLC
Chambersburg PA
CBHW060920060326
40690CB00041B/2789